A. REYNOSO

PRINCIPLES

of

MAGNETIC PARTICLE TESTING

by
CARL E. BETZ

Chemical Engineer
Vice President and Director (Retired), Magnaflux Corporation.
Honorary Life Member, Society for Nondestructive Testing
Member
American Society for Metals
American Society for Testing and Materials
American Association for the Advancement of Science

First Edition

Published by Magnaflux Corporation
Chicago, Illinois
February 1, 1967

PRINCIPLES OF MAGNETIC PARTICLE TESTING

LIST OF CHAPTERS

TABLE OF CONTENTS

CHAPTER 1
HISTORY OF THE MAGNETIC PARTICLE METHOD

CHAPTER 2
FUNDAMENTAL CONCEPTS OF THE METHOD

CHAPTER 3
SOURCES OF DEFECTS

7

CHAPTER 4
HOW AND WHY METALS FAIL

CHAPTER 5

DEFINITIONS OF SOME TERMS USED
IN MAGNETIC PARTICLE TESTING

CHAPTER 6

CHARACTERISTICS OF MAGNETIC FIELDS

CHAPTER 7

METHODS AND MEANS FOR
GENERATING MAGNETIC FIELDS

CHAPTER 8

DETERMINATION OF FIELD STRENGTH
AND DISTRIBUTION

CHAPTER 9

FIELD STRENGTH AND DISTRIBUTION
IN SYMMETRICAL OBJECTS

CHAPTER 10

FIELD DISTRIBUTION IN LARGE
OR IRREGULAR-SHAPED BODIES

CHAPTER 11

MAGNETIC PARTICLES—
THEIR NATURE AND PROPERTIES

Chapter 12

BASIC VARIATIONS IN TECHNIQUE

Chapter 13

THE DRY METHOD—
MATERIALS AND TECHNIQUES

CHAPTER 14
THE WET METHOD—
MATERIALS AND TECHNIQUES

CHAPTER 15
FLUORESCENT MAGNETIC PARTICLES—
THEIR NATURE AND USE

CHAPTER 16

BLACK LIGHT—ITS NATURE, SOURCES AND
REQUIREMENTS

CHAPTER 20
DETECTABLE DEFECTS

CHAPTER 21
NON-RELEVANT INDICATIONS

CHAPTER 22
INTERPRETATION, EVALUATION
AND RECORDING OF RESULTS

CHAPTER 23

INDUSTRIAL APPLICATIONS

LIST OF TABLES

LIST OF ILLUSTRATIONS

Frontispiece—Testing of Steel Billets for Seams, Using Fluorescent Magnetic Particles. The Process Discriminates Between Shallow and Deep Seams. *(Courtesy Youngstown Sheet and Tube Company.)*

CHAPTER 4

CHAPTER 9

CHAPTER 10

CHAPTER 11 *Page*

CHAPTER 12

CHAPTER 13

CHAPTER 14

CHAPTER 20

CHAPTER 21

CHAPTER 23

Frontispiece   (Courtesy Youngstown Sheet and Tube Company)

Testing Steel Billets for Seams, Using Fluorescent Magnetic Particles.
The Process Discriminates Between Shallow and Deep Seams.

PREFACE

When the first edition of the book "Principles of Magnaflux" was published in 1939, the magnetic particle method was still relatively unknown in industry. Certainly it was unappreciated. This was ten years after the inception of the method in 1929. Those were ten years of intensive work on the part of A.V. de Forest and F. B. Doane—and later of the staff of Magnaflux Corporation—to educate potential users in the value and usefulness of this testing method. To us, of Magnaflux Corporation, the need for the method appeared self-evident, but the natural resistance of human nature to trying out new ideas or any departure from time-honored customs, made potential users reluctant to adopt this novel testing procedure, and progress was slow indeed. At times our efforts took on the aspect of a parent patiently forcing a child to take unpleasant medicine, because we (the parents) knew it would be good for the recipients. The book, "Principles of Magnaflux", was written by F. B. Doane in 1939 with the purpose to make available the knowledge and experience in the method accumulated up to that time, to aid in this educational program.

Re-reading this first edition now—almost thirty years later—one is impressed by two points: first, how little the fundamental principles and guidelines of the method have changed; and second, how great has been the advance in the techniques of applying these principles and how widely the method has come to be used in areas not even visualized in 1939.

When the second revision of the book was issued in 1948 the growth in technique variations and expansion of fields of use which had occurred during World War II were so great that the book was almost doubled in size and contained a great deal of new matter relating to new materials, new knowledge and new understanding of the conditions which controlled performance, and many details of new applications.

Now, after eighteen more years, the 1948 version of the book is completely out-dated, and the author feels keen regret that this present effort could not have been presented five or more years earlier. As a compensating factor, however, much has occurred in those five years—such as the steel billet testing development, and applications in the inspection of rockets and aero-space vehicles— which could not have been included in a 1958 revision.

The present new volume, "Principles of Magnetic Particle Testing", is completely rewritten. The format and organization follows that of its companion volume, "Principles of Penetrants". A considerable amount of new material, much of it heretofore unpublished, has been incorporated. The earlier books were oriented toward the era of steam locomotives and piston aircraft engines— two uses which, major at the time, have passed out of the light of first importance in industrial uses in the United States (or are about to). Instead, this volume is oriented toward diesel locomotives, automotive and rolling mill applications, and space age testing problems.

There is much new detail in the information given regarding materials and equipment. A whole chapter (19) is devoted to the design and use of automatic and special-purpose units—a field in which there has been a tremendous advance in the past ten years. Chapters 8, 9 and 10 have expanded the discussion of field distribution in various shaped objects. These latter three chapters include some original work by the research laboratories of Magnaflux Corporation, and were prepared by Dr. G. O. McClurg, Research Director. Chapter 11, on materials, was prepared by Bruce Graham, Assistant Director of Materials Engineering. This department is constantly working to improve materials used in magnetic particle testing. A. E. Christensen, Manager of Engineering, is responsible for the preparation of Chapter 19 on special and automatic equipment. The important development in this approach to more effective and economical testing is the result of work during the past fifteen years by Mr. Christensen and his staff of experienced design and development engineering specialists.

In this connection a tribute is due to the late A. K. (Ding) Saltis, who pioneered the whole idea of magnetic particle testing of billets; and, after some five years of effort, succeeded in convincing the steel industry of the value of this important and now accepted steel mill practice.

In presenting a book on a subject as fluid and growing as is the magnetic particle method, the author is aware that what has been written regarding techniques and materials will probably no longer be accurate some few years hence. The reader is asked to appreciate this point, and regard the statements made here as being an expression of our knowledge and experience as of the present date, recognizing that in a few years new developments may make some present practices obsolete.

If the book is considered as a whole, the reader will find a considerable number of repetitions of various facts, discussions and illustrations. These repetitions are intentional, in order to present a topic completely in its own context, and to avoid requiring the reader to be constantly thumbing pages to locate cross-references. A considerable number of cross-references are, of course, inevitably included.

In addition to those persons mentioned above as having made contributions to the text, the author wishes to thank the many who have aided in this work in one way or another. Especially, thanks go to D. T. O'Connor, Dr. G. O. McClurg, and A. E. Christensen, who have patiently read and re-read the text. F. S. Catlin, Ronald Marks, Kenneth Schroeder and Bruce Tyler have also checked and re-checked the entire manuscript. These men have given what at times seemed merciless criticism of statements made and phrases used by the author. For this the author is grateful, since it has made possible the publishing of this volume with confidence that it is technically correct in all details, and that grammar and word usage have not been flagrantly abused. The corrections and suggestions made have added materially to the accuracy and style of the work. Arthur Debb has been helpful in the preparation of the sections on steel mill testing and practices.

F. S. Catlin and Ronald Marks, with the help of their able staff in the Market Development Department, have spearheaded the search for and the selection of illustrations, and have assumed the burden of the many details that are a part of the publication of any book. Mr. Catlin is also responsible for the design of the jacket cover.

The line drawings, with a few exceptions, are the work of the Ed Adams Studio of Indianapolis, Indiana. The photographs are to a large extent by Gordon Carson Studio, Franklin Park, Illinois.

Lastly, my thanks go to Mr. W. E. Thomas, who has in good-natured patience borne all the frustrating delays in the completion of this work, and has contributed never-failing suggestions, encouragement and support.

Carl E. Betz.
Greentree Farm,
Martinsville, Indiana.
September 1, 1966.

INTRODUCTION

by

W. E. THOMAS

This is a book which covers in considerable detail the many technical and practical aspects of Magnetic Particle Inspection. While it follows and replaces the various editions of "Principles of Magnaflux" originally written by the late Mr. F. B. Doane and later revised and up-dated by Mr. Carl E. Betz, it is a completely new work both in its organization and its detail, and illustrative material.

As covered by the author in the first chapter, this nondestructive method was first conceived by Major Wm. E. Hoke, but was developed into a practical industrial tool by Professor A. V. de Forest in 1929. Following the formation of a partnership between him and Mr. F. B. Doane in that year (which partnership was the predecessor to Magnaflux Corporation) they both contributed extensively to its initial development. A few years later Mr. Betz joined them and over the past 31 years has made numerous significant personal contributions.

The method is inherently so readily applied and its results so essential to the proving of the integrity of structural materials and parts, that it grew in use to a point where it is probably accurate to say that it is *the* most widely-used nondestructive testing method and is employed on more parts than any other method in use today.

Nondestructive testing is a term used to describe a variety of methods which may determine either the physical soundness or other characteristics of materials without injury to them in any way. This is the point of separation between such methods and the many other tests for strength or soundness which are used throughout industry, but which destroy or damage the part beyond its point of usefulness. The oldest of these commonly used "scientific" NDT methods is radiography, and of course its applications are many and its utilization is very widespread. In point of time, the magnetic particle test was the second to come into general use, and in the past 35 years we have seen a number of other test methods developed and a great variety of instrumentation designed and used.

The result of these nondestructive tests has been to help make possible the manufacture and use of lighter and stronger equipment of all kinds, and the use of that equipment with a high degree of safety to its operators and to the general public. All of us in this field of work feel a sense of personal satisfaction in having contributed to the fine safety record of railroads, aircraft, automobiles and trucks, turbines, bridges and structures of all kinds. Yet safety alone is not the sole reason for the employment of these tests. They are powerful factors in economical manufacture as well. The fabrication of defective materials is very expensive in both time and money and the assembly of defective components into a machine or structure usually requires expensive rework and replacement.

Magnaflux Corporation's policies have been based upon the high technical standards of men such as de Forest, Doane and Betz, as well as their personal and business integrity. Its people have considered that a primary part of their responsibility is to insure that the magnetic particle method be thoroughly understood and intelligently and economically used. To this end, a great many educational activities have been sponsored and conducted by the Company, and in this work Carl Betz has had a leading part.

It is a great pleasure for me to acknowledge his contributions to the technical development of the subject of the book and his personal leadership in the education and training of literally many thousands of people involved in its specification, use, and evaluation, for the benefit of all.

W. E. Thomas
President
Magnaflux Corporation

CHAPTER 1

HISTORY OF THE MAGNETIC PARTICLE METHOD

1. EARLY TESTING METHODS. In the years prior to 1920 the term "nondestructive testing" had not yet acquired a meaning or found a place in the language of Engineering. There were many testing methods, of course, but nondestructive testing as such, was not in existence at that time. None of the methods so important today had even been thought of. Although radiography had been born, its industrial application had not yet been visualized. Testing of materials was considered necessary to provide data for use in design, or to check the overall physical properties of the materials that were actually used. But such tests were pretty much limited to destructive tests of samples selected from large lots. These tests included chemical analysis; tensile, compressive or impact tests of steel and other metals; and similar efforts intended to reassure the engineer that he was actually getting the properties that he had assumed in his design. Proof tests of completed parts or assemblies gave assurance of *initial* strength, but gave no information as to what might be expected later in service.

But, as a matter of fact, the *need* for nondestructive testing had not yet become very urgent. Machinery was for the most part heavy and relatively slow-moving, and large factors of safety were the rule. Of course, fatigue failures did occur in spite of the efforts to design strength into moving parts by making them bigger than they really needed to be. Large crankshafts, railroad axles and similar stressed members broke all too frequently. But no concerted attack had yet been made on the real cause of such failures or on specific methods to avoid them. The part played by small notch-like defects in initiating fatigue failures was not yet fully recognized—so the urge to find and eliminate such flaws as a means to reduce failures in service had not yet come into being.

In the days following the end of World War I, two developments marked the birth of modern nondestructive testing. Work at the Bureau of Standards and at the General Electric Company had shown that X-rays could be used to obtain pictures of metallic articles, revealing internal conditions: and the alert observation of

47

a man—William E. Hoke—also at the Bureau of Standards had discovered the principle of the use of magnetic fields and ferromagnetic particles to locate surface cracks in magnetic metals. He found that the metallic grindings from hard steel parts which were being ground while held on a magnetic chuck, often formed patterns on the face of the parts which corresponded to cracks in that surface. This was the essential basis of longitudinal magnetization for the location of transverse cracks by the magnetic particle method.

It is true that other roots or seeds of modern nondestructive testing also existed at that time. The "oil and whiting" (or oil and chalk) method was in use, in the railroad field particularly, to locate cracks in heavy locomotive parts; and in the casting industry to locate gross cracks. This was the forerunner of penetrant testing, to be developed twenty years later.* Ringing methods to test castings, ceramics and chinaware for soundness were in use, and could be considered as foreshadowing sonic and ultrasonic methods for finding flaws. Magnetic and electrical methods were in use to determine certain physical properties, though they were not used for flaw detection at that time.

2. BEGINNINGS OF INDUSTRIAL RADIOGRAPHY. In the decade from 1920 to 1930 modern nondestructive testing began to show stirrings of life. In 1922 Dr. H. H. Lester set up in the Watertown Arsenal laboratory the X-ray equipment that started industrial radiography on its way. His work demonstrated that penetration of thick sections of steel was practical, and that the result revealed defects such as cavities, cracks, porosity and non-metallic inclusions where the defect thickness was as small as 2% of the metal thickness. This demonstration fired the imagination of engineers and scientists. The idea of nondestructively finding flaws in the interior of solid metal by literally looking through it was an exciting prospect. From that time on, developments and improvements have been continuous up to the present day, and have made radiography one of the outstanding nondestructive test methods.

3. THE MAGNETIC PARTICLE METHOD DEVELOPMENT. Toward the end of this decade in 1928 and 1929, A. V. de Forest did his initial work which eventually resulted in the magnetic particle method as we now know it. It is reported that some abortive efforts were made to apply Hoke's discovery in the middle 1920's, but the method did not "catch on". It remained for de Forest to envision the great

---

*See "Principles of Penetrants" by C. E. Betz, Magnaflux Corporation, 1963.

Fig. 1—Dr. H. H. Lester's Pioneer X-ray Laboratory
at the Watertown Arsenal. 1922.

possibilities which the method held. He realized that if cracks in any direction were to be located reliably, the direction of the magnetic field in the part could not be left to chance, nor be only longitudinal in direction. Rather, it was necessary to have some means for generating a field in *any* desired direction. This could not be done successfully in all cases with external magnets or coils carrying current—the only magnetizing means then available. To solve this need, he proposed magnetizing by passing a sufficiently large current through the piece being magnetized. This was the first use of circular magnetization, the method now so widely employed. He also insisted on the use of magnetic powder of controlled size, shape and magnetic properties, as being essential to consistent and reliable results. These contributions provided the impetus for the development of a truly practical and useful nondestructive test.

Magnaflux Corporation was formed by A. V. de Forest and F. B. Doane, first as a partnership known as A. V. de Forest Associates

Fig. 2—Prof. A. V. de Forest.

and subsequently incorporated in December 1934 as Magnaflux Corporation. F. B. Doane was associated with Pittsburgh Testing Laboratory in Pittsburgh, Pennsylvania at that time and did some of the early work on the development of magnetic powders there. Later, much of the work of developing techniques and the use of alternating current directly for magnetizing, was done with a "breadboard" set-up of equipment in Doane's basement at his home.

Fig. 3—F. B. Doane.

4. PROGRESS BETWEEN 1930 AND 1940. The decade from 1930 to 1940 saw large developments in radiography and magnetic particle testing, and also saw the start of other nondestructive testing methods. Figure 5 is a drawing which shows at a glance the progress and subsequent growth of various methods up into the 1960-1970 decade. The real proliferation of new methods did not start until the beginning of World War II, but has continued at an accelerating

Fig. 4—Early Experimental Equipment used by F. B. Doane. 1930.

rate from that time down to the present writing. There is every reason to believe that in the future many new methods will be devised to solve as yet unknown testing problems.

But during the 1930's the progress seems, from our present viewpoint, to have been painfully slow. For the magnetic particle method, techniques had to be developed and explored, equipment devised and materials improved. Above all, users had to be found to give industrial life to this promising testing procedure. It was during this period that the author became actively associated with the Corporation (1935), and contributed to the work of developing materials, equipment and techniques.

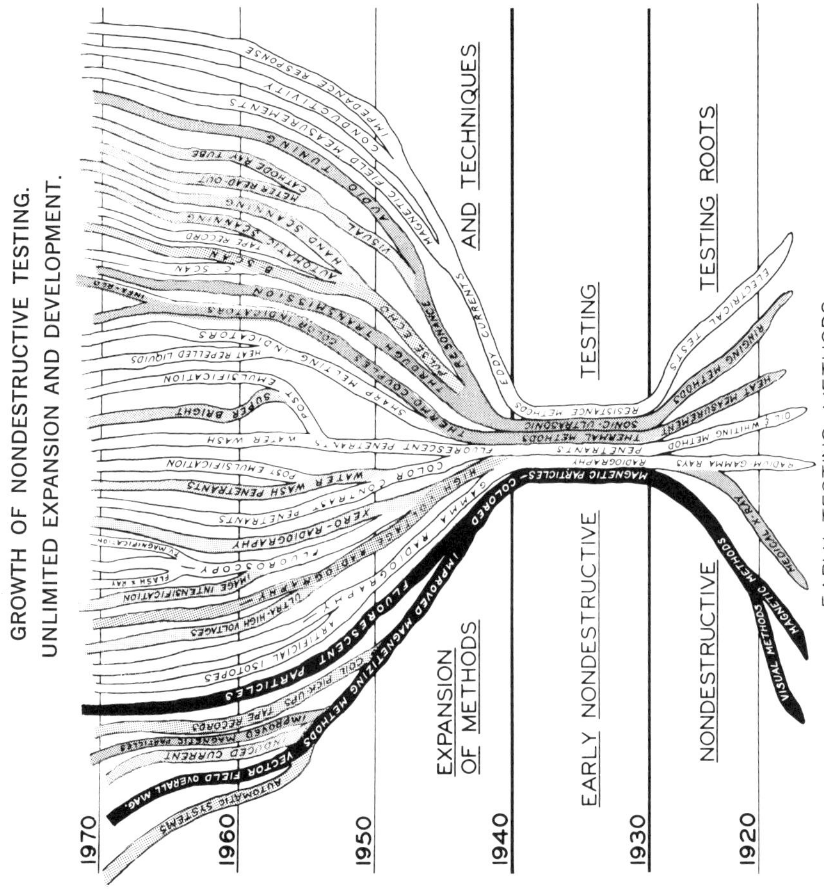

Fig. 5—"Tree of Growth" of Nondestructive Testing.

Fig. 6—Early A.C. Magnetizing Assembly. 1933.

5. EARLY EQUIPMENT. Early applications were crude and early equipment simple and not yet production minded. Figure 6 shows an assembly of electrical components—transformer, switch, meter, etc.—such as was put together in the plant for early users. Such an installation employed alternating current for magnetizing, and was first used for the testing of tool-steel bars. But most of the equipment at this time employed direct current for magnetizing, derived from storage batteries. Figure 7 shows a unit extensively used in the aircraft industry in the late 1930's, which provided both circular and longitudinal magnetization from storage battery power. The railroads also were early in their use of the new method, mainly for the location of fatigue cracks in axles, and motion parts of steam locomotives. The automotive industry, too, rapidly became interested and by 1940 many units were in use, both for the inspection of new parts such as front wheel spindles, crankshafts and connecting rods, and for the location during overhaul, of fatigue cracks on bus and truck engine parts.

During these years radiography also made tremendous strides; and some forms of eddy-current testing were put into successful use, notably for the inspection of steel bars for seams.

6. DEVELOPMENTS OF THE 1940'S. The advent of World War II brought a rapid increase in the demand for all nondestructive test-

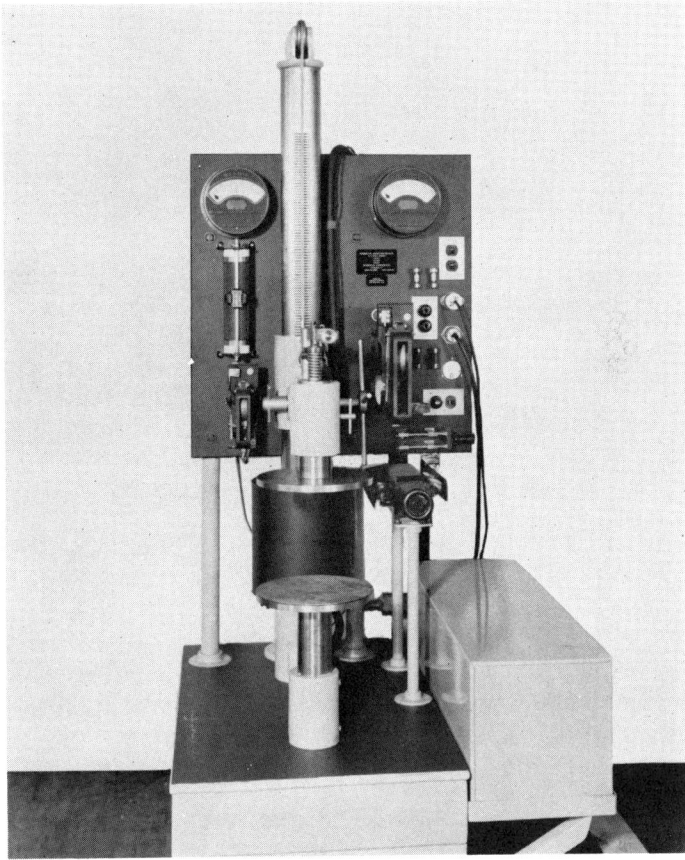

Fig. 7—Storage Battery Unit Used by the Aircraft Industry. 1932-1940.

ing methods. The magnetic particle method was ready to hand and found innumerable new applications. The ability of nondestructive testing techniques to assure the quality and integrity of all components of an assembly, gave confidence that equipment, armament and ammunition would perform reliably in the field. The need to speed production of war materials initiated the design and use of automatic testing units. Figure 8 shows an automatic unit for testing armor-piercing projectiles, which was put into service in 1943.

Fig. 8—Automatic Unit for Testing Armor-Piercing Projectiles. 1943.

Fluorescent penetrant methods appeared in mid-1942 and found immediate application. Penetrant methods were badly needed, to do for non-ferrous materials what the magnetic particle method was doing in the testing of iron and steel parts. Fluorescent penetrants were put to instant use. Fluorescent magnetic particles, too, were introduced about the same time. They found many applications at once, although their great value was not realized until the war was over.

Ultrasonic flaw detection, both by the resonance and the pulse-echo methods, was worked out and used to some extent during the war, though the large development of this useful technique did not take place until after the war.

7. PROBLEMS DUE TO RAPID EXPANSION OF USE. The rapid expansion of the use of magnetic particle testing during the war at first created many problems, mostly due to lack of experience with the method. Government requirements for inspection by means of magnetic particles often seemed unreasonable, but were applied widely. Many manufacturers who had never heard of the method resented the added trouble and expense imposed upon them. Inexperienced Government inspectors rejected parts that showed

indications, which rejections later experience proved to have been unjustified. It took many years to accumulate enough experience and data to interpret magnetic particle indications correctly, so that this tendency toward over-inspection did not cause waste of good material.

The Air Force standardized on the wet method with direct current, although in many cases the dry method with alternating current was better adapted for a particular inspection problem. Later in the war this position was modified and use of A.C. for magnetizing was approved by the Air Force under certain circumstances. But in a number of cases it was found that flaws known to exist were not indicated by either method and improved techniques had to be devised to insure complete reliability for the tests.

8. POST-WAR DEVELOPMENTS. In the years since the close of World War II a rapidly growing list of new test methods has come into being as the result of new testing problems. The magnetic

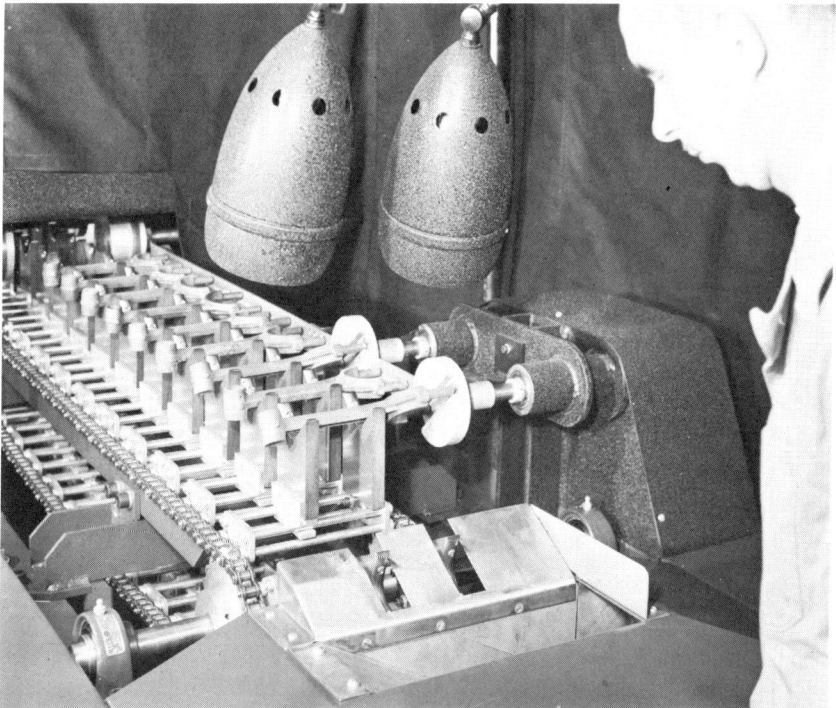

Fig. 9—Automatic Unit Including Part Rotation for Testing Automotive Connecting Rods, Using Fluorescent Magnetic Particles. 1948.

particle method of nondestructive testing has, however, retained its pre-eminent position for the detection of surface and near-surface flaws in ferromagnetic materials. Constant improvement of materials and techniques has kept the method abreast of the demands made by advances in metals and alloys, and their utilization in new designs and new applications.

Outstanding has been the improvement in the magnetic particles themselves, and especially the full realization of the value of the fluorescent method for rapid and reliable inspection. Of equal importance has been the application of the method in new fields, made possible by the recent introduction of new magnetizing techniques, and accumulated experience in designing larger and more automated equipment. Among these developments are multi-directional magnetization; the magnetization of large castings and other parts by the use of very high amperage currents (the "overall" method); water-borne suspensions of magnetic particles; and, in the area of equipment, the construction of large automatic equipment for the inspection of steel tube rounds, billets and blooms of considerable length and cross-section.

(Courtesy American Steel and Wire Division, U.S. Steel Corporation)
Fig. 10—Automatic Unit for Testing Steel Billets for Seams. 1956.

58

9. Nuclear and Space-Age Requirements. The need for absolute reliability and failure-free performance of missiles and rockets, and nuclear power reactors, has placed a new responsibility on all nondestructive testing procedures. Many new methods, and new techniques for old methods, have appeared to solve some of the new testing problems arising from the use of new materials and new construction ideas.

The magnetic particle method has found increasing usefulness in the solution of various of these new testing problems. No other nondestructive testing method is its equal for the detection of very fine and very shallow surface cracks in ferromagnetic materials. Penetrant methods have this same pre-eminence for non-ferrous metals and other non-magnetic materials such as plastics, ceramics, etc. Although both methods have certain limitations, and are excelled in certain applications by other kinds of nondestructive tests, these two methods have continually enhanced their position of being more widely used than any other methods.

10. Future of Magnetic Particle Testing. The acceptance of nondestructive testing by Industry is today practically universal. Its value, not only in assuring flaw-free products, but also in effecting material cost savings in the manufacture of almost every sort of product, has placed it in the position of being an essential tool in the manufacturing process. There is no question as to the continued usefulness of the magnetic particle method in such a climate. Experience with the method is now so extensive that confidence in its indications and in the ability of experienced inspectors to interpret them, is the rule in those industries where the method is applicable. And the record already established of adaptability to the solution of new testing problems leaves little doubt that when new situations arise in its field, new problems will be solved.

CHAPTER 2

FUNDAMENTAL CONCEPTS OF THE METHOD

1. WHAT IS THE MAGNETIC PARTICLE METHOD? The magnetic particle method of nondestructive testing is a method for locating surface and subsurface discontinuities in ferromagnetic material. It depends for its operation on the fact that when the material or part under test is magnetized, discontinuities which lie in a direction generally transverse to the direction of the magnetic field will cause a leakage field to be formed at and above the surface of the part. The presence of this leakage field, and therefore the presence of the discontinuity, is detected by the use of finely divided ferromagnetic particles applied over the surface, some of these particles being gathered and held by the leakage field. This magnetically held collection of particles forms an outline of the discontinuity and indicates its location, size, shape and extent.

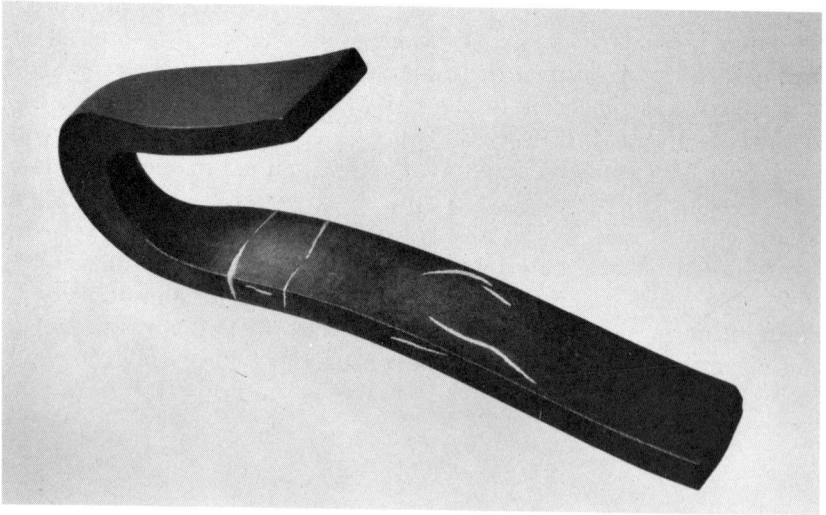

Fig. 11—Typical Dry Powder Pattern. This is the Original Demonstration Piece Used by A. V. de Forest, and Later by F. B. Doane.

There are many factors that affect the formation and appearance of this powder pattern. Some of these are:

(a) Direction and strength of the magnetic field.

(b) Method of magnetization employed.

(c) Size, shape and direction of the discontinuity.

(d) Character of the magnetic powder and the method of applying it.

(e) Magnetic characteristics of the part being tested.

(f) Shape of the part, which affects the distribution of the magnetic field.

(g) Character of the surface of the part—whether smooth or rough, or plated or painted, etc.

The effect of these various factors on the obtaining of optimum results in the detection of discontinuities will be discussed in detail in subsequent chapters.

2. HOW DOES THE METHOD WORK? The method involves three essential steps:

(a) Magnetizing the material or part under test.

(b) Applying the ferromagnetic particles over the surface.

(c) Examining the surface for powder patterns or indications.

3. MAGNETIZATION. The magnetic field inside the part must run in a direction such that it is intercepted by the discontinuity. A number of techniques are in use to accomplish this result, and will be discussed in later chapters.

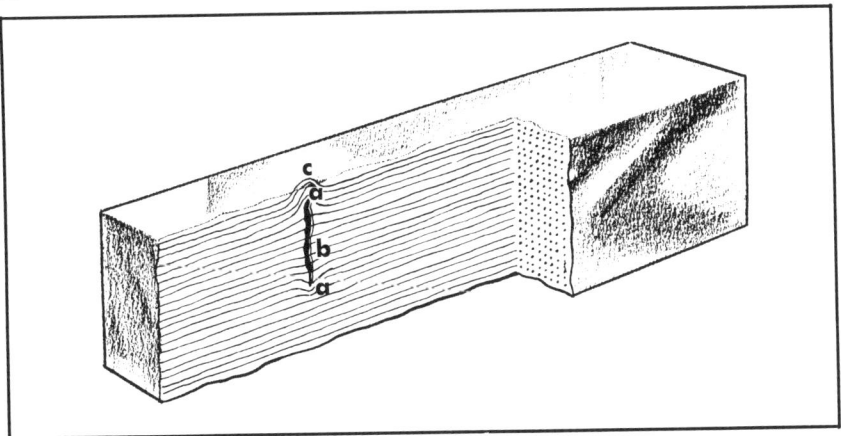

Fig. 12—Field Distortion at a Discontinuity Lying Wholly Below the Surface.

When the field is intercepted by the discontinuity, some of the field is forced out into the air above the discontinuity, to form a "leakage field". Figure 12 illustrates the path of flux lines intercepted by a crack-like discontinuity lying wholly below the surface. In this case the flux follows three alternative paths:

(a) Around the discontinuity on both edges, causing an increase in flux density in these areas, in the interior of the material.

(b) Across the discontinuity. This path offers a high reluctance or resistance to the passage of the flux, since the permeability of the gap—usually air-filled—is much lower than that of the surrounding metal.

(c) Through the air above the discontinuity, causing a leakage field. The strength and intensity of this leakage field is determined to a large extent by the amount of metal between the discontinuity and the surface, the strength of the field inside the metal, and the dimensions of the discontinuity in a direction at right angles to the surface.

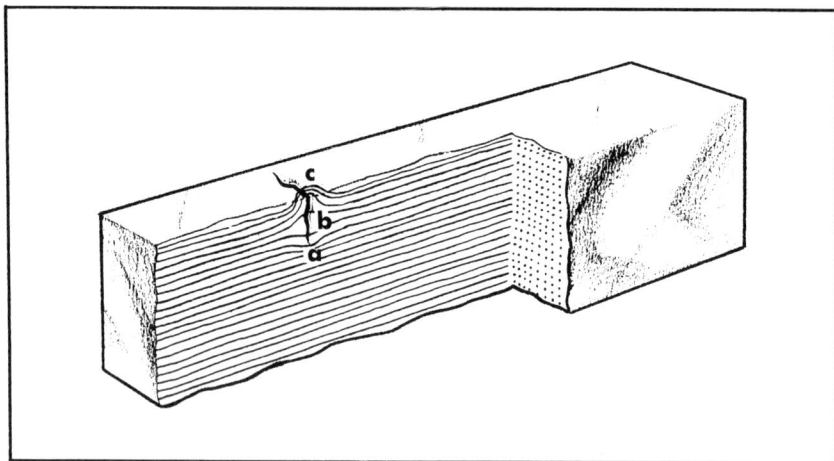

Fig. 13—Field Distortion at a Discontinuity which is Open to the Surface.

The closer the discontinuity is to the surface, the stronger and more concentrated will be the leakage field. If the discontinuity actually breaks the surface, path (a) between the discontinuity and the surface no longer exists, and the leakage flux jumps the gap created by the discontinuity. In this case the field at the surface will be strong and sharply concentrated, if the discontinuity is narrow and crack-like. If it is shallow, and more like a scratch, there may be no leakage field at all. The flux will simply stream-line below the scratch in the body of the metal. Figure 14 illustrates this case.

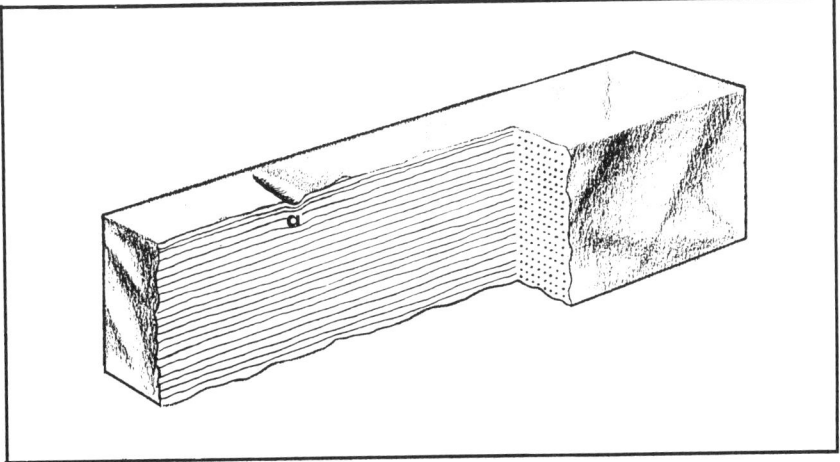

Fig. 14—Field Distortion at a Surface Scratch.

4. APPLYING THE FERROMAGNETIC PARTICLES. After the part under test has been properly magnetized, the ferromagnetic particles are applied. These may be dusted on as a dry powder, or flowed on as a suspension in some liquid. Those particles which come under

Fig. 15—Typical Discontinuity Pattern as Indicated by the Wet Method. (Seamy Wrist Pin)

the influence of the leakage field produced by a discontinuity offer a lower reluctance path to the magnetic flux than does the air, and they are consequently drawn and held by the leakage field. These magnetically held particles then provide the visible evidence of the presence of a leakage field, and therefore of the presence of some kind of discontinuity.

Since leakage fields can be produced by conditions other than metallic discontinuities, a magnetic particle pattern is not *prima facie* evidence of the presence of a "defect". Conditions such as those caused by two components of an assembly in contact, or a sharp change in section will also produce leakage fields and patterns of magnetic particles. Sharp changes in permeability from any cause will do likewise. These circumstances will be described in detail in later chapters.

Strength and concentration of the leakage field is a controlling factor in determining the character and appearance of the indication. It must be strong enough to overcome gravity, surface irregularities, blowing off of excess powder or draining away of excess liquid suspension, and other forces tending to prevent the gathering and holding of particles to form an indication.

5. EXAMINATION OF THE SURFACE FOR MAGNETIC PARTICLE PATTERNS. Visual inspection of the surface of the part after it has been magnetized and magnetic particles applied requires only good light, good eyesight, and close attention and alertness on the part of the inspector. The magnetic particles are available in several colors to increase the visibility of indications under any given conditions. Fluorescent magnetic particles which glow in darkness or dim light when exposed to near-ultraviolet radiation ("black light"), offer the ultimate in contrast and visibility, and are now very widely used in many applications.

The inspector usually marks the location of the indications which he sees, and may or may not, depending on his experience and the nature and importance of the inspection, accept or reject the part at that time. In some cases he may segregate those parts showing indications, and hold them for later disposition. Sometimes this may require the judgment of quality control or design personnel. With increasing experience in working with the method however, the inspector himself acquires, in most cases, the judgment necessary for making the accept-reject decision himself. In highly repetitive tests, as in mass production, all the variables of the test can be controlled, so that *only* those defects considered important are indicated.

6. WHAT CAN THE MAGNETIC PARTICLE METHOD FIND? The method operates to indicate the presence of surface leakage fields, and any such fields from whatever cause will be indicated, provided that:

(a) The field is strong enough to hold the particles applied. Leakage fields at very shallow discontinuities may be very weak, and the fields at the surface formed by discontinuities wholly below the surface may be weak and diffuse.

(b) The magnetic particles used are suitable for the purpose—that is, are fine enough and have high enough permeability to be held.

Under proper circumstances—that is, with proper powder and proper magnetization—exceedingly fine discontinuities can be detected. Even grain boundaries in steel and the outlines of magnetic domains can be shown by using special techniques. The magnetic particle method is the *most sensitive means available* for locating *very fine* and *very shallow* surface cracks in ferromagnetic materials. At the other end of the scale, indications may be produced at cracks that are large enough to be seen by the naked eye. In this case magnetic particle testing is still worth while, because the presence of a prominent and easily seen powder pattern makes for more rapid inspection, and assures that the crack will not be missed by the inspector. Exceedingly wide surface cracks will not produce a powder pattern at all if the surface opening is too wide for the particles to bridge.

Discontinuities which do not actually break through the surface are also indicated in many instances by this method, though certain limitations in this case must be recognized and understood.

If the discontinuity is fine and sharp and close to the surface—as, for example, a long stringer of non-metallic inclusions—a good sharp indication can be produced. As the discontinuity lies deeper, the indication becomes fainter. Put another way, the deeper the discontinuity lies below the surface, the larger it must be to give a readable indication. (See the discussion of Detectable Defects, in Chapter 20.)

The general statement can therefore be made that the magnetic particle method of nondestructive testing is pre-eminent for finding all sizes and depths of cracks that break the surface, but runs into more and more difficulty in detecting internal discontinuities as they lie deeper and deeper below the surface.

7. ON WHAT KINDS OF MATERIAL DOES IT WORK? The method can be used on any ferromagnetic material, though not in all cases with equal effectiveness. It works best on steels and alloys that have a high permeability. Discontinuities lying wholly below the surface are more likely to be located in soft steels having high permeability, than in hardened steels and alloys which in nearly all cases have a lower permeability. This difference is less critical if surface defects only are being sought.

In the case of gray or malleable iron castings, surface cracks are easily located. The method also works quite well on metallic nickel and cobalt. On the other hand, stainless steel and other alloys which are in the austenitic state cannot be tested with magnetic particles at all, since iron in this state is non-magnetic.

8. WHAT ARE THE ADVANTAGES OF THE METHOD? The magnetic particle method has a number of outstanding advantages within its field of usefulness—that is, on ferromagnetic materials. Some of these are the following:

(a) It is the best and most reliable method available for finding surface cracks, especially very fine and shallow ones.

(b) It is rapid and simple to operate.

(c) The indications are produced directly on the surface of the part, and are a magnetic picture of the actual discontinuity. There is no electric circuitry or electronic read-out to be calibrated or kept in proper operating condition.

(d) Operators can learn the method easily, without lengthy or highly technical training.

(e) There is little or no limitation due to size or shape of the part being tested.

(f) It will detect cracks filled with foreign material.

(g) No elaborate pre-cleaning is ordinarily necessary.

(h) It will work well through thin coatings of paint, or other non-magnetic coverings such as plating.

(i) Skilled operators can judge crack depth quite accurately with suitable powders and proper technique.

(j) It lends itself well to automation.

(k) It is relatively inexpensive to operate.

(l) It utilizes electro-mechanical equipment that can be ruggedly built for plant environment, and adequately maintained by existing plant personnel in most instances.

9. WHAT ARE THE GENERAL LIMITATIONS OF THE METHOD? Although the method has many desirable and attractive advantages, it has, as does every method, certain limitations. These the operator must be aware of, and take into account by observing the precautions which they dictate. Some of these are:

(a) It will work only on ferromagnetic materials.

(b) It is not in all cases reliable for locating discontinuities which lie wholly below the surface.

(c) The magnetic field *must* be in a direction that will intercept the principal plane of the discontinuity, a requirement that the operator *must* be aware of and know how to meet. Sometimes this requires two or more sequential inspections with different magnetizations.

(d) A second step—demagnetization—following inspection is often necessary.

(e) Post-cleaning, to remove remnants of the magnetic particles clinging to the surface, may sometimes be required after testing and demagnetizing.

(f) Odd-shaped parts sometimes present a problem with respect to how to apply the magnetizing force to produce a field in the proper direction.

(g) Exceedingly heavy currents are sometimes required for the testing of very large castings and forgings.

(h) Care is required to avoid local heating and burning of highly finished parts or surfaces at the points of electrical contact.

(i) Individual handling of parts for magnetization is usually necessary, a disadvantage particularly with large numbers of small parts.

(j) Although the indications are easily seen, experience and skill in interpreting the significance of the indication is sometimes needed. Compared to some other methods such as radiography, ultrasonics and eddy-currents, however, interpretation of magnetic particle indications is far simpler in most cases.

10. COMPARISON WITH OTHER METHODS. There are four other major nondestructive testing methods commonly in use for the detection of surface and subsurface discontinuities in ferromagnetic materials. In determining which of them to use in any given case, the inspector should know the relative merits and limitations of the several methods. The following brief comparison will be helpful in arriving at a proper choice when such a selection is required:

(a) *Radiography* is superior to magnetic particle testing in most cases for the location of discontinuities which lie wholly below the surface. It is not nearly so effective, however, for locating surface cracks, in which *magnetic particles excel all other methods,* on magnetic materials. The magnetic particle method is usually faster and less costly to apply than is radiography, especially when 100% inspection of numerous articles is required, and the results are immediate. It is also less hampered by the shape of the part than is radiography.

(b) *Penetrants* are the equal of magnetic particles for the location of surface cracks, provided the cracks are clean and are not already filled with foreign substances. Because of this limitation in the case of the penetrant method, and the fact that it is often slower, the magnetic particle method, which produces immediate results, is to be preferred on magnetic material for locating surface discontinuities in most cases. For locating discontinuities on the interior of the metal, penetrants are of course completely unsuitable, whereas on magnetic materials, magnetic particles can do a remarkably good job under favorable circumstances and controlled conditions.

(c) *Ultrasound* is inferior to the magnetic particle method for the location of exceedingly shallow surface cracks. There is *no known minimum limit* to the depth of cracks magnetic particles will detect, provided proper materials and techniques are used. Ultrasound, on the other hand, requires a certain minimum depth—although this is very small—in order to get sound wave reflection. In the case of below-the-surface discontinuities ultrasound is in general superior, as it will locate under favorable conditions, discontinuities at great depths below the surface. However, magnetic particles are less handicapped by irregular shapes than is ultrasound.

(d) *Eddy-currents* are subject to limitations similar to ultrasound for the location of surface cracks, since a finite depth is required to unbalance the impedance bridge and secure readable indications. Eddy current methods in general are better adapted to the testing of non-magnetic than magnetic materials. Eddy currents, in conjunction with magnetic saturating bias fields, are sometimes better for the location of below-the-surface discontinuities than magnetic particles, but they are more limited in the depth they can penetrate. They can usually indicate only discontinuities lying close to the surface. Magnetic particles are less hampered by size and shape than are eddy currents.

Because the magnetic particle method excels all others for finding very shallow surface cracks in ferromagnetic materials, it is *the best method* to use for location of fatigue cracks in parts made from such materials. Fatigue cracks, which occur in parts highly stressed and subject to stress variations in service, almost invariably start at the surface and propagate inwards through the section. If these fatigue cracks can be located early in their progress, a breakdown can be avoided and the part can often be salvaged. Because of its tremendous ability in locating such fine shallow surface fatigue cracks, the magnetic particle method is used in preference to all others for overhaul or maintenance inspection of machinery of all kinds.

CHAPTER 3

SOURCES OF DEFECTS

1. GENERAL. Nondestructive testing processes produce indications of conditions in metals and other materials by indirect means, and such indications must be interpreted to determine the condition which is causing them. The problem of the testing engineer is, in this respect, something like that of the physician in medical diagnosis. He must decide correctly from the symptoms what the disease (condition) is that fits the symptoms. To do this with certainty, the physician or surgeon must be thoroughly familiar with the human body, with its processes and with the diseases it is subject to; and he must also know what the symptoms are that these diseases produce.

Similarly, the nondestructive testing engineer must know the nature of metals (or other materials he may be called upon to test); how they are made, how they are worked into finished form, and the discontinuities that may be produced by each step in their manufacture. Such knowledge and familiarity with manufacturing processes gives the inspector, in advance, information very helpful in alerting him as to what sort of defects may or may not be present, and where they are likely to occur. As an aid to successful inspection by nondestructive testing means, there is no substitute for the knowledge of "what to look for and where to look for it". Magnetic particle testing is no exception.

This and the following chapter are devoted to the subject of defects in iron and steel and how they are produced; and to a discussion of how metals fail and how such defects contribute to failure. It seems proper to present this information as a preliminary to the description of how defects are actually found with magnetic particle testing.

2. SOME DEFINITIONS. There are several synonyms for the word "defect", and in nondestructive testing these words are not at all interchangeable. On the contrary they have specific connotations that must be recognized by those working in this field. These words are "blemish", "flaw" and "defect". The dictionary definitions are:

70

Blemish. Imperfection. Blemish suggests something superficial such as a blot or a stain.

Flaw. A small defect in continuity or cohesion, such as a crack or break—or a departure from perfection.

Defect. The lack or want (not always visible) of something essential to completeness or perfection.

In nondestructive testing these definitions acquire some specific meanings not clearly conveyed by the dictionary language.

A part may have a blemish or a flaw in it and still not be "defective" in the sense that its usefulness for its intended purpose will be impaired. A spot or stain would have no effect on the service performance of a connecting rod of an automobile engine and would certainly not make it defective; but on a mirror, for example, such a spot might make the mirror unuseable and would then unquestionably make the mirror defective.

3. WHAT IS A DEFECT? In nondestructive testing language the word "defect" is correctly applied only to a condition which will interfere with the safe or satisfactory service of the particular part in question.

To avoid confusion, we use a fourth word—"discontinuity". This word covers the condition before it is determined whether it is a defect or not. The cause of magnetic particle indications is in all cases a discontinuity—whether physical or magnetic. And if we exclude those discontinuities that are present by design and consider only those present in the metal by accident or as the result of some manufacturing process, these may still not in all cases make the part defective in the sense that its service performance will be affected unfavorably.

We come, therefore, to the conclusion that a discontinuity is not necessarily a defect. It is a defect only when it will interfere with the performance of the part or material in its intended service. And a discontinuity which may make one part defective may be entirely harmless in another part designed for a different service. Further — as in the case of the mirror — what would be only a blemish in most cases may become a defect in a product in which appearance is a major factor in acceptable service use.

So we should be careful to refer to a discontinuity as a defect only when it makes the specific part in which it occurs unsuitable for the purpose for which it was designed and manufactured.

It is difficult to adhere strictly to this precept, but its significance should never be lost sight of. "Discontinuity" is a long word and difficult to say—whereas "defect" is short and easy. Even in this book "defect" may be used in a loose sense on occasions when "discontinuity" would be the correct term.

But the nondestructive testing inspector should always understand that a "discontinuity is a defect only when it interferes with the service for which the part was intended".

4. MAGNETIC DISCONTINUITIES. Magnetic particle indications are produced by *magnetic* discontinuities in the metal which is being examined. These discontinuities may not always be actual physical breaks in the continuity of the metal. Magnetic discontinuities may be produced by causes other than actual breaks or flaws in the metal. Therefore a magnetic particle indication does not necessarily show the presence of a defect. *Magnetic* discontinuities may be caused by:

(a) An actual *metallic* discontinuity or rupture at or near the surface of a part, which may have been present in the original metal, or may have been produced by subsequent forming, heating or finishing processes.

(b) Actual *metallic* discontinuities which are, however, present by design—as for example, a forced fit between two members of an assembly.

Fig. 16—Typical Magnetic Particle Indication of Cracks.

72

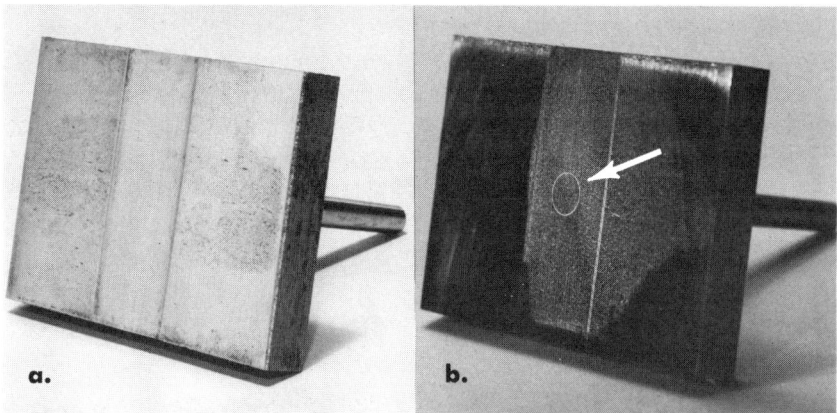

Fig. 17—Magnetic Particle Indication of a Forced Fit. a) White Light View.
b) Fluorescent Particle Indication.

(c) The junction between two dissimilar ferromagnetic metals
having different permeabilities; or between a ferromagnetic
metal and a non-magnetic material. Indications may be
produced at such a point even though the joint is perfectly
sound. Such an indication may be produced in a friction or
flash weld of two dissimilar metals.

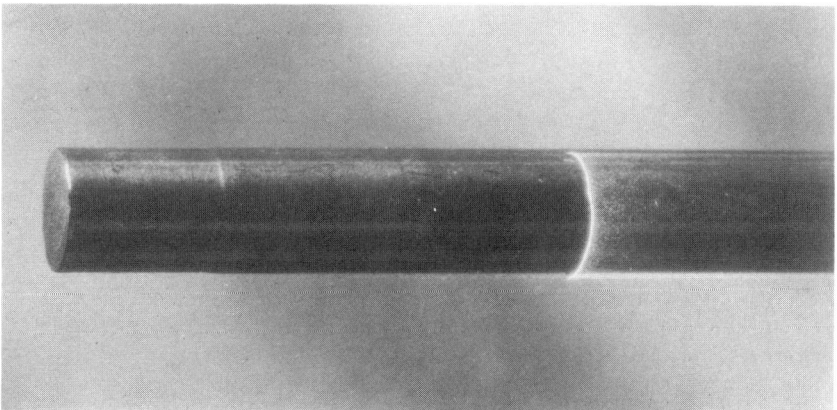

Fig. 18—Magnetic Particle Indication at the Weld Between
a Soft and a Hard Steel Rod.

(d) The junction between two ferromagnetic metals by means
of non-magnetic bonding materials, as in a brazed joint.
An indication will be produced though the joint itself may
be perfectly sound.

73

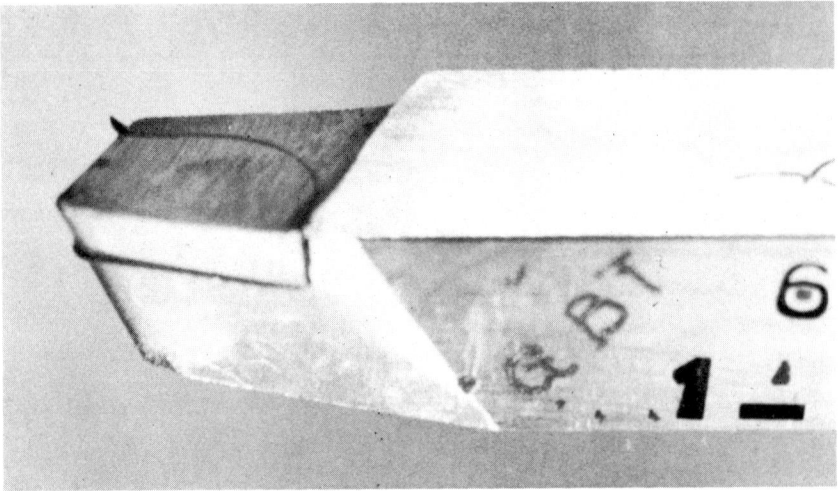

Fig. 19—Magnetic Particle Indication of the Braze Line of a Brazed Tool Bit.

(e) Segregation of the constituents of the metal, where these have different permeabilities—as, for example, low carbon

Fig. 20—Magnetic Particle Indications of Segregations.

74

areas in a high carbon steel, or areas of ferrite, which is magnetic, in a matrix of stainless steel which is austenitic and therefore non-magnetic.

5. CLASSES OF DISCONTINUITIES. There are a number of possible ways of classifying discontinuities that occur in ferromagnetic materials and parts. One broad grouping which is useful in Penetrant testing is that based on location—whether surface or subsurface. (See Chapter 20 for a full discussion). This grouping is also useful in magnetic particle testing, since the ability of this method to find members of these two groups varies sharply. But beyond this, the classification is too broad to be very useful.

Another possible system is to classify discontinuities by the processes which produce them. Although such a system is too specific to be suitable for all purposes, it is used extensively. We speak of forging defects, welding defects, heat-treating cracks, grinding cracks, etc. Practically every process, from the original production of metal from its ore, down to the last finishing operation, can and does introduce discontinuities which magnetic particle testing can find. It is therefore important that the nondestructive testing engineer or inspector be aware of all of these potential sources of defects.

6. CONVENTIONAL CLASSIFICATION SYSTEM. For many years it has been customary to classify discontinuities for the purpose of magnetic particle testing (as well as for other nondestructive testing methods) according to their source or origin in the various stages of production of the metal, its fabrication and its use.

Broadly these are referred to as follows:

(a) Inherent—Produced during solidification from the liquid state.

(b) Processing—Primary.

(c) Processing—Secondary, or finishing.

(d) Service.

7. INHERENT DISCONTINUITIES. This group of discontinuities is present in metal as the result of its initial solidification from the molten state, before any of the operations to forge or roll it into useful sizes and shapes have begun. The names of these inherent discontinuities are given and their sources described below.

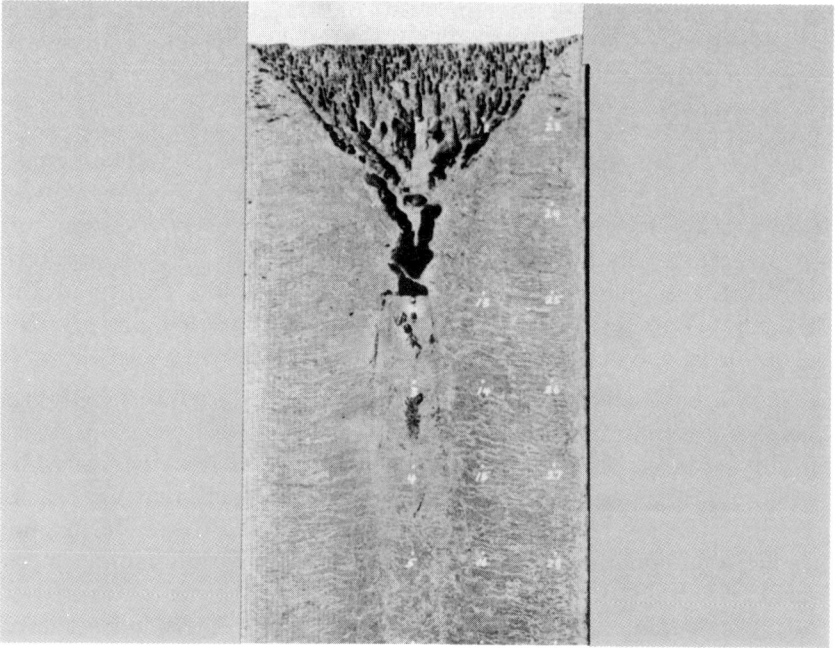

Fig. 21—Cross-section of Ingot Showing Shrink Cavity.

(a) *Pipe.* As the molten steel which has been poured into the ingot mold cools, it solidifies first at the bottom and walls of the mold. Solidification progresses gradually upward and inward. The solidified metal occupies a somewhat smaller volume than the liquid, so that there is a progressive shrinkage of volume as solidification goes on. The last metal to solidify is at the top of the mold, but due to shrinkage there is not enough metal to fill the mold completely, and a depression or cavity is formed. This may extend quite deeply into the ingot. See Fig. 21. After early breakdown of the ingot into a bloom, this shrink cavity is cut away or cropped. If this is not done completely before final rolling or forging into shape, the unsound metal will show up as voids called "pipe" in the finished product. Such internal discontinuities, or pipe, are obviously undesirable for most uses and constitute a true defect. Special devices ("hot tops") and special handling of the ingot during pouring and solidifying can, to a large degree, control the formation of these shrink cavities.

(b) *Blowholes.* As the molten metal in the ingot mold solidifies there is an evolution of various gases. These gas bubbles rise through the liquid and many escape. Many, however, are trapped as the metal freezes. Some, usually small, will appear near the surface of the ingot; and some, often large, will be deeper in the metal, especially near the top of the ingot.

Many of these blowholes are clean on the interior and are welded shut again into sound metal during the first rolling or forging of the ingot; but some, near the surface, may have become oxidized and do not weld. These may appear as seams in the rolled product. Those deeper in the interior, if not welded shut in the rolling, may appear as laminations.

(c) *Segregation.* Another action that takes place during the solidification of the molten metal is the tendency for certain elements in the metal to concentrate in the last-to-solidify liquid, resulting in an uneven distribution of some of the chemical constituents of the steel as between the outside and the center of the ingot.

Various means have been developed to minimize this tendency, but if for any reason severe segregation does occur, the difference in permeability of the segregated areas may produce magnetic particle indications. Unless severe, such segregation is generally not deleterious.

(d) *Non-metallic Inclusions.* All steel contains more or less matter of a non-metallic nature. The origin of such matter is chiefly the de-oxidizing materials added to the molten steel in the furnace, the ladle or the ingot mold. These additions are easily oxidizable metals such as aluminum, silicon, manganese and others. The oxides and sulphides of these additions constitute the bulk of the non-metallic inclusions.

When finely divided and uniformly distributed, such non-metallic matter does not usually injure the steel. However, it sometimes gathers into large clumps which, when rolled out, become long "stringers." These stringers in some cases are objectionable. When such a string occurs at the surface or just under the surface of a highly stressed part or a bearing surface, it may lead to fatigue cracking. In general, non-metallic inclusions in steel seldom constitute a real defect, though they are often indicated with magnetic particles. Non-metallic inclusions are sometimes added to steel intentionally. The addition of lead or sulphur to steels for the purpose of improving their machinability is common practice. Such steels will show excessive amounts of non-metallic inclusions, which serve to break up the chips when the metal is turned or otherwise machined. Machining time is reduced and tool life is lengthened.

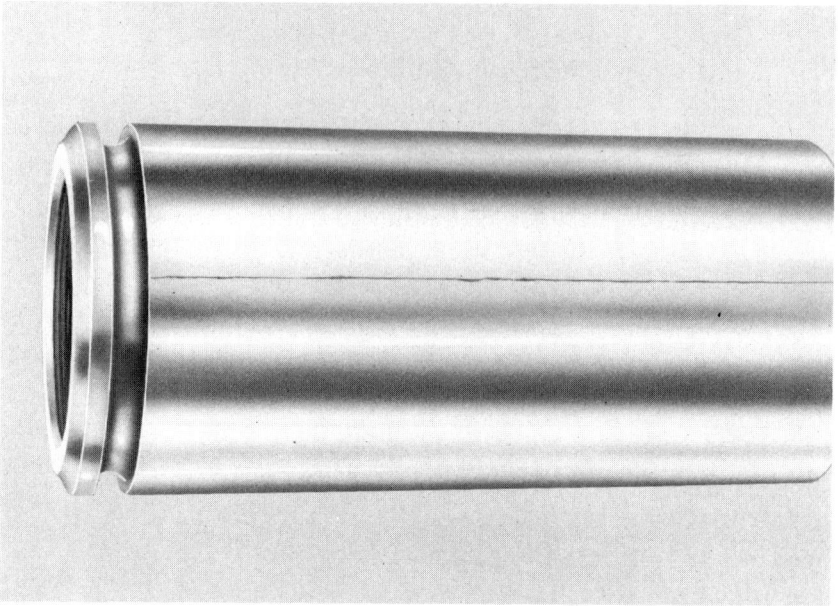

Fig. 22—Magnetic Particle Indication of a Sub-surface Stringer of Non-Metallic Inclusions.

Such steels, if tested with magnetic particles, may show alarming looking patterns, which have no significance as defects. The magnetic particle operator must be familiar with this type of steel. Though serving a useful purpose in their proper field, these steels should never be used for critical or highly stressed parts, or parts subject to fatigue in service.

(e) *Internal Fissures.* Because of the stresses set up in the ingot as the result of shrinkage during cooling, internal ruptures may occur which may be quite large. Since no air normally reaches the surfaces of these internal bursts, they may be completely welded shut during rolling or other working, and leave no discontinuity. If there is an opening from the fissure to the surface, however, air will enter and oxidize the surfaces. In such a case welding does not occur, and they will remain in the finished product as discontinuities.

(f) *Scabs.* When liquid steel is first poured into the ingot mold there is considerable splashing or spattering up and against the cool walls of the mold. These splashes solidify at once and become oxidized. As the molten steel rises and the mold becomes filled, these splashes will be reabsorbed to a large extent into the metal. But in some cases they will remain as scabs of oxidized metal adhering to the surface of the ingot. These may remain and appear on the surface of the rolled product. If they do not go deeply into the surface they may not constitute a defect, since they may be removed on machining. Figure 23 illustrates this condition on a rolled bloom.

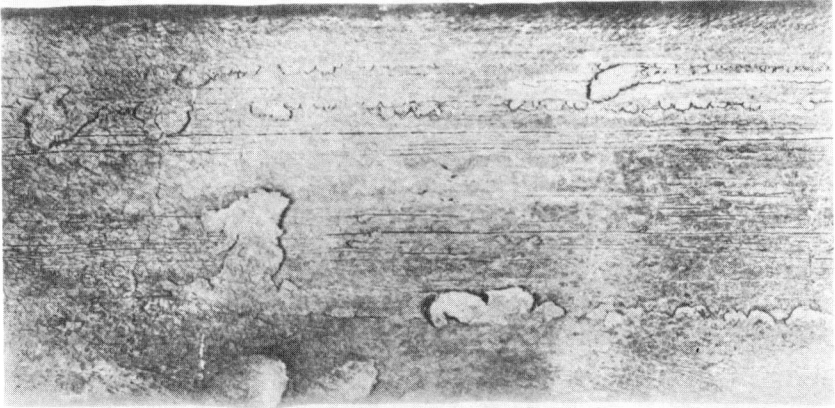

(Courtesy U.S. Steel Corporation)
Fig. 23—Scabs on the Surface of a Rolled Bloom.

(g) *Ingot Cracks.* Surface cracking of ingots occurs due to surface stresses generated during cooling of the ingot. They may be either longitudinal or transverse, or both types may occur together. As the ingot is worked into billets by rolling, these cracks form long seams. Inspection of quality-product billets for seams of this type with magnetic particles is now common practice in modern mills. This permits removal of the seams by flame scarfing, chipping or grinding without waste of good metal. If not removed before further rolling these deeper seams appear, greatly elongated, on finished bars and shapes, often making them unsuitable for many purposes.

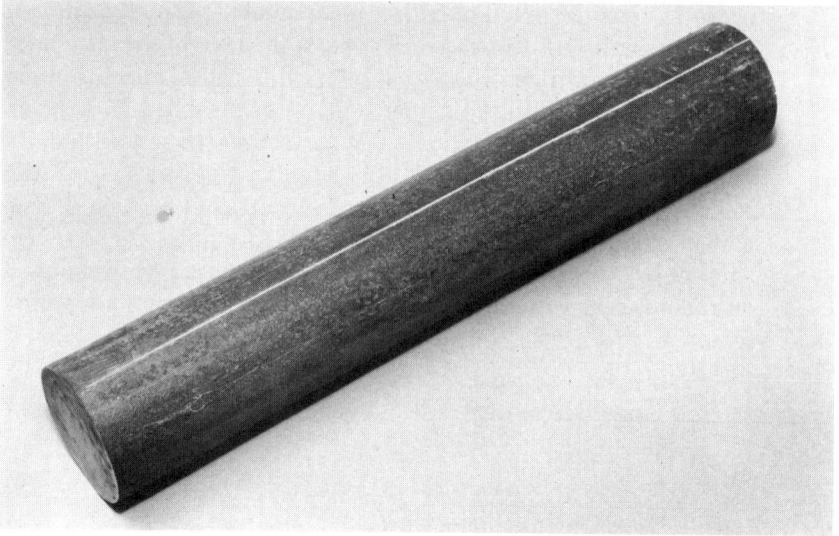

Fig. 24—Seam on a Bar Shown by Magnetic Particles.

8. PRIMARY PROCESSING DISCONTINUITIES. When steel ingots are worked down into usable sizes and shapes such as billets and forging blanks, some of the above described inherent defects may appear. But the rolling and forging processes may themselves introduce discontinuities which in many cases constitute defects. Primary processes as here considered, are those which work the metal down, by either hot or cold deformation, into useful forms such as bars, rod and wire, and forged shapes. Casting is another process usually included in this group since, though it starts with molten metal, it results in a semi-finished product. Welding is similarly included for similar reasons.

A description of the discontinuities introduced by these primary processes follows:

(a) *Seams.* Seams in rolled bars or drawn wire are usually highly objectionable, and often down-grade the product and make it unusable for first quality purposes. As has just been

80

described, severe seams may originate from ingot cracks, but by proper cleaning up (conditioning) of the billets by scarfing, grinding or chipping these can be eliminated before final rolling. Conditioning is now usually aided by use of magnetic particle testing to indicate the length and severity of the seams. See Chapter 19. If properly conditioned at the billet stage, seams from *this* source need not appear in the final rolled product.

But seams can be introduced by the rolling or drawing processes themselves. Laps can occur in the rolling of the ingot into billets as the result of over-filling of the rolls. This produces projecting fins, which on subsequent passes are rolled into the surface of the billet or bar. When severe, the billet often cannot be salvaged and is downgraded.

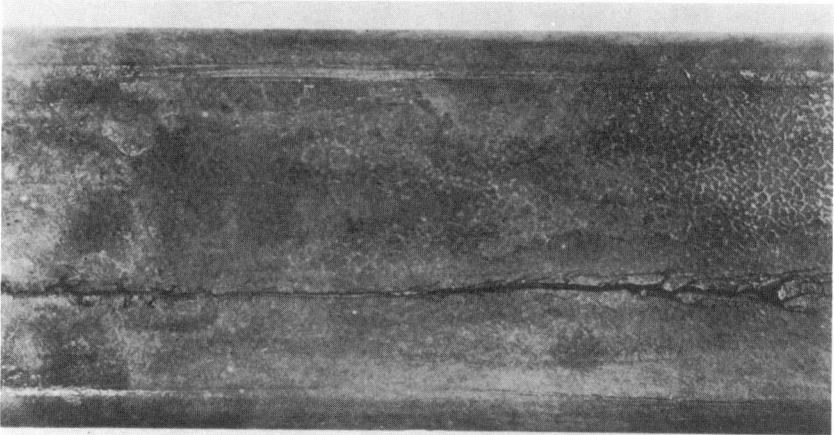

(Courtesy U.S. Steel Corporation)
Fig. 25—Surface of a Steel Billet Showing a Lap.

Similarly, even when billets which have been conditioned, and are free of seams, are rolled into bars or rod, laps resulting from over-filled rolls can occur, producing long and often very deep seams in the finished product. In similar fashion, under-fills in the rolling process may, on subsequent passes be squeezed to form a seam, which often runs the full length of the bar. Seams derived from laps will usually emerge to the surface of the bar at an acute angle. Seams derived from the folds produced by an under-filled pass are likely to be more nearly normal to the surface of the bar. Seams or die marks may also be introduced in the drawing process due to defective dies. Such seams may or may not make the product defective. For some purposes, such as springs or bars for heavy upsetting, the most minute surface imperfections (or discontinuities) are cause for rejection. For others, where for example machining operations are expected to remove the outer layers of metal, seams which are not too deep will be machined away.

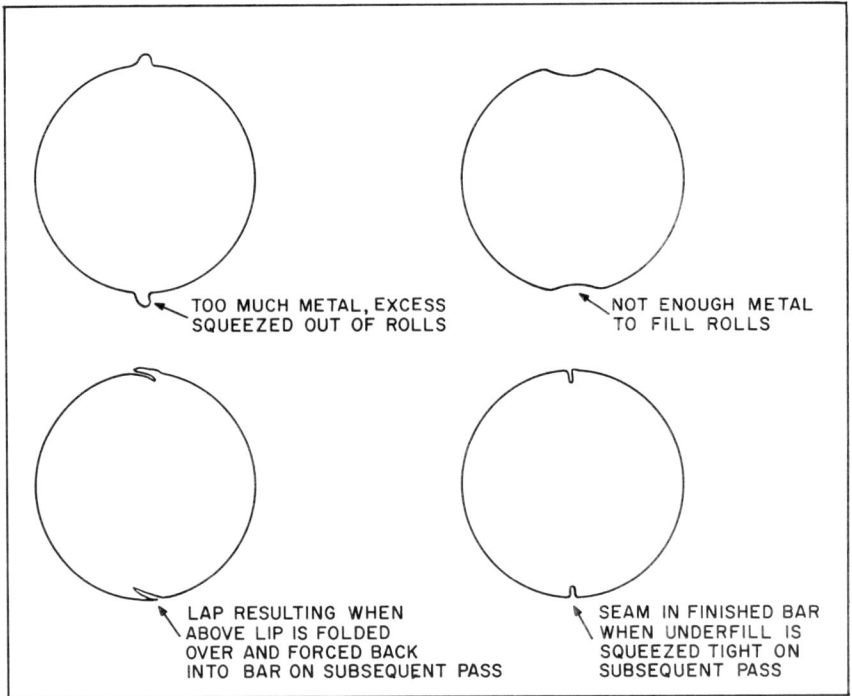

TOO MUCH METAL, EXCESS
SQUEEZED OUT OF ROLLS

NOT ENOUGH METAL
TO FILL ROLLS

LAP RESULTING WHEN
ABOVE LIP IS FOLDED
OVER AND FORCED BACK
INTO BAR ON SUBSEQUENT PASS

SEAM IN FINISHED BAR
WHEN UNDERFILL IS
SQUEEZED TIGHT ON
SUBSEQUENT PASS

Fig. 26—How Laps and Seams are Produced by the Rolls. Over-fills and Under-fills.

(b) *Laminations.* Laminations in rolled plate or strip are formed when blowholes or internal fissures are not welded tight during rolling, but are enlarged and flattened into sometimes quite large areas of horizontal discontinuities. Laminations may be detected by magnetic particle testing on the cut edges of plate, but do not give indications on plate or strip surfaces, since these discontinuities are internal and lie in a plane parallel to the surface. Ultrasonic mapping techniques are used to define them.

CHAPTER 3
SOURCES OF DEFECTS

Fig. 27—Magnetic Particle Indications of Laminations Shown on Flame
Cut Edge of Thick Steel Plate.

(c) *Cupping.* This is a condition created when, in drawing or
extruding a bar or shape, the interior of the metal does not
flow as rapidly as the surface. Segregation in the center of
the bar usually contributes to the occurrence. The result is a
series of internal ruptures which are severe defects whenever
they occur. They may be indicated with magnetic particles,
but only if the ruptures are large and approach the surface
of the bar. The cupping problem can be minimized by
changing die angles.

(d) *Cooling Cracks.* When alloy and tool steel bars are rolled
and subsequently run out onto a bed or table for cooling,
stresses may be set up due to uneven cooling which can be
severe enough to crack the bars. Such cracks are generally
longitudinal, but not necessarily straight. They may be quite
long, and usually vary in depth along their length. Figure
29 shows the magnetic particle indications of such a crack,
and also sections through the crack at three points where the
depth of the crack is different. The magnetic particle indication varies in intensity, being heavier at points where the
crack is deepest.

Fig. 28—a) Fluorescent Magnetic Particle Indications of Severe Cupping in Drawn Spring Stock, Ground into the Ruptures at One End. b) Section Through Severe Cupping in 1⅜-inch Bar.

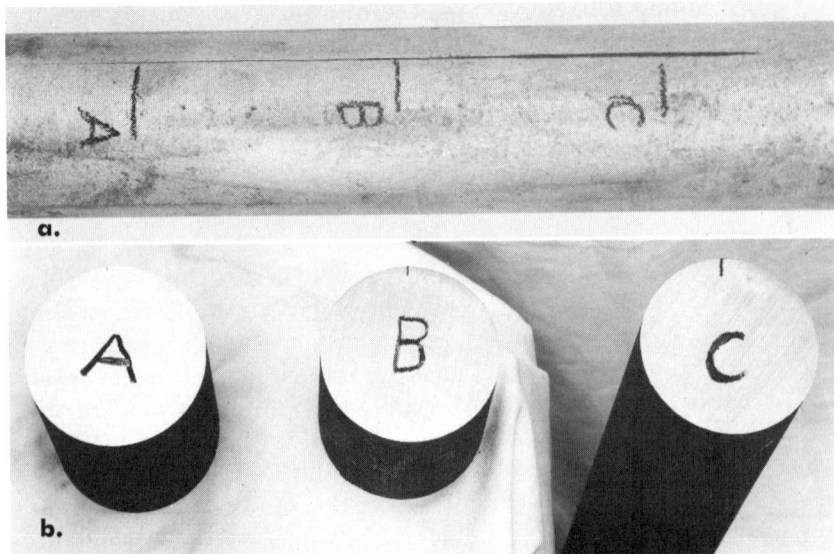

Fig. 29—Magnetic Particle Indications of Cooling Cracks in an Alloy Steel Bar. a) Surface Indications. b) Cross Sections Showing Depth.

84

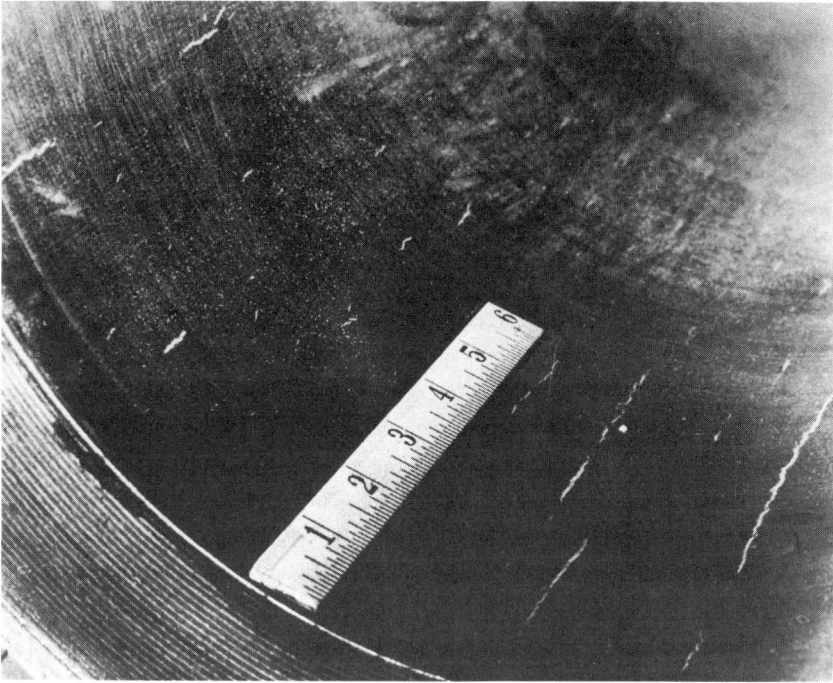

Fig. 30—Magnetic Particle Indications of Flakes in the Bore of a
Large Hollow Shaft.

(e) *Flakes.* Flakes are internal ruptures that may occur in steel
as the result of too rapid cooling. It is believed that the re-
lease of dissolved hydrogen gas during the cooling process
causes these ruptures, and that controlled slow cooling after
forging or otherwise hot-working the metal will reduce their
occurance. Flaking usually occurs in fairly heavy sections
and certain alloys are more susceptible than others. Figure
30 shows magnetic particle indications of flakes which have
been exposed on a machined surface. Since these ruptures
are deep in the metal—usually half way and more from the
surface to the center of the section—they will not be shown
by magnetic particle testing on the original surface of the
part.

(f) *Forging Bursts.* When steel is worked at improper temperatures it is subject to cracking or rupturing. Too rapid or too
severe a reduction of section can also cause bursts or cracks.
Such ruptures may be internal bursts, or they may be cracks

85

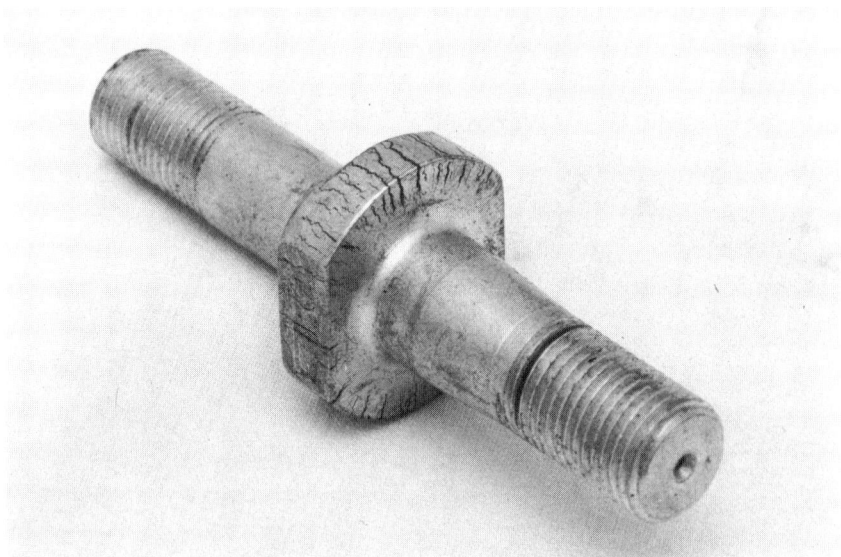

Fig. 31—Magnetic Particle Indications of Forging Cracks or Bursts
in an Upset Section. Severe Case.

on the surface. When on the surface they are readily found
by magnetic particle testing. If interior, they are usually
not shown except when they have been exposed by machining.

(g) *Forging Laps.* As the name implies, forging laps or folds
are formed when, in the forging operation, improper handling of the blank in the die causes the metal to flow so as
to form a lap which is later squeezed tight. Since it is on the
surface and is oxidized, this lap does not weld shut. This

Fig. 32—Cross Section of a Forging Lap. Magnified 100X.

86

type of discontinuity is sometimes difficult to locate, because it may be open at the surface and fairly shallow, and often may lie at only a very slight angle to the surface. In some unusual cases it also may be solidly filled with magnetic oxides.

(h) *Burning.* Overheating of forgings, to the point of incipient fusion, results in a condition which renders the forging unusable in most cases, and is referred to as burning. Actual oxidation is however not the real source of the damage, but rather the partial liquidation due to heat, of material at the grain boundaries. Burning is a serious defect but is not generally shown by magnetic particle testing.

(i) *Flash Line Tears.* Cracks or tears along the flash line of forgings are usually caused by improper trimming of the flash. If shallow, they may "clean up" if machined, and do not make the part defective; or, they may be too deep to clean up and in such cases the forging cannot be salvaged. Such cracks or tears can easily be found by magnetic particles.

Fig. 33—Magnetic Particle Indication of Flash Line Tear in an Automotive Spindle Forging. Partially Machined.

(j) *Castings.* Steel and iron castings are subject to a number of defects which magnetic particle testing can easily detect. Surface discontinuities are formed in castings due to stresses resulting from cooling, and are often associated with changes in the cross section of the part. These may be hot tears or they may be shrinkage cracks which occur as the metal cools down. Sand from the mold, trapped by the hot metal, may form sand inclusions on or near the surface of castings. Gray

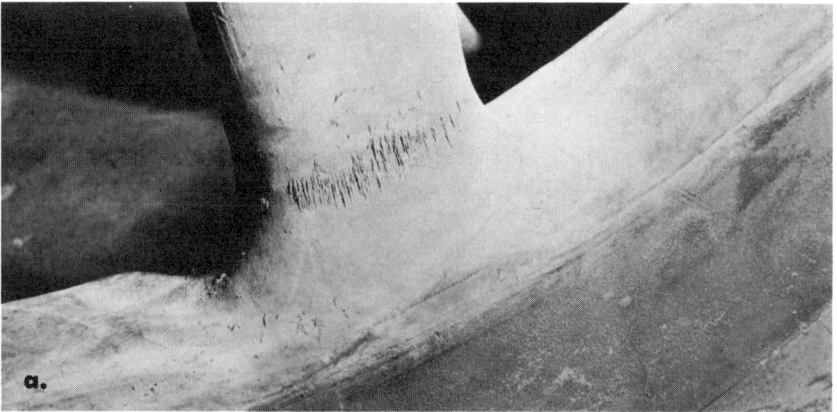

Fig. 34 — Magnetic Particle Indications of Defects in Castings. a) Surface Indications of an Internal Shrink Cavity. b) Handling Crack in Gray Iron Casting.

iron castings may be quite brittle, and are often cracked—usually at thin sections—during the "shake-out" or by rough handling during sorting.

(k) *Weldments.* A number of kinds of discontinuities may be formed during welding of both thin and heavy sections. Some are at the surface and some are in the interior of the metal. Some of the defects peculiar to weldments are lack of penetration, lack of fusion, undercutting, cracks in the weld metal, crater cracks, cracks in the heat affected zone, etc. These defects and their detection will be discussed in detail, along with castings, in Chapter 24.

9. SECONDARY PROCESSING OR FINISHING DISCONTINUITIES. In this group are those discontinuities associated with the various finishing operations, after the part has been rough-formed by rolling, forging, casting or welding. Discontinuities may be introduced by machining, heat treating, grinding and similar processes. These are described below:

(a) *Machining Tears.* These are caused by dragging of the metal under the tool when it is not cutting cleanly. Soft and ductile low carbon steels are more susceptible to this kind of damage than are the harder, higher carbon or alloy types. Machining tears are surface discontinuities and are readily found with magnetic particles.

(b) *Heat Treating Cracks.* When steels are heated and quenched to harden them, or are otherwise heat treated to produce desired properties for strength or wear, cracking may occur if the operation is not correctly suited to the material and the shape of the part. Most common are quench cracks, caused when parts are heated to high temperatures and then suddenly cooled by immersing them in some cool medium, which may be water, oil or even air. Such cracks often occur at locations where the part changes section—from light to heavy—or at fillets or notches in the part. The edges of keyways and the roots of splines or threads are likely spots to watch for quenching cracks. Cracks may also result from too rapid heating of the part, which may cause uneven expansion at changes of cross-section, or at corners where heat is absorbed—from three sides—more rapidly than in the body of the piece. Corner cracking may also occur during quenching, because of more rapid heat loss at such locations.

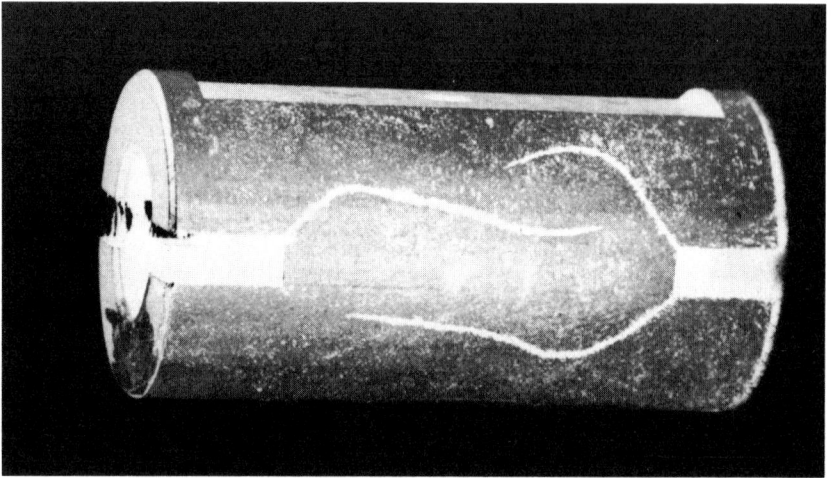

Fig. 35—Magnetic Particle Indications of Quenching Cracks, Shown with Dry Powder.

Heat treating cycles can be designed to minimize or eliminate such cracking, but for critical parts, testing with magnetic particles is a safety measure usually applied, since such cracks are serious and their detection presents no difficulty.

(c) *Straightening Cracks.* The process of heat treating often causes some warping of the part due to slight unevenness in the cooling during quenching. A hardened shaft, for example, may come from the heat treat operation not quite straight. In many cases these can be straightened in a press, but if the amount of bend required is too great, or if the shaft is very hard, cracks may be formed. These, again, are very readily found with magnetic particles.

(d) *Grinding Cracks.* Surface cracking of hardened parts as the result of improper grinding is frequently a source of trouble. Grinding cracks are essentially thermal cracks and are related to quenching cracks in more ways than one. They are caused by stresses set up by local heating under the grinding wheel. They are in nearly all cases avoidable if proper wheels, proper cuts and proper coolants are used, and if wheels are properly dressed when required. But since proper grinding requires constant attention and care which is not always provided in practice, these defects do occur. Since they are sharp surface cracks they are easily located with magnetic

90

particles, even if shallow, and locating them is usually of vital importance. Salvage of the cracked part is seldom possible since grinding is usually a final precision finishing operation. The best use of magnetic particle testing in this case is to conduct sampling tests to monitor the grinding operation, and thereby control it to avoid the formation of grinding cracks.

Fig. 36—Fluorescent Magnetic Particle Indications of Typical Grinding Cracks.

Hardened surfaces often retain internal stresses from the quenching operation which are not severe enough to cause cracking at the time of quenching. During grinding however, the relatively small increment of stress set up by local heating under the grinding wheel may cause rupture when added to the residual stress already present. Such surfaces usually crack severely and extensively, as illustrated in Fig. 37.

(e) *Etching and Pickling Cracks.* Hardened surfaces which contain residual stresses may be cracked if they are pickled or etched in acid. Attack by the acid of the surface fibres of the metal gives the internal stress a chance to be relieved by the formation of a crack. Before this action was fully understood, the heat treatment of the part was often blamed for the cracking, when such cracking actually occurred during acid cleaning for plating or for other purposes. The heat treat operation did, however, deserve some of the blame, by leaving the part with locked up internal residual stresses.

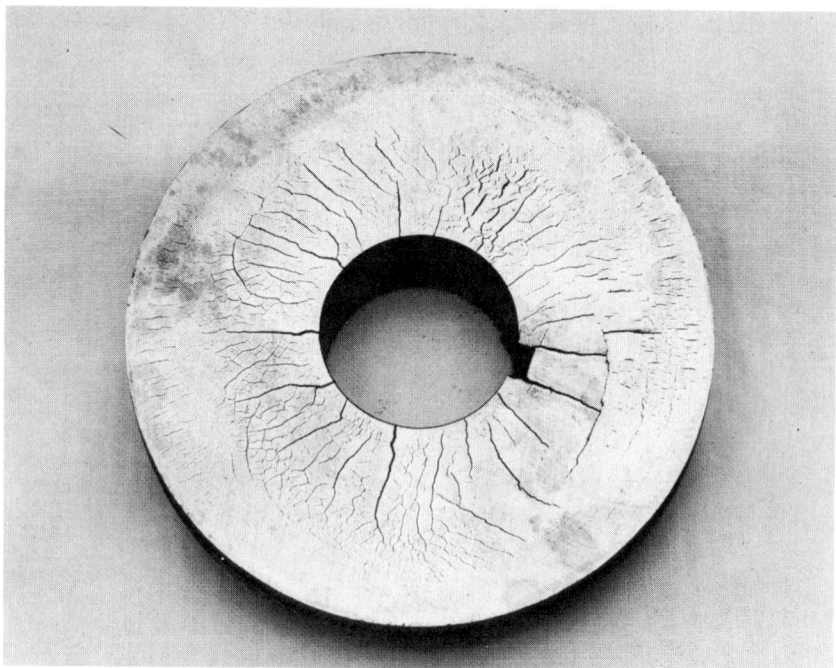

Fig. 37—Magnetic Particle Indications of Grinding Cracks in a
Stress-Sensitive, Hardened Surface.

Fig. 38—Magnetic Particle Indications of Plating Cracks.

(f) *Plating Cracks.* When hardened surfaces are to be electroplated, care must be taken to ensure that pickling (or other cleaning operations preparatory to plating) does not produce cracks. Sometimes cracks are formed during the plating operation itself.

Residual stresses leading to etching or plating cracks may also be the result of cold work. Spiral springs, cold wound, then pickled for plating or hot galvanizing, have also shown such cracks. The hot galvanizing process itself may also produce cracks in surfaces containing residual stresses. Penetration of the hot zinc between the grain boundaries during the hot dip process provides points for relief of such stresses by the formation of cracks. Copper penetration during brazing may result in similar cracking if the parts contain residual stresses. Molten alloys from the bearing of a railroad axle journal during a "hot box" will penetrate the surface of the heated journal and provide the starting point for a fatigue crack and axle failure.

10. SERVICE CRACKS. The fourth major classification of discontinuities comprises those which are formed or produced after all fabrication has been completed and the part has gone into service. The objective of magnetic particle testing to locate and eliminate discontinuities during fabrication, is to put the part into service free from defects. However, even when this is fully accomplished,

Fig. 39—Magnetic Particle Indication of a Typical Fatigue Crack.

failures in service still occur as a result of cracking caused by service conditions.

(a) *Fatigue Cracks.* The subject of fatigue cracking will be thoroughly discussed in the following Chapter, 4. As a source of discontinuities, the phenomenon of fatigue is a prolific one. Metals which are subjected to alternating or fluctuating stresses above a certain critical level (the fatigue strength) will eventually develop cracks, and finally fracture. Fatigue cracks, even very shallow ones, can readily be found with magnetic particles, and the part often can be salvaged.

(b) *Corrosion.* Parts which are under tension stress in service and are at the same time exposed to corrosion from whatever cause, may develop cracks at the surface, referred to as stress corrosion cracking. Such cracks, under continuing corrosion and stress (whether reversing, or fluctuating or not) will progress through the section until failure occurs. When corrosion is added to fatigue-producing service conditions, this type of service failure is called corrosion fatigue.

(c) *Overstressing.* Parts of an assembly in service that are stressed beyond the level for which they were designed are very likely to crack or break. Such over-stressing may occur as the result of an accident; or a part may become over-

Fig. 40—Fluorescent Magnetic Particle Indications of Cracks in Crankshaft of Small Aircraft Engine, Damaged in Plane Accident.

94

loaded due to some unusual or emergency condition not anticipated by the designer; or a part may be loaded beyond its strength because of the failure of some related member of the structure.

After complete failure has occurred magnetic particle testing obviously has no application as regards the fractured part. But other parts of the assembly which may appear undamaged may have been overstressed during the accident or overloading from other causes. Examination by magnetic particle testing is usually carried out in such cases to determine whether any cracks have actually been formed. With this precaution salvage of good parts after the assembly failure is often possible.

11. OTHER SOURCES OF DEFECTS. In this chapter an attempt has been made to familiarize the reader with most of the common sources of defects (discontinuities) which occur in iron and steel. Actually the list as given here is nowhere near complete. But the inspector working with magnetic particle testing will encounter these discontinuities which have been described more frequently than those from less common or more obscure conditions. He will often have the metallurgical laboratory of the plant available for consultation, and the metallurgist will usually be able to assign a cause to an indicated discontinuity and assess its importance.

CHAPTER 4

HOW AND WHY METALS FAIL

1. GENERAL. One of the principal reasons for the existence of nondestructive testing is the need to find and eliminate defects in parts or assemblies so as to prevent their failure after they have been fabricated and put into service. In order to apply these methods effectively, the nondestructive testing engineer or the inspector should know how and why metals fail. Those who must accept and reject parts that show magnetic particle indications of discontinuities (or indications by any other nondestructive testing method) would be handicapped if they did not know the metallic weaknesses that can lead to the several types of metal failure.

This knowledge is essential to an accurate appraisal of the effect which the detected condition is likely to have on the performance of the part in service. Two other factors are equally essential—an accurate knowledge of what the service requirements are going to be, and an accurate interpretation of the actual condition present, which the nondestructive test had indicated.

2. METAL FAILURE. Ever since metals were discovered and useful implements made from them they have been subject to failure. Metal axes broke under impact, swords broke off at the hilt in combat, and when metal axles were used for wheels instead of wood, they too often broke under heavy loads. Naturally such failures would always occur at the most inconvenient times—that is, when a little extra load was being imposed during some critical operation or service incident. A sword that broke in battle was likely to result in serious consequences for the wielder!

It has been the problem of the ages to try to make metal parts that would not break in service; and in every age, the advances in design have been pushing the frontiers of knowledge for this purpose.

3. EARLY ATTEMPTS TO AVOID FAILURE. The natural and obvious remedy, when a part broke, was to conclude that the part must be weak and must be "beefed up"—made bigger and therefore (it was assumed) stronger. Someone then invented the term "safety factor"

96

—more recently called with some justification the "factor of ignorance"—to describe this beefing-up process.

Knowledge of the distribution of the stresses within a part, both dynamic and static, was not very extensive or even accurate. The means for studying the effects which stress distribution had on the service performance of a part had not yet been devised. Today we have several highly accurate methods of experimental stress analysis, and these have helped immeasurably in studies that have lead to an understanding of metal failure. The most important bit of knowledge coming out of these studies was the realization that "bigger" is not necessarily "stronger".

Today the factor of safety has acquired a new significance. The designer can now use a much higher proportion of the ultimate strength of metals with assurance, since by using the methods of stress analysis, he can know in advance what and where the maximum stresses in the part will be. In his design he can then avoid concentrations above those that his selected factor of safety will permit. Today, factors of less than two are often employed instead of factors of 3, 4 and 5 which were formerly the rule. Even factors of less than one have been used when load cycles are very few in number. Furthermore, this low safety allowance is as much a cushion against the possibility of unusual over-stresses in the service life of the structure or part, as it is an allowance for unknown metallic defects that may occur. This is because nondestructive testing methods, if intelligently applied and properly interpreted can almost entirely eliminate the danger of defective components being allowed to go into service.

4. METALLURGY. Metallurgy in the early days was in the field of the Alchemist, and his rituals and strange formulae for making metals strong were shrouded in mystery. But the early fabricator of metals certainly had learned a thing or two about metallurgy, even if he did not know what he was doing. Witness the fine steel that was forged into the weapons and armor of the middle ages—heat treated, hardened and tempered to exactly the proper point.

The addition of the methods of metallography (the study of the structure of metals with the aid of the microscope) sixty or more years ago, gave metallurgists the tool they needed to learn about and understand the part played by metallic microstructure in the strength and working characteristics of metals. This technique was first brought to a high state of usefulness by Dr. Albert Sauveur, working at Harvard University, about 1910. Many, if not most, of the famous names appearing in the story of the intensive expansion of our knowledge of metals which has occurred during the succeeding years, are those of men who studied under Dr. Sauveur.

5. STRENGTH VERSUS FAILURE IN METALS. Studies of the internal construction of steels and other metals quickly brought understanding of the basic factors involved in strength or weakness, ductility or brittleness, softness or hardness. And with this understanding, a new approach to the problem of how—and why—metals fail became possible. It was quickly realized that strength under dynamic loading was quite different from strength under static conditions. Discontinuities or other flaws that were apparently harmless in static service were found to be major factors leading to failure under service conditions that involved stress variations. Merely making metals stronger by heat treatment or by addition of alloying elements did not avoid service failures, and in fact often had the opposite effect.

Today, the study of the various modes of failure and the conditions leading to their occurrence has given us the understanding that is needed to design parts that will give failure-free performance in service. This result is not only possible, but realizable, provided the importance of the service warrants the cost involved to accomplish it. Required is the choice of the right metal or alloy and its proper heat treatment for the expected service conditions; complete stress analysis studies to make certain that no undue stress concentrations occur in any part if unexpected loads occur; and full use of the methods of nondestructive testing to give assurance that each part is free of defects which would introduce weaknesses and stress concentrations not allowed for by careful design.

6. HOW METALS FAIL. Two major modes of fracture of steel are recognized. These are:
  (a) Cleavage types, usually called brittle fractures, in which failure occurs without plastic flow.
  (b) Shear types, usually referred to as ductile fractures, in which plastic flow occurs before fracture.

In both these types of fracture the failure line cuts across grain boundaries—that is, follows slip planes within the grains of the metal. However, these slip planes are on different axes of the crystal in the cases of the two types of fracture.

Fig. 41—Examples of Silky and Crystalline Fractures.
a) Ductile or Silky. b) Brittle or Cleavage.

The appearance of the surface of the fracture in these two modes is entirely different. The fracture surface of a shear or ductile fracture is fine-grained or "silky", whereas in the case of cleavage fractures the surface is crystalline in appearance. Fracture by propagation along grain boundaries also occurs but is less common and involves rather special conditions. (See Section 23 and 24, this Chapter.)

7. CONDITIONS LEADING TO FAILURE. The following list enumerates the most common of the recognized causes of failure in iron and steel:

(a) Simple over-stressing, due to overloading.
(b) Impact, induced by sudden and rapid stress rise.
(c) Fatigue, induced by stress variations below the elastic limit.
(d) Corrosion, induced by corrosion of stressed metal.
(e) Creep, induced by prolonged exposure to high temperature under static loading.

These will be discussed in the following sections.

8. OVERSTRESSING. It is self-evident that if steel is stressed beyond its strength it will pull apart and break. Overloads sufficient to produce this result may occur in service from several causes.

Unusual conditions—such as an aircraft suddenly encountering excessively violent clear air turbulance—can induce stresses which are severe enough to pull the plane's structural members apart. Or, one member of an assembly may break, due to fatigue or some other cause, and throw additional load onto other members, creating stress too high for the latter to withstand. Sudden stoppage of a vehicle, as in an accident, may also overstress some member of the frame or the engine beyond its strength, and cause parts to break.

Failures caused by overstressing are usually of the ductile type, although brittle fractures can also occur from simple over-stressing in the case of very hard metals, or at low temperatures.

The point at which the actual fracture originates may be determined by the presence of some local stress-raiser, such as a nick or a scratch in the surface of the part, even though this condition is not the primary cause of the failure.

9. IMPACT. Impact is actually another form of over-stressing. The difference is principally in the rate at which the stress is applied. A sudden blow or impact can impose a very high stress at an exceedingly rapid rate, and failure occurs in the brittle mode. Such a brittle fracture may occur even in normally ductile materials, if the rate of rise of stress is fast enough.

10. EFFECT OF TEMPERATURE. Brittle fractures at ordinary temperatures are most likely to occur only in hard materials which normally break with little or no plastic flow preceding fracture. However, temperature has a great deal to do with the mode of fracture of nearly all steels and alloys.

Fig. 42—Brittle Fracture Which Occurred in a Tank Holding Gas at Low Temperature.

At low temperatures soft steels which are normally ductile will fail in the brittle mode, even without fast-rising stress application. And at certain elevated temperatures, such steels may become "hot short" and fail in the brittle mode on over-stressing. In both cases, however, failure is more apt to occur when the stress is applied rapidly, as by impact.

The composition of steels and alloys has a great influence on their behavior under low temperature loading conditions. The degree of "notch sensitivity" of the steel is a major factor. Notch sensitivity is defined as "a measure of the reduction in strength of a metal caused by the presence of stress concentration."* Steels having a low notch sensitivity will withstand conditions favorable for brittle failure better than those having high notch sensitivity. Alloys

---

*A.S.M. Metals Handbook, 8th Edition, Vol. I, Page 26.

normally in the austenitic state at ordinary temperatures or temperatures below normal—stainless steels of many types are austenitic—have a very low notch sensitivity and are not as likely to fail in the brittle mode, even at very low temperatures, as are ordinary carbon or alloy steels.

11. FATIGUE FAILURES. In the early efforts to understand why metals broke in service, fatigue failures were very puzzling. It was observed that a part in service when subjected to reversing or widely varying stresses often broke suddenly, even though these stresses were well below the elastic limit of the metal. Examination of the surface of the fracture showed a portion of it to be smooth, with peculiar concentric markings, while the remainder was crystalline. This immediately led to the assumption that when metal became "tired" after long service, it "crystallized" and then broke with a brittle type fracture.

After metallographic methods were employed, studies under the microscope showed that there was no readily discernable difference between the structure of the metal at the break and of metal some distance away. And it was not long before the true explanation of this partly crystalline fracture was arrived at—that is, the

(Courtesy Sydney Walker, ARO Engineering Corp.)
Fig. 43—Face of a Typical Fatigue Crack.

fatigue crack first started and spread very slowly. During this period the faces of the crack "worked" against each other to produce the smooth part of the fracture. Progress of the crack came in small increments, each advance leaving its circular mark to give the "oyster shell" appearance to the surface. When the remaining sound part of the section finally broke it broke suddenly, producing the crystalline portion.

12. STRESS-RAISERS. Examination of the smooth portion of the fracture face almost always showed these ring-like markings, which appeared to have a common center at some point at the original surface of the part. This point was clearly the point at which the fatigue crack actually started its (usually) slow progress across the section. It was further observable that this point of origin could in most cases be associated with some small surface defect, such as a nick, scratch or tool mark.

Experimental stress analysis methods showed that the tension stress across such a surface mark could be several times greater than the nominal average stress over the surface. Even though the defect was small, this multiplication of the local stress level across the defect was high enough to start the crack. For example, as a loaded railroad car axle rotates, the surface stress is reversed with each turn, from highly positive to highly negative. Presence of a nick or circumferential scratch will furnish the stress concentration necessary to start a fatigue crack. The stress reversals as the axle rotates causes the faces of a starting fatigue crack to work against each other as the crack advances. Eventually the remaining sound portion of the axle breaks.

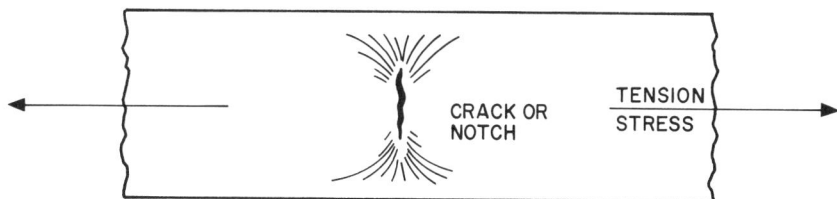

Fig. 44—Stress Concentration at a Surface Notch or Crack.

13. DESIGN STRESS-RAISERS. Further stress analysis studies showed that various features of design could also act as stress-raisers, producing stress concentrations far above the average for the surface. Stresses at a hole, for example, can be three times the average tension stress for the surface. Stiffening members intended

Fig. 45—Stress Concentration at a Hole in a Surface Stressed in Tension, Shown by the Brittle Coating Method.

to strengthen the part often turn out to be stress-raisers of high magnitude, defeating their purpose entirely. Fatigue cracks were often observed starting at such locations, due to these stress concentrations.

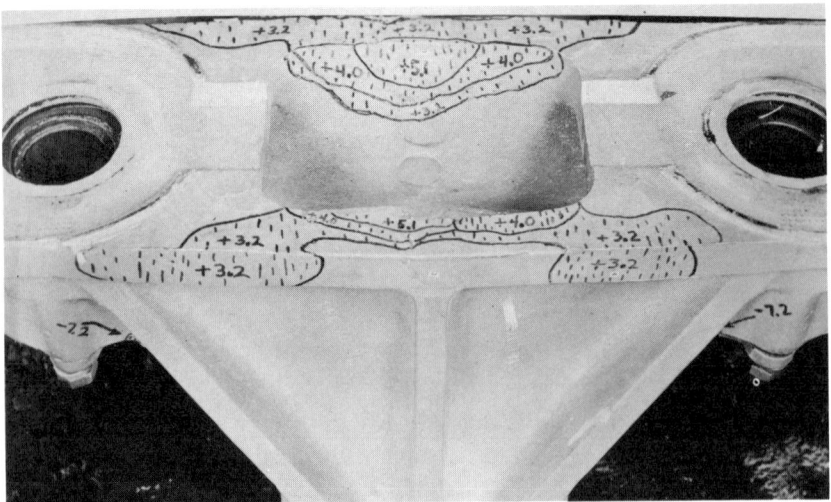

Fig. 46—Stress Concentration Pattern on a Ribbed Casting, Shown by the Brittle Coating Method.

CHAPTER 4
HOW AND WHY METALS FAIL

14. FATIGUE OF METALS. Research and experience thus gradually explained the "how" of fatigue. The "why" is still not altogether clear, but new methods of studying what happens inside the grains of steel under fluctuating or reversing stress is today by way of giving this answer also. From the point of view of the nondestructive testing engineer, however, the "how" is of more immediate importance than the "why".

Tremendous numbers of tests were made to produce failures in the laboratory under controlled conditions. Testing machines were devised which permitted test specimens to be subjected to large numbers of stress reversals under various levels of maximum stress. It was learned that for many metals there was a definite relation between stress level and the number of cycles before a fatigue crack was formed (or before fracture occurred). The R. R. Moore rotating-beam type of fatigue testing machine was at first the most common means used for such testing, since with it the number of stress reversals could easily be run into the millions in a relatively short period of time. Also, it reproduced the conditions of shafts and axles which so frequently were the victims of fatigue failures. Other types of testing machines which produced tension stress variations

(Courtesy Wiedeman Division, Warner & Swasey Co.)
Fig. 47—The R. R. Moore Rotating-Beam Fatigue Testing Machine.

or reversals in other ways, often on actual structural components, gave added information.

15. FATIGUE STRENGTH. From this mass of data, certain relationships emerged which gave experimental means for predicting the fatigue life of various types of steels and alloys. These relationships are called "Fatigue Strength", "Fatigue or Endurance Limit", "Fatigue Ratio" and "Fatigue Life".

Fatigue Strength* is defined as "the maximum stress that can be sustained for a specified number of cycles without failure, the stress being completely reversed within each cycle, unless otherwise stated".

Fatigue or Endurance Limit is "the maximum stress below which a material can presumably endure an infinite number of stress cycles".

Fatigue Life is "the number of stress cycles that can be sustained prior to failure for a stated test condition".

Fatigue Ratio is "the ratio of the fatigue limit for cycles of reversed flexural stress, to the tensile strength".

16. FATIGUE TESTING. To determine the values of fatigue strength and fatigue limit for any given steel requires running many tests, often going to many millions of cycles. To get the true values for the steel under test, it is necessary that the test specimen be free from flaws that would act as stress raisers and thereby would reduce the fatigue values below what they would be for a flaw-free specimen. Even the corrosive effect of the water vapor or oxygen in the air can affect the results, and some tests have been run in a vacuum or an atmosphere of inert gas. Acceptable values for the fatigue properties had, therefore, to be the average of many runs made on test specimens as free from all stress raisers or flaws as careful selection made possible.

Today, fatigue limits for most of the ordinary steels and alloys are readily available to designers.

17. DESIGNING FOR FATIGUE. With known values for fatigue strength and fatigue limits it is theoretically possible to produce a part for a given service that will not fail in fatigue. Knowing the number of cycles expected during the service life of the part, the

---

*These definitions are quoted from the A.S.M. Metals Handbook, Vol. I, 8th Edition, Page 16.

designer has merely to set a maximum stress limit which is below the fatigue limit of the steel he intends to use.

As a practical matter, however, this is extremely difficult to achieve. Even though the use of experimental stress analysis methods indicates that no *design* stress-raisers are present in the part as designed, some of the various types of discontinuities discussed in Chapter 3 are almost certain to be present. Cracks, scratches, nicks, seams, near-to-surface stringers of non-metallic inclusions or large interior inclusions—all these are stress raising hazards which cannot be satisfactorily allowed for in advance, even by the use of a large factor of safety. This is because the size, location and stress-raising severity of these probably-occurring defects cannot be known in advance.

The methods of nondestructive testing are the means by which such stress-raising conditions can be eliminated. This requires the careful inspection of each part by one or more nondestructive tests suitable for the conditions involved. For iron, steel and other ferromagnetic materials, the magnetic particle method is the simplest and most reliable for locating and evaluating most of these discontinuities which would act as damaging stress-raisers.

18. INSPECTING FOR FATIGUE CRACKS. Fatigue cracks almost invariably start at the surface, induced by the presence of some sort of stress-raiser. They are most likely to occur in areas of high tension stress. Such areas include holes, notches, sharp changes of section, stiffening members, etc. The most likely places to look for fatigue cracks, therefore, are at sharp changes of section such as fillets, keyways, splines, roots of screw threads and gear teeth, etc. On crankshafts, the most common point for fatigue cracks to appear is at the fillets where the main or throw bearings join the cheeks of the crank.

Fatigue cracks just starting may be very shallow and very short, but such indications must not be ignored, since in most cases the conditions which caused the crack to start will cause it to propagate to failure. Depth of an isolated crack can often be estimated by its length, since propagation tends to be uniform in all directions from the point of origin. The "oyster shell" markings on a fatigue fracture usually consist of uniform circular concentric lines and clearly indicate the point of origin. Multiple fatigue cracks are often encountered on shafts and railroad axles. In these cases, many shallow cracks have started and later "joined up", sometimes making a

Fig. 48—Fatigue Cracks Starting Out of a Sharp Fillet on a Racing
Stock Car Safety Hub.

(Courtesy Sydney Walker, ARO Engineering Corp.)
Fig. 49—Fracture Through a Small Fatigue Crack Showing Circular Shape of the
Smooth Portion.

ring of cracks completely around the axle or shaft. In this case the apparent length of the crack obviously does not indicate the depth.

19. FATIGUE IN TORSION. When a part is being stressed in torsion, the maximum tension stresses occur in a direction 45° to the torsional axis. The most common machine part in this category is the helical wire spring. Fatigue cracks in such springs will have a direction 45° to the axis of the *wire*. Longitudinal seams in the wire from which the spring was wound will therefore lie at 45° to the direction of *maximum tension stress*, and invariably form a nucleus for the start of fatigue cracks. Wire (or the rod from which the wire is drawn) intended for coiling into springs, such as valve springs, the service load variations of which are severe and rapid, is usually inspected with extreme care to eliminate wire with even very fine surface seams, pits or other stress raisers.

Fig. 50—Fatigue Cracks at 45° to the Axis of the Wire on a Helical Spring.

20. EXPERIMENTAL STRESS ANALYSIS. The methods of experimental stress analysis have been repeatedly referred to in the preceding discussions. There are several useful procedures in this field. A brittle coating (Stresscoat*) applied to the surface before stressing gives, when the part is stressed, a pattern of stress location and distribution, and the direction of maximum stress. Photo-elastic methods employ models made from certain plastics which, when stressed and examined with polarized light, show stress distribution and concentrations visually. Electric resistance strain gauges provide means for measuring surface stresses at points which the shape of the part and the nature of the service may have indicated. The maximum stresses, the location of which brittle coating or photoelastic tests have determined, can then also be measured with great accuracy by use of these strain gauges.

Fig. 51—Brittle Coating Stress Analysis Kit.

By making a proto-type of the first design of a new casting and subjecting it to stress analysis by the brittle coating method, the areas of stress concentration can be mapped and changes made in the design until the stress distribution approaches uniformity, or at least is free of serious concentrations. Figure 53 shows an example of redesign by this means. The final design, at right, is lighter, stronger, and much easier to cast.

---

*Stresscoat. Registered in the United States Patent Office. Property of Magnaflux Corporation.

110

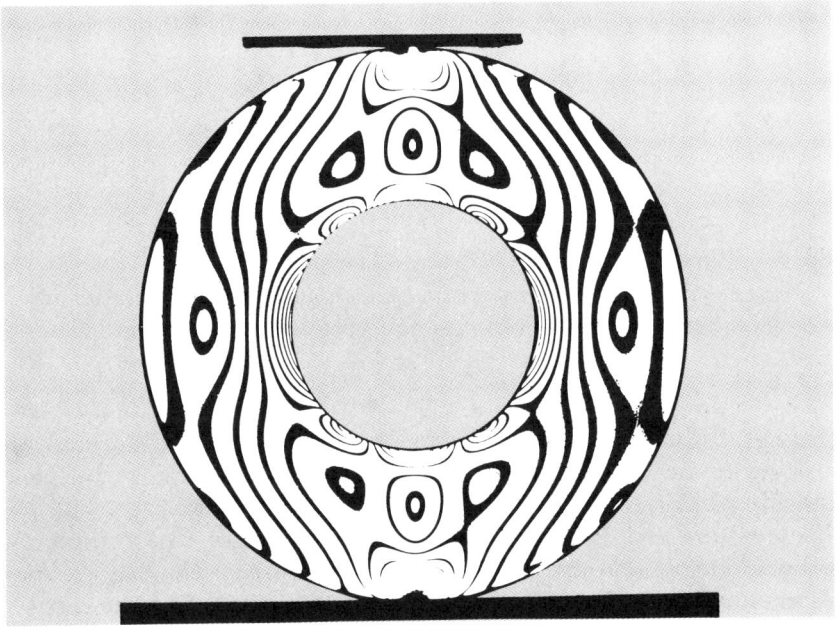

(Courtesy Prof. Wm. H. Murray, M.I.T.)
Fig. 52—Stress Pattern Shown in Small Ring Made of a Photo-Elastic Resin. The Ring is Stressed in Compression and Photographed with Transmitted Polarized Light.

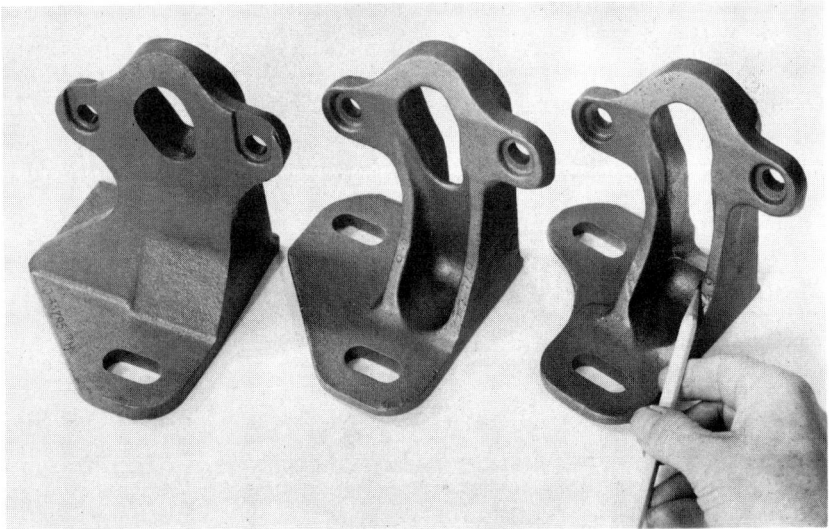

Fig. 53—Progressive Improvement of Casting Design Resulting from Use of Brittle Coating Stress Analysis.

21. RATE OF PROPAGATION OF FATIGUE CRACKS. Once a fatigue crack has started, it will usually continue to progress through the section until failure occurs. Fortunately, the rate of propagation through steel is usually relatively slow, especially if stress levels do not rise far above design limits. It is therefore usually possible to detect such cracks in their early stages by inspection at intervals that may be separated by several months or even longer. Failure in service can thereby be prevented.

Fatigue cracks in general propagate much more rapidly in non-ferrous metals such as aluminum and bronze, than they do in most steels.

22. SALVAGE OF PARTS SHOWING FATIGUE CRACKS. Regular inspection of machine or other parts liable to fatigue failure makes possible the early detection of cracks when still very shallow, as has just been said. In such cases the cracks can often be ground or "blended" out and the part safely returned to service. The amount of metal which it is necessary to remove to eliminate the crack completely must not, of course, go beyond the minimum dimension permitted for the section involved.

23. CORROSION. When metal is under tension stress and is at the same time subject to corrosion, cracks develop and progress through the section until failure occurs. The cracking is due to corrosive attack on the grain boundary material, so that propagation of this type of cracking is *inter*-granular, as opposed to fatigue cracking, which is *trans*-granular.

Stress corrosion cracking will proceed in stressed members even under static loading conditions. The stresses favoring such corrosive

Fig. 54—Corrosion Fatigue Cracks on an Oil Well Sucker Rod.

attack need not be externally applied, but may be residual stresses from some previous operation such as cold-forming, heat treating or welding.

The rate of propagation of stress corrosion cracking may be increased if the stress is variable. Under such circumstances the phenomenon is called corrosion fatigue. Parts operating under high reversing stresses and subject to corrosion at the same time will fail much more rapidly than under simple fatigue. All values for fatigue limits are useless when corrosion enters the picture. There is no predictable endurance limit in the case of corrosion fatigue.

Magnetic particle testing will readily locate the presence of corrosion fatigue cracks.

24. CREEP. Steel under tensile load and at high temperatures undergoes slow changes that may eventually end in failure. The metal elongates in the direction of the tensile stresses. This phenomenon is called "creep". The greater the load and the higher the temperature, the greater is the amount and rate of creep.

Creep data is available for many steels for design purposes, although the time required for making a series of creep tests at various stress levels and at various temperatures can be extremely long. Progress of creep is very slow indeed, especially at the lower temperatures and stresses. Such tests have shown, however, that there is no "creep limit" corresponding to the fatigue endurance limit, and in design, stresses and temperatures must be adjusted so that failure will not occur during the service life of the part or structure.

When cracking occurs under creep conditions the cracks progress along the grain boundaries as in corrosion cracking. Cracks due to creep can be readily detected with magnetic particle testing.

25. SUMMARY. The discussion in this chapter is not presented as a complete coverage of the subject of metal failure. It is intended merely to give the magnetic particle testing operator or inspector some knowledge of the more common ways in which metals fail, to aid him in his examination of parts that have been in service, and to increase his awareness of the importance of defect detection by nondestructive testing means. The information given in this and the preceding chapter will also be of help in the problem of "what to look for and where to look for it".

CHAPTER 5

DEFINITIONS OF SOME TERMS USED IN MAGNETIC PARTICLE TESTING

1. NEED FOR DEFINITIONS. For the individual who undertakes to apply magnetic particle testing for the detection of flaws, certain terms commonly used in connection with the method should be defined, and be understood by him. He need not understand the vast amount of theory, nor be familiar in detail with all the areas of physics and electrical engineering with which many of these terms are concerned. But he must understand their significance as they apply to, and are used in connection with, magnetic particle testing.

The definitions as given are not put in highly technical terms, but are worded with particular reference to their use in the language of magnetic particle testing.

2. GROUPS OF TERMS. The terms defined fall into four groups. These are:

(a) Terms relating to magnetism.
(b) Terms relating to electricity.
(c) Terms relating to electro-magnetism.
(d) Terms relating specifically to magnetic particle testing.

The definitions are arranged in alphabetical order under each group.

3. TERMS RELATING TO MAGNETISM.

(1) *Magnetism.* Magnetism is a property of matter that was known to the ancients. The iron oxide mineral, magnetite, (or lodestone) exhibited, in its natural state, the property of attracting bits of iron to itself. Fragments also had the ability of aligning themselves in a north and south direction, and were the first compasses used for navigation. Later scientific investigators showed that all matter, including liquids and gasses, was affected by magnetism in one way or another, but to widely varying degrees. Some materials such as iron were strongly attracted to a magnet—others much less so. The ability of matter to attract other matter to itself in this manner is called magnetism.

(2) *Coercive Force.* See Hysteresis.

(3) *Demagnetization.* The process of removing the magnetism existing in a part.

(4) *Ferromagnetic materials.* With most metals and other materials, the attraction or repulsion when under the influence of a magnet is very slight. A few materials, notably iron and steel, and cobalt and nickel, are attracted strongly. These materials are said to be ferromagnetic. In magnetic particle testing we are concerned only with ferromagnetic materials.

(5) *Flux Density.* By this term is meant the number of flux lines per unit of area, taken at right angles to the direction of the flux. It is the measure of magnetic field strength.

(6) *Gauss.* This is the unit of flux density or induction. The strength of field induced in a ferromagnetic body is described as being so many Gausses. It is usually designated by the letter "B". Numerically, one Gauss is one line of flux per square centimeter of area.

(7) *Hall Effect.* An effect used in the measurement of magnetic fields. Assume an electric current being passed through a thin rectangular metal plate. The lines of flow are parallel to the edges of the rectangle. Then pairs of symmetrical points, one on each of these edges, will be at the same potential. (See Fig. 55) Now let a uniform magnetic field be applied

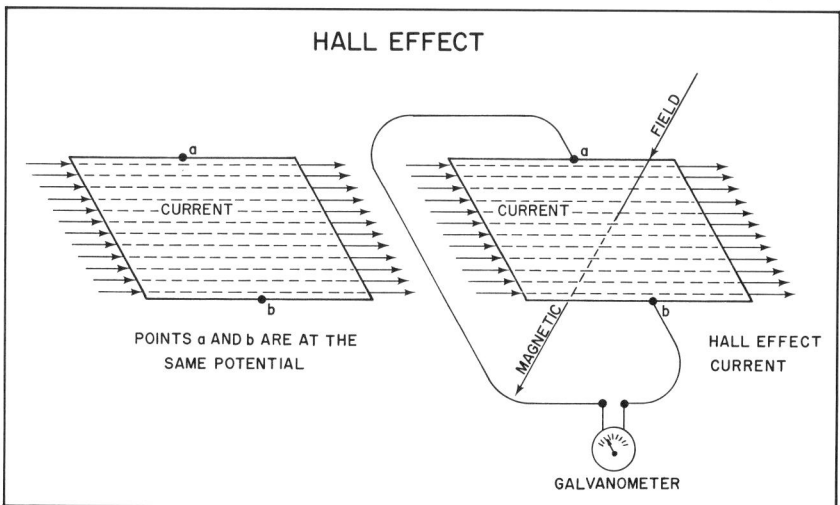

Fig. 55—The Hall Effect.

115

at right angles to the plane of the plate. These pairs of points will no longer be at the same potential, and if any pair be connected through a galvanometer, a transverse current will flow. The amount and direction of current so set up is determined by the strength and direction of the applied magnetic field. Transverse difference of potential is directly proportional to the longitudinal current and to the magnetic field, and inversely to the thickness of the plate.

By controlling all the other variables, the strength and direction of the magnetic field can be measured with accuracy.

(8) *Horse-shoe Magnet.* A bar magnet, bent into the shape of a horse-shoe so that the two poles are adjacent. Usually the term applies to a permanent magnet.

(9) *Hysteresis.* When an unmagnetized piece of iron is exposed to a gradually increasing magnetizing force, and the strength of the induced field in the iron, "B", is plotted against the magnetizing force, "H", a curve like that shown in Fig. 56

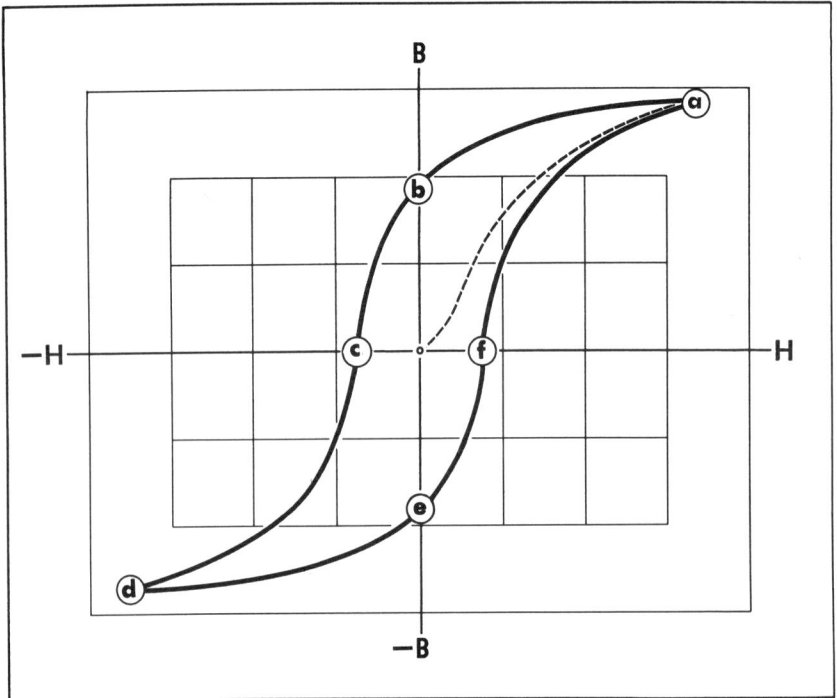

Fig. 56—Hysteresis Curve.

is produced. As H is increased, B also increases along the line "oa" to a point at which further increase in the magnetizing force, H, produces no further increase in the flux density other than that which would occur in non-magnetic material or air. This point, "a", on the curve is called the *Saturation Point* and the iron is said to be magnetically *saturated* at this point.

As the magnetizing force is then reduced, the flux density does not decrease to zero by the same curve, but lags behind, so that when H reaches zero there is still some flux in the piece of iron. This field is called the *Residual Field*, and on the curve of Fig. 56 is the value of B at which the curve crosses the B axis. This is point "b" on the curve.

As the magnetizing force is further decreased, now in the negative direction, the value of B decreases till it crosses the H axis at point "c". The value of negative or reverse magnetizing force necessary to bring the flux density back to zero after saturation, and demagnetize the piece—distance "co" on the H axis—is called the *Coercive Force*.

As the magnetizing force is further decreased in the negative direction the iron reaches a point of negative saturation ("d"), and as H is then increased in the positive direction the curve passes the point "e" on the B axis, the negative residual field. The point "f" on the H axis is equal and opposite to the coercive force "c". Further increase of H brings the field strength B back again to the point of saturation, "a", thus completing the curve.

The curve "abcdefa" is called the *Hysteresis Curve,* or *Hysteresis Loop.* It gives much information about the magnetic characteristics of a ferromagnetic material, and the terms just defined will be much used in this discussion.

(10) *Induction.* Magnetic induction is the magnetism induced in a ferromagnetic body by some outside magnetizing force.

(11) *Lines of Force.* When a piece of paper is laid over a magnet, and iron filings or other iron powder is sprinkled over the paper the powder arranges itself into a pattern as shown in Fig. 57. This pattern is called a magnetograph. It appears to consist of a series of curved lines and suggests that the magnetic force of the field flows along these lines. Although

there is actually no known movement, it is most convenient to think of the field as flowing through the magnet and through the air around it, following the path of these lines. They are called *Lines of Force.*

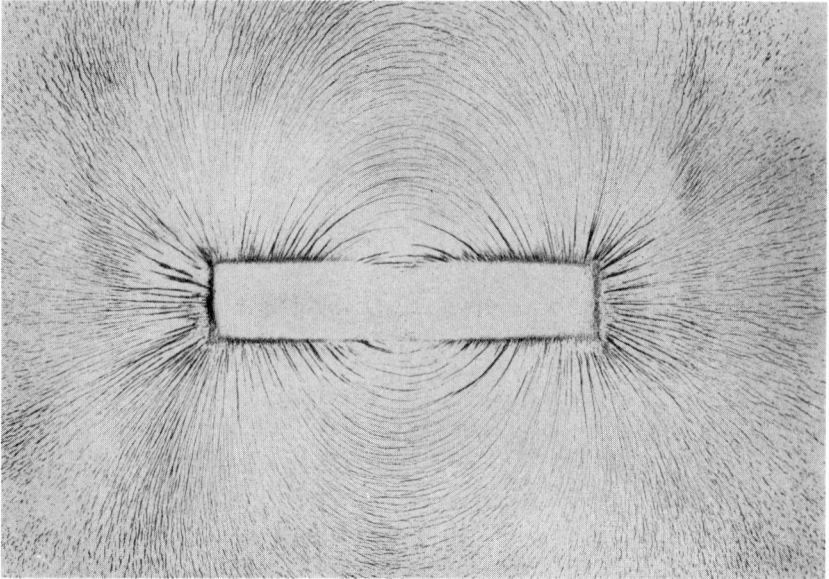

Fig. 57—Magnetograph of the Field Around a Magnet.

(12) *Magnet.* Materials that show the power to attract iron and other substances to themselves, and that exhibit polarity, are called *Magnets.*

(13) *Magnetic Field.* The space around a magnet within which ferromagnetic materials are attracted is called a *Magnetic Field.*

(14) *Magnetic Flux.* The concept that the magnetic field is flowing along the lines of force suggests that these lines are therefore "flux" lines, and they are called *Magnetic Flux.* The strength of the field is defined by the number of flux lines crossing a unit area taken at right angles to the direction of the lines.

(15) *Magnetizing Force.* For the purpose of this discussion, magnetizing force is considered to be the total force tending to set up a magnetic flux in a magnetic circuit. It is usually designated by the letter "H" and the unit is the *"Oersted".*

118

(16) *Magnetic Pole*. The flux lines have direction, and as their path passes from the iron of the magnet into the air, or from the air into the iron, magnetic poles are set up. Ferromagnetic materials are attracted to these poles. The point where the flux leaves the magnet is called the *north pole*, and where the flux enters the magnet is called the *south pole*. These names are derived from the compass. One end of the compass needle always points to the earth's north magnetic pole, and is called the "north-seeking" pole. The other end which points south is called the "south-seeking" pole. For purposes of magnetic language the terms are shortened to simply *North Pole* and *South Pole*.

(17) *Magnetograph*. A magnetograph is a picture of a magnetic field made by the use of iron powder under conditions that allow it to arrange itself into the pattern of the field. See "Lines of Force" (11).

(18) *Oersted*. This is the unit of field strength which produces magnetic induction. It is designated by the letter "H". The Oersted and the Gauss are numberically equal in air or in a vacuum. Oersted (H) refers to the magnetizing force tending to magnetize an unmagnetized body, and Gauss refers to the field (B) so induced in the body.

(19) *Paramagnetic and Diamagnetic*. All materials are affected by magnetic fields. Those which are attracted are called Paramagnetic. Those which are repelled are called Diamagnetic. The reaction to a magnetic field of these two classes of substances is very slight indeed. The few materials that are strongly attracted by magnetic fields are called ferromagnetic.

(20) *Permanent Magnet*. Materials such as Lodestone exhibit the property of magnetism at all times and with little or no reduction in strength when measured against time. Such materials are called *permanent magnets*. Hard steels and many alloys, when magnetized, retain much of their magnetism and become permanent magnets.

(21) *Permeability*. This is the term used to refer to the ease with which a magnetic field or flux can be set up in a magnetic circuit. It is not a constant value for a given material, but is a ratio. At any given value of magnetizing force, permea-

bility is B/H, the ratio of flux density, B, to magnetizing force, H.

(22) *Reluctance.* Reluctance is the opposition to the establishment of magnetic flux in a magnetic circuit presented by the path through which the flux must pass, and is that which determines the magnitude of the flux produced by a given magnetizing force. It is analogous to resistance in an electric circuit.

(23) *Remanent Magnetism.* This is the term applied to the magnetism remaining in a magnetic circuit after the magnetizing force has been removed.

(24) *Residual Field.* This is the field left in a piece of ferromagnetic material when the magnetizing force has been reduced to zero. It is represented by point "b" on the hysteresis curve (Fig. 56).

(25) *Retentivity.* Retentivity is the property of a given material of retaining, to a greater or lesser degree, some amount of residual magnetism.

(26) *Saturation.* This term refers to that degree of magnetization where a further increase in H produces no further increase in the field in a given material other than is produced in air. This is point "a" on the hysteresis curve.

(27) *Temporary Magnet.* A temporary magnet is a piece of soft iron or steel which has little or no retentivity, so that it returns substantially to the unmagnetized state when the magnetizing force is withdrawn.

4. TERMS RELATING TO ELECTRICITY. The following listed electrical terms are those most frequently used in connection with magnetic particle testing:

(1) *Alternating Current or A.C.* Alternating current is current that reverses its direction of flow at regular intervals. Such current is frequently referred to as A.C.

(2) *Ampere.* This is the unit of electrical current. One ampere is the current which flows through a conductor having a resistance of one ohm, at a potential of one volt.

(3) *Conductivity.* This is the inverse of resistance, and refers to the ability of a conductor to carry current.

(4) *Decay.* As used in connection with electricity, decay is the falling off to zero of the current in an electrical circuit. Magnetic fields and electrical potentials can also decay in a similar sense.

(5) *Direct Current or D.C.* As the name implies, this term refers to an electric current flowing continually in one direction through a conductor. Such current is frequently referred to as D.C.

(6) *Full Wave Rectified A.C.—Single Phase.* This is rectified alternating current for which the rectifier is so connected that the reverse half of the cycle is "turned around", and fed into the circuit flowing in the same direction as the first half

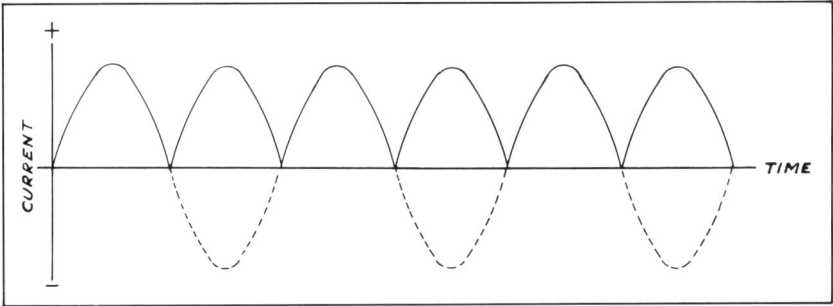

Fig. 58—Wave Form of Full Wave Rectified Single Phase A.C.

of the cycle. This produces pulsating D.C., but with no interval between the pulses. Such current is also referred to as single phase full wave D.C.

(7) *Full Wave Rectified Three Phase A.C.* When three phase alternating current is rectified the full wave rectification system is used. The result is D.C. with very little pulsation—

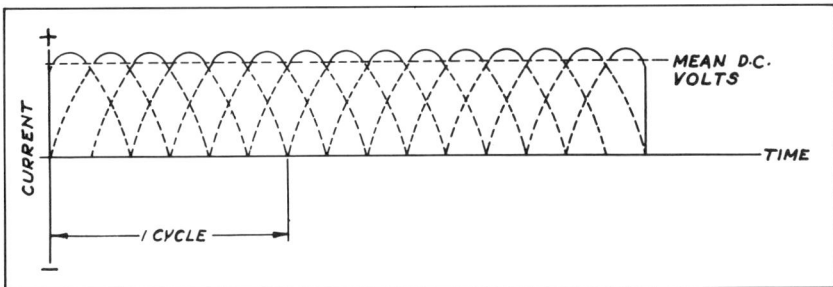

Fig. 59—Wave Form of Three Phase Rectified A.C.

121

in fact only a ripple of varying voltage distinguishes it from straight D.C. (Fig. 59).

(8) *Half Wave Rectified A.C.* When a single phase alternating current is rectified in the simplest manner, the reverse half of the cycle is blocked out entirely. The result is a pulsating uni-directional current with intervals when no current at all is flowing. This is often referred to as "half wave" or as pulsating direct current. This type of current is illustrated in Fig. 60.

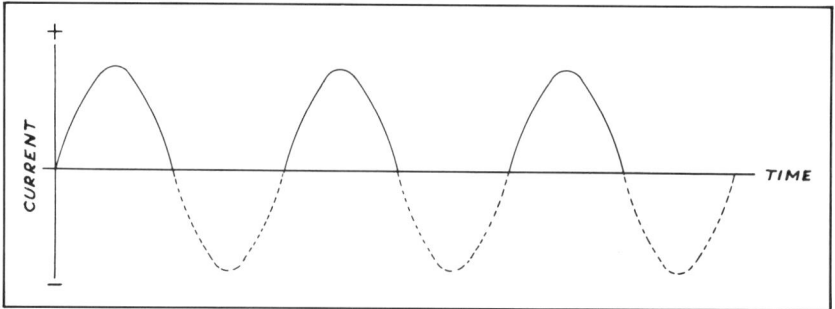

Fig. 60—Wave Form of Half Wave Rectified Single Phase A.C.

(9) *Impedance.* This term is used to refer to the total opposition to the flow of current represented by the combined effect of resistance, inductance and capacitance of a circuit.

(10) *Inductive Reactance.* This is the opposition, independent of resistance, of a coil to the flow of an alternating current.

(11) *Ohm.* The ohm is the unit of electrical resistance. It is the value of a resistance that will pass one ampere of current at a potential of one volt.

(12) *Rectified Alternating Current.* By means of a device called a rectifier, which permits current to flow in one direction only, alternating current can be converted to direct or uni-directional current. This differs from direct current in that the current value varies from a steady level. This variation may be extreme, as in the case of half wave rectified single phase A.C. (see definition #8) or slight, as in the case of three phase rectified A.C. (see definition #7).

(13) *Resistance.* Resistance is the opposition to the flow of an electrical current through a conductor. Its unit is the ohm.

(14) *Single Phase Alternating Current.* This term refers to a

simple current, alternating in direction. Commercial single phase current follows a sine wave, illustrated in Fig. 61. Such a current requires only two conductors for its circuit. Most common commercial frequencies are 25, 50 and 60 cycles per second.

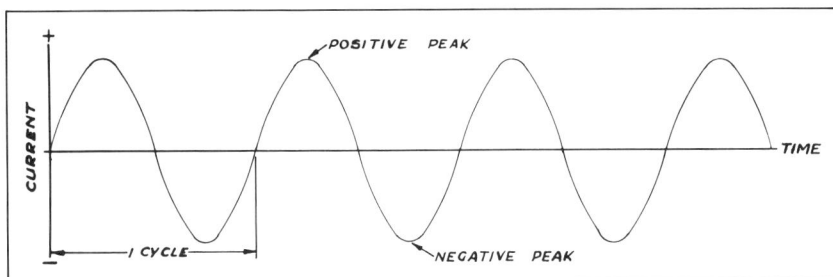

Fig. 61—Wave Form of Single Phase Alternating Current.

(15) *Three Phase Alternating Current.* Commercial electricity is commonly transmitted as three single phase currents— that is, three separate currents following separate sine curves, each at 60 cycles (or other frequency) per second, but with the peaks of their individual curves one-third of a cycle apart. This set of curves is illustrated in Fig. 62. At least three (sometimes four) conductors are required for three phase alternating current.

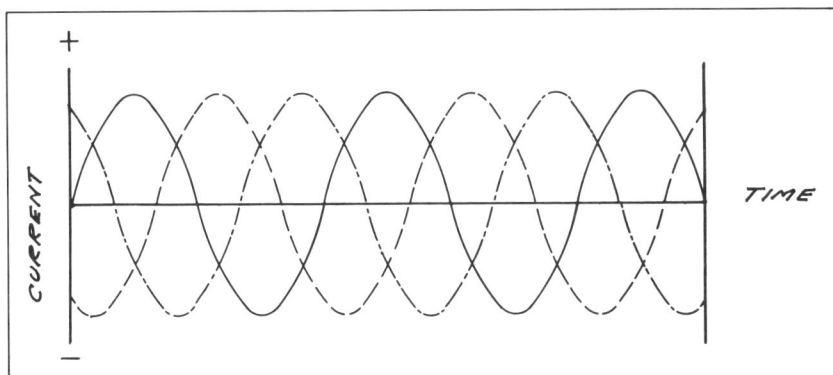

Fig. 62—Set of Wave Forms of Three Phase Alternating Currents.

(16) *Transient Currents.* These currents are of short duration, generated by sudden changes in the electrical or magnetic conditions existing in an electrical or magnetic circuit.

(17) *Volt.* The volt is the unit of electromotive force which tends to cause an electric current to flow through a conductor.

5. TERMS RELATING TO ELECTROMAGNETISM. The following group of terms has to do with magnetism induced by the electric current:

(1) *Ampere Turns.* This term refers to the product of the number of turns in a coil and the number of amperes of current flowing through it. This is a measure of the magnetizing or demagnetizing strength of the coil. For example: 800 amperes in a 6 turn coil = 800 × 6 = 4800 ampere turns.

(2) *Electro-magnet.* When ferromagnetic material is surrounded by a coil carrying current it becomes magnetized and is called an electro-magnet.

(3) *Induced Current.* Two fundamental electrical principles must be understood in order to define the term induced current.

(a) When current flows through a conductor, it sets up a magnetic field at right angles to the direction of current flow. The strength of this field varies directly with the current.

(b) When a conductor in the form of a closed loop moves through a magnetic field, a current will flow in the conductor. This is the basis of transformer action.

Passing an alternating current through a conductor will set up a fluctuating magnetic field. If a second conductor in the form of a closed loop is placed in this field, the action of the fluctuating field moving across the conductor will set up a second alternating current of the same frequency. This is an induced current.

When the moving field is produced by the sudden discontinuance of a direct current (a "collapsing field") the induced current is of only very short duration. Induced currents such as this are used as a method of magnetizing ring-shaped parts.

(4) *Inductance.* When the current in a circuit is varied, the change in the magnetic field surrounding the conductor generates an electromotive force (voltage) in the circuit itself.

If another circuit is adjacent to the first the change in the magnetic field of the first will generate an electromotive force in the second. The phenomenon is known as inductance—either self-inductance or mutual inductance. Inductance is measured by the electromotive force produced in a conductor by unit rate of change of the current. The unit of inductance is the Henry which is that inductance in which an electromotive force of one volt is produced when the inducing current is changed at the rate of one ampere per second.

(5) *Loop*. This term refers to a single turn of wire or cable used to carry electric current. It is used for magnetizing and demagnetizing purposes.

(6) *Solenoid*. A solenoid is a coil consisting of a number of loops of wire or cable to carry electric current. It may be used for both magnetizing and demagnetizing purposes.

6. TERMS RELATING TO MAGNETIC PARTICLE TESTING. The following group consists of terms that are directly in reference to magnetic particle testing.

(1) *Air Gap*. When a magnetic circuit contains a small gap which the magnetic flux must cross, the space is referred to as an *air gap*. Cracks produce small air gaps on the surface of a part.

(2) *Black Light*. Black light is near-ultraviolet light having a wave length of 3650 Angstrom Units. It is used in fluorescent magnetic particle testing to cause the particles to fluoresce and give off visible light.

(3) *Central Conductor*. A central conductor is a conductor that is passed through the opening in a ring or tube, or any hole in a part, for the purpose of creating a circular or circumferential field in the tube or ring, or around the hole.

(4) *Circular Magnetization*. Magnetization of a piece of ferromagnetic material in such a way that the flux path is completely contained within the article, is called circular magnetization. There are usually no external poles. It is usually produced by passing current directly through the piece.

(5) *Coil Shot*. A "shot" of magnetizing current passed through a solenoid or coil surrounding a part, for the purpose of

longitudinal magnetization is called a "coil shot." Duration of the passage of the current is usually very short—often only a fraction of a second.

(6) *Defect.* A condition, of whatever kind, that renders a part unsuitable for its intended service.

(7) *Discontinuity.* This is the general term used to refer to a break in the metallic continuity of the part being tested—a break that may be a crack, a seam or other interruption to the continuity of the material. It is such a break that produces a magnetic particle indication.

(8) *Distorted Field.* The direction of a magnetic field in a symmetrical object will be substantially uniform if produced by a uniformly applied magnetizing force, as in the case of a bar magnetized in a solenoid. But if the piece being magnetized is irregular in shape, the field is distorted and does not follow a straight path or have a uniform distribution.

(9) *False Indication.* Any patch or pattern of magnetic particles not caused or held in place by a leakage field of any kind is called a false indication.

(10) *Fluorescence.* The light given off by certain materials while they are being irradiated with short wave ultraviolet or near-ultraviolet light.

(11) *Heads.* The clamping contacts on a stationary magnetizing unit.

(12) *Head Shot.* A "shot" of magnetizing current passed through a part or a central conductor while clamped between the head contacts of a stationary magnetizing unit, for the purpose of circular magnetization of the part is called a "head shot." Duration of the passage of the current is usually less than one second.

(13) *Indication.* This term refers to any magnetically-held magnetic particle pattern on the surface of a part being tested.

(14) *Leakage Field.* This is the field forced out into the air by the distortion of the field within a part caused by the presence of a discontinuity.

(15) *Longitudinal Magnetization.* Magnetization of a material in such a way that the magnetic flux runs substantially parallel

to the long axis of the part, the flux path completing itself through the air outside the material, is called longitudinal magnetization. It is sometimes called bi-polar magnetization, because *at least* two external poles exist in longitudinal magnetization.

(16) *Magnetic Discontinuity.* This refers to a break in the *magnetic* uniformity of the part—a sudden change in permeability. A magnetic discontinuity may not be related to any actual physical break in the metal, but it may produce a magnetic particle indication.

(17) *Metallic Discontinuity.* This terms refers to an actual break in the continuity of the metal of a part, and may be located on the surface—as for instance, a crack; or deep in the interior of the part—as for example, a gas pocket.

(18) *Multi-directional Magnetization.* Two separate fields, having different directions, cannot exist in a part at the same time. But two or more fields in different directions can be imposed upon a part sequentially in rapid succession. When this is done magnetic particle indications are formed when discontinuities are located favorably with respect to the directions of each of the fields, and will persist as long as the rapid alternations of field direction continue. This, *in effect,* does constitute two or more fields in different directions at the same time, and enables the detection of defects oriented in any direction in one operation. The method has also been called by the name of "Duovec".*

(19) *Non-Relevant Indications.* These are true indications produced by leakage fields. However the conditions causing them are present by design or accident, or other features of the part having no relation to the damaging flaws being sought. The term signifies that such an indication has no relation to discontinuities that might constitute defects.

(20) *Part.* Throughout discussions of the applications of magnetic particle testing it is necessary to refer to the article being tested. The term "part" is used quite generally for this purpose, as a quick and easy word to say. Often the article actually is a "part"—of an engine or other assembly—but

---

*Duovec. Trademark registered in the U.S. Patent Office. Property of Magnaflux Corporation.

the word "part" implies only that it is the piece of material, of whatever sort, that is being tested.

(21) *Prods.* Two hand-held electrodes which are pressed against the surface of a part to make contact for passing magnetizing current through the metal. The current passing between the two contacts creates a field suitable for finding defects with magnetic particles.

(22) *Quick-Break.* Sometimes called "Fast Break". The sudden breaking of a direct current causes a transient current to be induced in the part by the rapid collapse of the magnetic field. In magnetic particle testing, fast breaking of the magnetizing current is used to generate a transient current in a part which is favorable for finding transverse defects at the ends of longitudinally magnetized bars. Such defects are often concealed by the strong polarity at the bar ends. At such locations the lines of force of the longitudinal field are leaving the bar in a direction normal to the surface, which prevents them from intercepting transverse defects in those areas. The field induced by the transient current does intercept such discontinuities. See Chapter 7, Fig. 83.

(23) *Resultant or Vector Field.* When two or more magnetizing forces operating in different directions are simultaneously applied to a ferromagnetic material, a resultant field is produced, having a direction which is determined by the relative strengths and directions of the applied magnetizing forces. Such a field is also referred to as a vector field.

If either or both of the applied magnetizing forces are themselves varying in direction or amount, the resultant field is moving or swinging in direction and strength. Such a moving resultant field is sometimes referred to as a "swinging field".

(24) *Threshold.* In reference to currents or magnetic fields, the minimum strength necessary to create a looked-for effect is called the threshold value. For example, the minimum current necessary to produce a readable indication at a given defect, is the threshold value of current for that purpose.

7. GENERAL COMMENTS. The list of definitions that has been given in the foregoing sections is not a glossary of all the terms that are used in connection with magnetic particle testing. It is intended to give a sufficient background of understanding so that the text of later chapters need not be interrupted by the need to define such terms; and avoids the possibility that the following discussions assume too much previous knowledge on the part of the reader. In general other terms will be defined in the text when they occur.

CHAPTER 6

CHARACTERISTICS OF MAGNETIC FIELDS

1. INTRODUCTION. Thirty-seven years ago (1929) when Magnetic Particle Testing was first put to use, remarkably little published information existed on those aspects of magnetism and magnetic fields which are basic to the understanding and proper application of this nondestructive testing method. The magnetic properties of iron and steel and various alloys were well known, of course, but the use of magnets and magnetic fields was preponderantly in the area of power generation and use—generators, motors, transformers, switch gear, relays, etc. Little had been done to analyze the behavior of magnetic fields *inside* the ferromagnetic material, since all the above mentioned applications are mainly concerned with the paths of fields *external* to the magnets. Nor was there any very satisfactory theory of the nature of magnetism itself.

Since that time, our understanding of atomic structure has resulted in theories of magnetism which have a sound basis, and new interest in various uses and applications of magnetism have stimulated extensive studies. The powerful magnetic fields used in our mammoth "atom smashers" are on one end of this line of interest, and the very faint fields involved in magnetic guidance systems, and protection of vessels from magnetic mines and magnetic guided missiles are at the other end. And there is a tremendous volume of literature available today on these areas of the subject. (References on this phase of magnetism are listed in the Bibliography Appendix.)

However, little of the knowledge resulting from all this research and development is of much direct value in the uses to which magnetism is put for magnetic particle testing purposes, and there is no point in devoting space in this book to the general subject of magnetic theory. We are interested not so much in what magnetism is, but rather in how it behaves with respect to defect-detection with magnetic particles. We are further interested in the strength and direction of fields *within* ferromagnetic material, and how to create fields of proper strength and direction. And we also need to know

130

how fields distribute themselves, both as to strength and as to direction, inside parts of various sizes and irregular shapes.

Our interest stems from a single point of view—that is, to produce a leakage field at a discontinuity, strong enough to form a readable indication with magnetic particles. One guide in this interest is the basic fact that the field must intersect the discontinuity at some appreciable angle, preferably approaching 90°. Therefore *direction of field* must be determined and known for reliable inspection. Next in importance is the strength of the field at the point where the discontinuity lies. The field obviously must be strong enough to produce a leakage field above the threshold strength for readable indications. It can, on the other hand, be *too strong* and produce confusing side effects.

Therefore in this and the next four chapters we will consider these several elements from the point of view of defect detection— that is, how fields behave, how they are produced and how they distribute themselves in various ferromagnetic parts.

2. MAGNETIC FIELD AROUND A BAR MAGNET. If we consider a bar of ferromagnetic material which has been permanently magnetized, we conceive it to be surrounded by a field of force which we call a magnetic field. By laying a piece of paper over the mag-

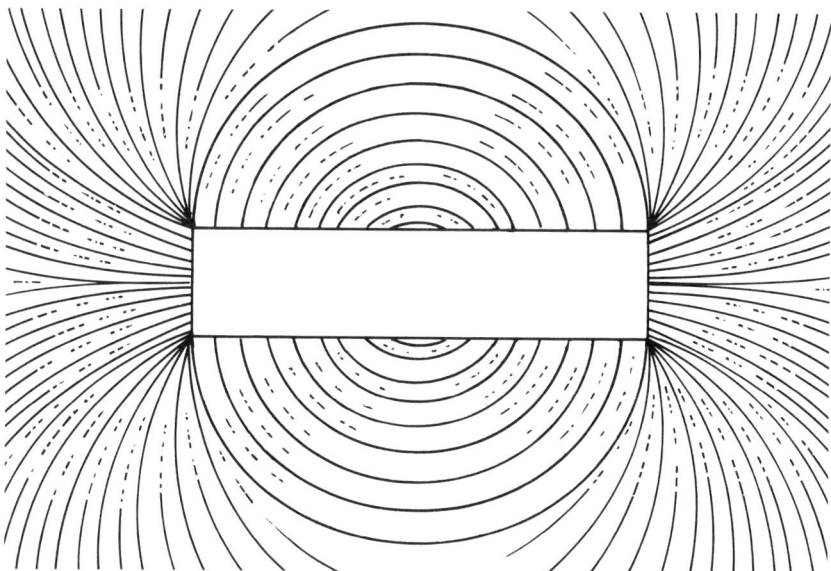

Fig. 63—Field Around a Bar Magnet.

netized bar and sprinkling "iron filings" on the paper, the fine iron particles will arrange themselves in what appear to be lines. Such a pattern is called a magnetograph. It should be emphasized that this pattern represents only the field which lies *outside* the bar. Nor is the field of force around the magnet confined to the plane of the paper on which the magnetograph is produced. It exists on all sides, surrounding the magnet completely. The magnetograph is a cross-section of the field in the plane of the magnet itself.

The appearance of the magnetograph suggests that this field is made up of lines, which extend, for the most part, in curved paths from one end of the magnet to the other. Some of the lines leave and re-enter the bar along its length, the number of such lines increasing toward the ends of the bar. If the bar is of uniform cross-section and composition throughout its length we can infer that the field passes through it with a practically uniform distribution, except near the ends. As the result of the appearance of the magnetograph, magnetic fields have long been thought of as being made up of "lines of force", and certain properties have been observed regarding them. Furthermore it is easy to think of these lines of force, or magnetic flux, as flowing through the bar, out into the air, and circling back to re-enter the bar at the other end. This concept of flow is most convenient in talking about the behavior of magnetic fields.

3. POLES. Although the flux is not proveably flowing, it does have directional properties. The compass, a magnetized needle, will

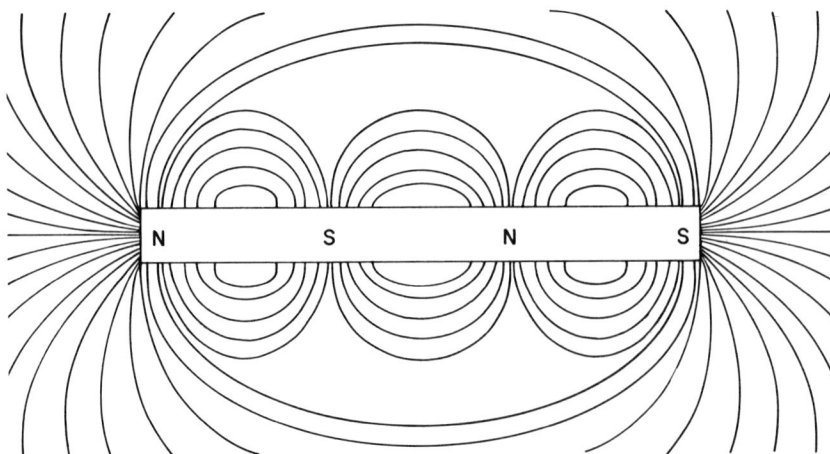

Fig. 64—Consequent Poles on a Bar Magnet.

swing so that the same end always points to the north. Similarly, compass needles will always point to the same end of a bar magnet. The two ends at which most of the flux lines leave and re-enter the bar are called poles. "The "north" pole is that end at which the lines are thought of as leaving the bar, and the "south" pole is that at which they re-enter. If the north poles of two magnets are brought together the magnets repel each other. If a north and a south pole are brought together they will attract each other and tend to be drawn together. The rule is: "like poles repel, unlike poles attract".

Normally a bar magnet has only two poles, one north and one south, located at opposite ends of the bar. However a magnet may have a number of poles, called "consequent poles"—some north and some south. Figure 64 gives some examples of consequent poles in magnetized parts.

4. MAGNETIC ATTRACTION. The concept of flux lines includes some other characteristics. Each line is considered to be a continuous loop, which is never broken, but must complete itself through

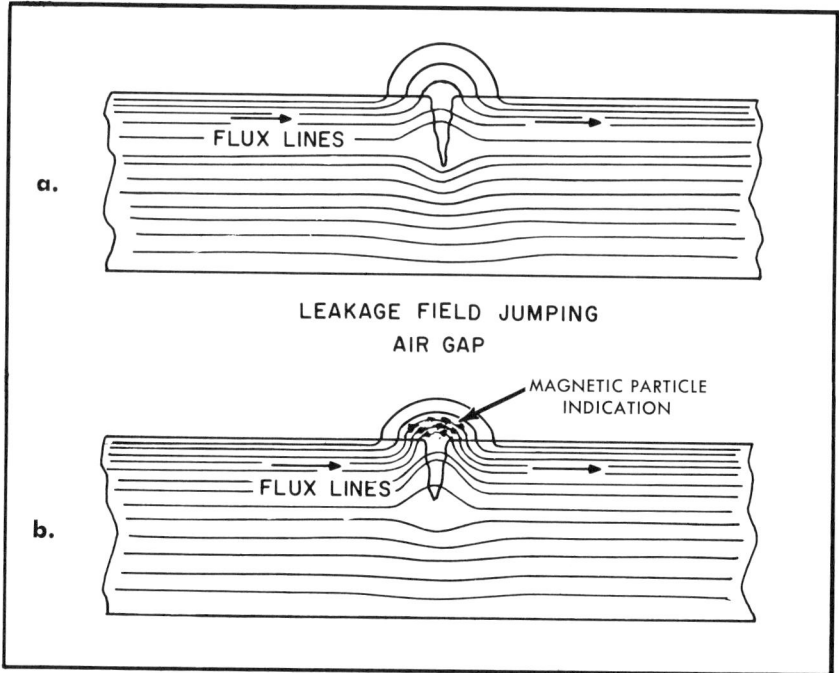

Fig. 65—Bridging of the Air Gap at a Crack.
a) Leakage Field. b) Magnetic Particle Indication.

some path. The lines always leave the magnet at right angles to the surface. They tend always to seek the path of lowest reluctance in completing their loop. A piece of soft iron, therefore, when placed in a magnetic field, will be drawn toward the magnet, so that more and more lines of force may traverse it, and the high reluctance path through the air will be as short as possible. This is the action which causes magnetic particles to gather at leakage fields at discontinuities. The leakage field is jumping across a relatively high reluctance air gap at a crack. The magnetic particles offer a lower reluctance path to the flux, and are therefore drawn to, and bridge, the air gap.

5. A CRACKED BAR MAGNET. If a bar magnet having two poles, N. and S., at opposite ends is broken in the center of its length, two complete bar magnets will result, each having a N. and S. pole. This process of breaking can go on till there are four, eight or any number of separate complete magnets.

If the magnet is cracked—not broken completely in two—a somewhat similar result occurs. A North and South pole will form at opposite edges of the crack, just as though the break had been complete. The *strength* of these poles will be different from that of the fully broken pieces, and will be a function of the *depth* of the crack and the width of the air gap at the surface.

The fields set up at cracks or other physical or magnetic discontinuities in the surface are called *leakage fields*. The strength of this leakage field determines the number of magnetic particles which will be gathered to form indications—strong indications at strong fields, weak indications at weak fields.

6. EFFECT OF FLUX DIRECTION. In the cases we have been considering—that is, a straight bar magnet—the flux lines are assumed to run lengthwise in the bar, from end to end. Such a bar is said to be longitudinally magnetized. A crack produced by partially breaking the bar transversely will lie at right angles to the long axis of the bar, and consequently will also be at 90° to the flux lines traversing the bar. Such a crack-orientation offers the greatest interruption to the lines of force, and therefore forms a strong leakage field.

Assume, however, a straight crack-like discontinuity lying in a direction parallel to the axis of the bar, and therefore parallel to the lines of force. No lines would be cut and there would be no leakage field. A fundamental requirement, therefore, if a discontinuity is to produce a leakage field and a readable magnetic particle pattern is that it must intercept the lines of force at some angle. The leakage field will be strongest if the angle is 90° and will become weaker as the angle the discontinuity makes with the lines of force becomes smaller.

It is, however, often true that a crack which has a general direction parallel to the axis of the bar, but which deviates somewhat from a straight line, may give quite strong magnetic particle indications. In such a case the segments of the crack not exactly parallel

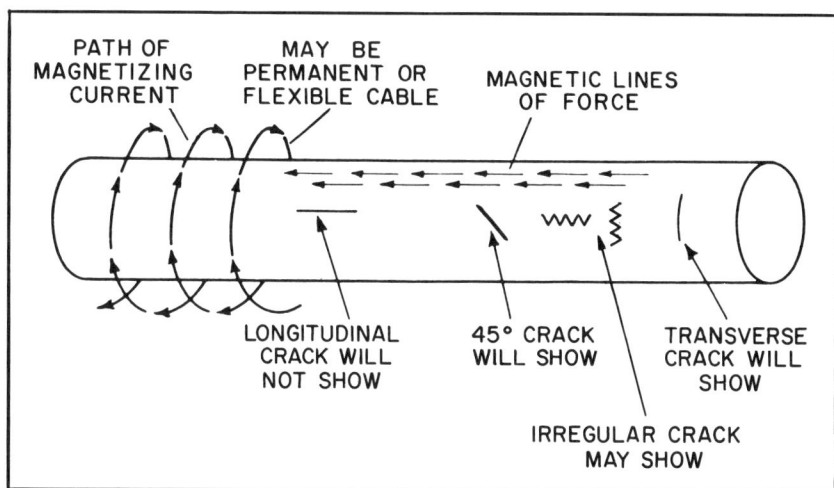

Fig. 66—Effect of Crack-Orientation in a Longitudinally Magnetized Bar.

to the axis do, locally, intercept the lines of force. Figure 66 illustrates these several situations.

7. CIRCULAR MAGNETIZATION. If the bar magnet we have been considering is bent into a circle so that the two ends are brought close together (see Fig. 67) the small air gap will have a strong flux jumping across it and will have a strong attraction for iron pieces or particles brought close to it. This is essentially the familiar horse-shoe magnet.

But if we think of this gap as being completely closed to form a ring as if by welding, so that there is no metallic discontinuity of any kind, there will be no leakage field or external poles. The field originally present in the bar will still exist in the ring although there will be no external evidence of its presence. In fact, the field

135

in the ring will be stronger than that which was in the bar, since the all-metallic circuit has a low reluctance, with no high reluctance air gap. Also, there is no self-demagnetizing effect which is present in the bar magnet and tends to limit the field in it. The ring is now said to be circularly magnetized.

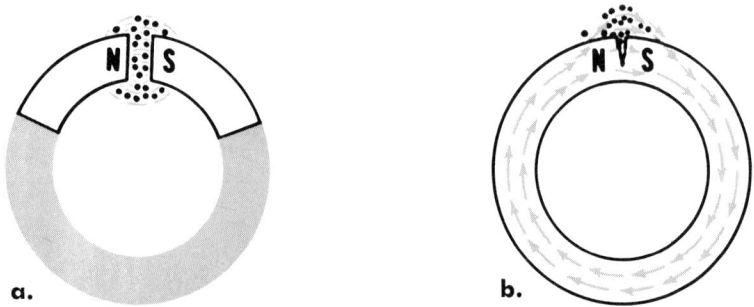

Fig. 67—Circular Magnetization.
a) Incomplete Ring with a Small Air Gap.
b) Complete Ring with no Air Gap, but with a Crack.

8. CIRCULAR MAGNETIZATION AND CRACKS. If the ring just described is cracked on the outer cylindrical surface in a direction parallel to the axis of the ring, the internal circular field would be intercepted and local poles would at once be formed. See Fig. 67. The leakage field would then form an indication of the crack if magnetic particles are applied. Here again, direction of the field with relation to the direction of the crack is of prime importance.

Instead of the ring, let us suppose that a bar or tube is circularly magnetized. See Fig. 68. In this case it is the crack lying parallel to the long axis of the bar or tube which intercepts the flux lines at 90°. Cracks at right angles to the axis would be parallel to the

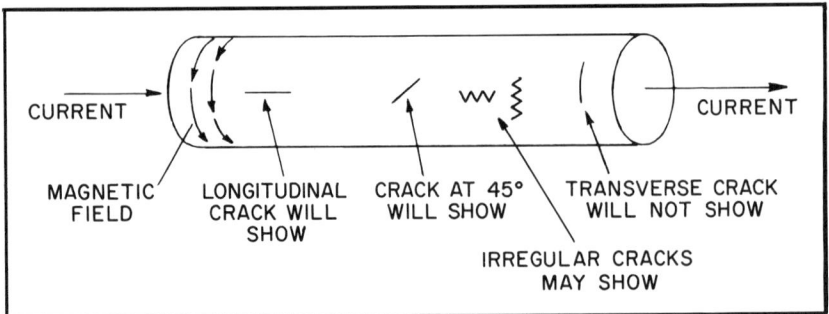

Fig. 68—Effect of Crack Orientation in a Circularly Magnetized Bar.

136

field and would create no leakage field. Cracks at intermediate angles would create leakage fields of strengths varying with the angle; and the crack, essentially transverse, which "wanders" from a straight line would probably give an indication.

9. DIFFICULTIES OF ESTABLISHING PROPER FIELDS. Since successful magnetic particle testing depends on establishing fields in parts in directions which will cut across the expected or possible cracks, some of these field characteristics must be thoroughly understood. For example, it must be remembered that the flux path will always go the route of lowest reluctance. This means it will follow ferromagnetic paths when it can, in preference to taking a path containing air gaps of high reluctance.

Thus, if it is desired to inspect a long bar by longitudinal magnetization in a short coil, it is not usually possible to secure a truly longitudinal field clear to the end of the bar when the coil is located at the mid-point of the bar's length. The flux tends to leave the bar at some distance from the coil, and return around the outside of the coil to the bar on the other side, leaving the ends of the bar not properly magnetized. When the bar is not much longer than the coil, it will be satisfactorily magnetized for its entire length. But if the bar is *very short*, so that its length is less than its diameter—as, for example a disc-shaped part such as a gear—a longitudinal field may not be set up at all.

Unless the disc is held exactly in the plane of the coil, it will be found that the field takes a path across a diameter, instead of passing through the disc from face to face. If such a piece is hand-held in the coil it is practically impossible to establish a longitudinal field in it to locate circumferential cracks on the cylindrical surface by this procedure. See Fig. 69. The proper manner in which to magnetize a disc to locate circumferential cracks is to pass current diametrically through it while clamped between the heads of a magnetizing unit in a manner similar to that used for ring-shaped parts as shown in Fig. 84, Chapter 7. The disc should be tested twice, with current passed across the diameters at 90° to each other.

Similarly, it is difficult to set up a truly transverse field at right angles to the axis of a bar. If the bar is placed crosswise in a short coil, with the bar positioned at nearly 90° to the coil's axis, the result will be a non-uniform but essentially longitudinal field following the length of the bar. Transverse fields can be set up in a bar by the use of yokes, when the yokes are positioned so that the

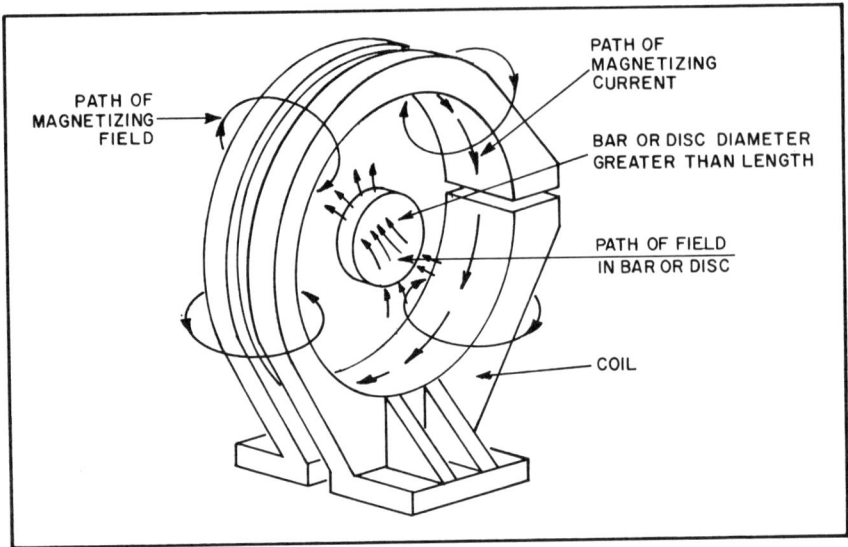

Fig. 69—Magnetic Flux Passing Diametrically Through a Disc Instead of from Face to Face, when the Disc is Placed at an Angle of <u>Nearly</u> 90° to a Longitudinal Field in a Coil.

two poles are diametrically opposite each other and close to the center of the bar.

10. DISTORTED FIELDS. We have been considering symmetrical objects for our magnetization discussion, in which the fields set up are likely to be symmetrical also, and approximately uniform. Fields may, however, be distorted and caused to depart from expected directions by various conditions, and by the shape of the part.

Variations in permeability in different sections of the part, or in different layers of the metal of an otherwise symmetrical part, may cause the internal field, whether circular or longitudinal, to depart from the path it would take in a homogeneous symmetrical object. Variations in hardness, grain size, composition (such as carbon content) all could cause considerable variation in permeability, and affect the distribution of the field.

A piece of un-magnetized soft iron in contact with the surface of a circularly magnetized part, will cause distortion of the circular field. See Fig. 70. Irregular shaped parts will also cause fields to be distorted and to follow paths which are often difficult to predict.

Chapters 8, 9, and 10 are devoted to a discussion of this important subject of field distribution.

138

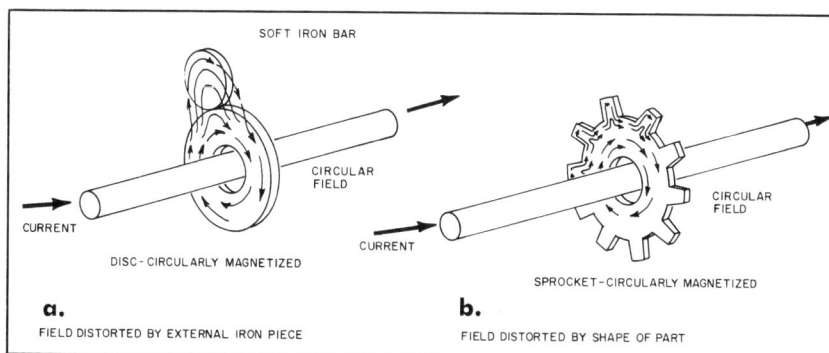

Fig. 70—Distortion of the Circular Field in a Part Caused by
a) Contact with a Piece of Soft Iron.
b) The Shape of the Part.

11. PARALLEL FIELDS. If a ferromagnetic bar is placed alongside, and parallel to, a conductor carrying current, a field will be set up in the bar which is more transverse than circular, and is of very little use for magnetic particle testing. Figure 71 shows the path of the field so produced. Operators have tried to use this method as a substitute for the head shot for the purpose of producing circular magnetization. Not only is the field produced not circular, but its essentially transverse direction is effective for finding seams etc. over only a very limited area of the bar. Furthermore, the external field around the conductor and the bar attract magnetic particles and produce confusing backgrounds. This method of magnetization should not be used for magnetic particle testing purposes.

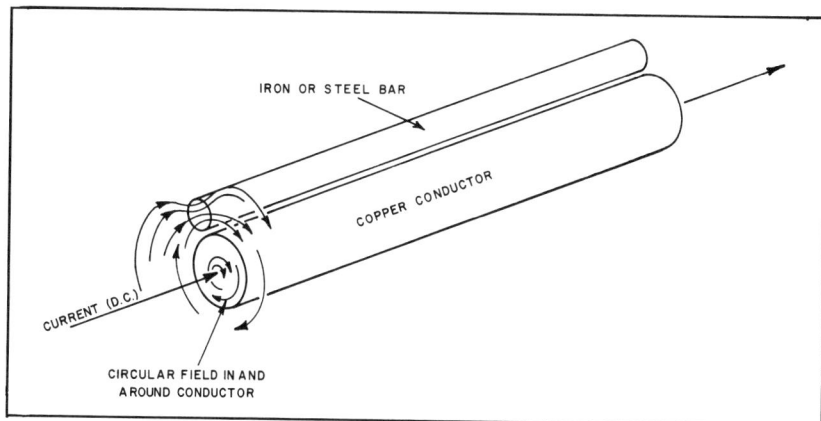

Fig. 71—Field Produced in a Bar by a "Parallel" Current.

139

12. PRODUCTION OF SUITABLE FIELDS. The characteristics of magnetic fields most important to the practice of magnetic particle testing have been discussed in this chapter, without reference to the methods of producing them. In the following chapter, methods and means for producing fields are discussed, together with the effect of the many variations in magnetizing methods on the results of the inspection.

CHAPTER 7

METHODS AND MEANS FOR GENERATING
MAGNETIC FIELDS

1. THE EARTH'S FIELD. One of the properties of permanent magnets is that they can magnetize other ferromagnetic parts. This may sometimes be done merely by putting the piece to be magnetized in the external field around a permanent magnet. A more effective way is to touch or stroke the piece to be magnetized with one pole of a permanent magnet. Under such conditions, soft iron will be temporarily magnetized while a piece of hard iron or steel will become a permanent magnet.

The Earth is in effect a huge permanent magnet having a north magnetic pole just north of Prince of Wales island in Northern Canada, and a south magnetic pole in the Antarctic Ocean south of Australia, just off the coast of the Antarctic Continent at Commonwealth Bay. The flux lines of the Earth's field circle the globe in a generally north and south direction and, though with many deviations on the way, concentrate at these magnetic poles. The Earth's field is quite weak compared to the strength of fields used in magnetic particle testing. However, long bars of iron or steel can be magnetized to an appreciable degree by placing them for a period of time in a north and south direction—that is, parallel to the lines of force of the Earth's field. Magnetization is hastened by striking the bars with a hammer to vibrate them. The natural lodestones derived their magnetism from their location in the Earth's field.

2. MAGNETIZATION WITH PERMANENT MAGNETS. For magnetic particle testing purposes, permanent magnets can be, and sometimes are used to magnetize parts. This method of magnetization has many limitations and is properly used only when these limitations do not prevent the formation of satisfactory leakage fields at discontinuities.

Permanent magnets in general are capable of setting up only what are essentially longitudinal fields. Invariably poles are present on the parts so magnetized, due to the contact of the poles of the permanent magnet used for magnetization. This results in confusing adherence of particles at such external poles. Control of field direction is possible over only limited areas. A single pole—north or south—of a permanent bar magnet set on end on the surface of a steel plate creates a sort of radial field in the plate around the pole. The flux of this field leaves the plate surface at some distance from the point of contact to return to the pole at the opposite end of the magnet. Cracks which happen to be crossed by such a field pattern will usually be indicated, provided the field produced in the plate is sufficiently strong. When the poles of a horse-shoe magnet or a permanent magnet yoke are placed upon the surface of a steel plate or part, the field travels through the plate or part from one pole of the magnet to the other. Along a straight line drawn between the poles, the flux will be relatively straight and will be strongest near the poles of the magnet or yoke. Field strength along this line will vary, and be weakest at the point midway between the poles. The actual strength at any point will

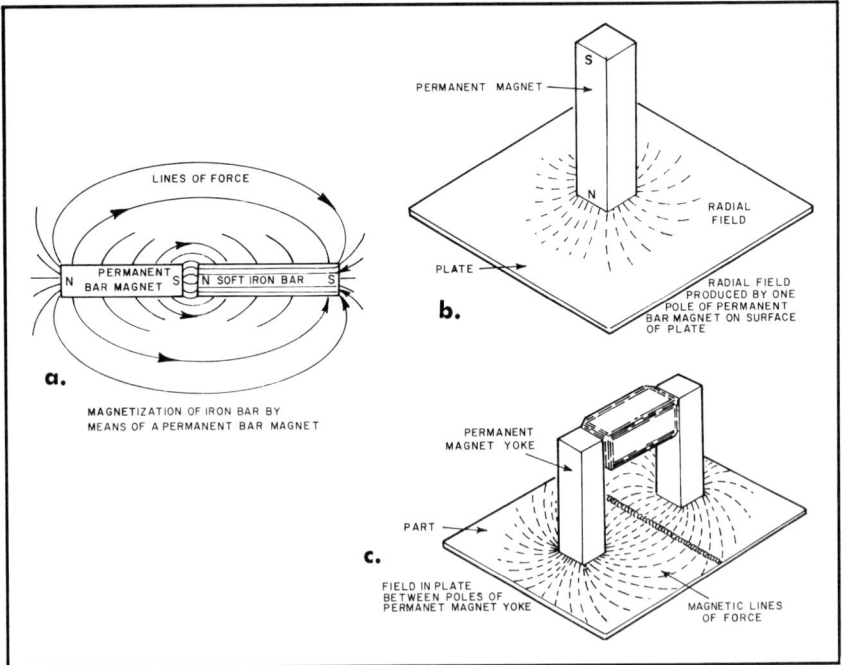

Fig. 72—Magnetization With a Permanent Magnet.
a) Magnetizing a Bar by Placing One Pole of a Permanent Magnet at One End.
b) Magnetizing a Plate by Setting a Permanent Bar Magnet on End on the Plate.
c) A Permanent Horse-shoe Magnet, or Yoke, Placed on the Surface of a Plate or Other Part.

depend on the strength of the magnet and the distance between the poles. Cracks at right angles (or nearly so) to this line will be indicated, provided field strength is adequate. Outside this limited area the field spreads out, and cracks favorably located with respect to field direction may or may not be shown, again depending on the place where they occur. Figure 72 illustrates the uses of permanent magnets for magnetization of parts. This method of magnetization should be used only by experienced operators who are aware of and understand the limitations of the method.

Some of the other drawbacks to the use of permanent magnets are 1) that it is not possible to vary the strength of the field at will, nor 2) can large areas or masses be magnetized with enough strength to produce satisfactory crack indications, and 3) if the magnet is very strong, it may be difficult to remove it from contact with the part.

3. ELECTRIC CURRENTS FOR MAGNETIZATION. Use of electric currents is by far the best means for magnetizing parts for magnetic particle testing purposes. Either longitudinal or circular fields can be set up easily; strength of field can be readily varied; and by use of the several types of current, useful variations in field strength and distribution can be accomplished. Electric current in one or another of its types, is used for magnetization for the purposes of magnetic particle testing in all but a very small percentage of cases.

4. FIELD IN AND AROUND A CONDUCTOR. When an electric current is passed through a conductor, such as a rod or wire, a magnetic field is set up in the conductor and in the space surrounding it. If the conductor is uniform and straight, the density of the field, or the number of magnetic lines of force, will be uniform along its length, and will decrease with increasing distance out from the conductor. The strength of the field will be directly proportional to the strength of the current flowing (that is, the number of amperes). The lines of force take concentric circular paths around the rod or wire, so that the field is circular and at 90° to the axis of the conductor.

The field has directional properties, which depend on the direction in which the current is flowing. A simple rule for determining the direction of the field is to grasp the conductor (not necessarily literally while current is flowing!) with the right hand so that the thumb points in the direction of current flow. The fingers will then point in the direction of the field. Another way is to bring a compass

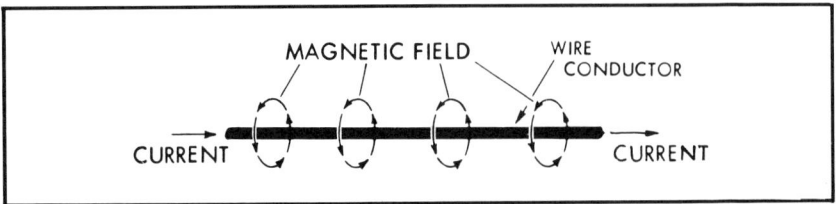

Fig. 73—Field Around a Conductor Carrying Direct Current, and Its Direction.

needle into the field. The north pole of the compass will point in the direction of the lines of force. Fig. 73 illustrates diagrammatically the field around a conductor and its direction.

5. LOOP. If the conductor carrying current is bent into a single loop, the lines of force surrounding the conductor will pass through the loop, all in one direction. The field within the loop then has definite direction, corresponding to the direction of the lines running

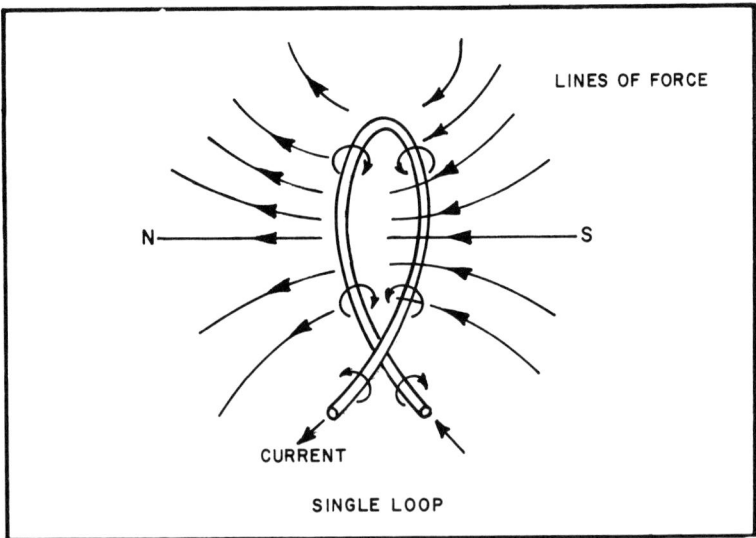

Fig. 74—Field In and Around a Loop Carrying Direct Current, Showing Polarity.

through it. One side of the loop will be a north pole and the other a south pole. See Fig. 74.

6. SOLENOID. If instead of only one turn, the conductor carrying current is looped a number of times, the coil or solenoid will similarly be longitudinally magnetized—one end will be of north polarity and the other south. The strength of the field passing through the interior of the solenoid will be proportional to the

product of the current in amperes and the number of turns of wire in the solenoid—that is, the ampere turns. Thus the magnetizing force of such a coil can be varied either by varying the current or the

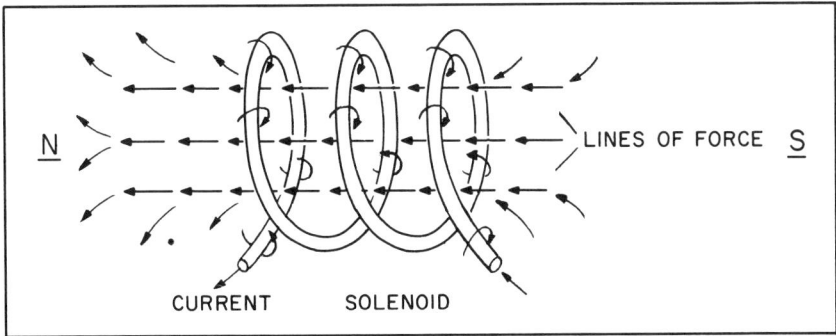

Fig. 75—Field In and Around a Solenoid Carrying Direct
Current and Its Direction.

number of turns in the coil. The measure of the magnetizing strength of a coil is thus expressed in terms of *Ampere Turns.*

7. YOKES. Electromagnetic yokes are U-shaped cores of soft iron with a coil wound around the base of the U. When direct current is passing through the coil the two ends of the core are magnetized with opposite polarity, and the combination is an electromagnetic yoke, similar to a permanent horseshoe magnet. Yokes are sometimes used for creating longitudinal fields in the same way as the permanent magnet yokes. D.C. yokes have some advantage over the latter in that their strength can be varied by varying the current in the coil; and also that they can be put on and removed from the part in the unmagnetized condition when no current is flowing. Yokes magnetized with alternating current also have numerous applications, and have the additional advantage that they can be used for demagnetizing as well as magnetizing.

8. SOLENOIDS FOR MAGNETIZATION. As has been said, solenoids carrying current are the preferred means for setting up longitudinal fields in ferromagnetic parts. For large parts the solenoid may be wound directly around the part, using several turns of flexible cable. For smaller parts, coils wound on fixed frames are most often used.

9. EFFECT OF COIL DIAMETER. Diameter of the coil with relation to the dimensions and shape of the part being magnetized is a large factor in securing proper magnetization. In particular, the length and diameter of the part must be considered in relation to the length and diameter of the coil, when deciding on the amount of current to use for proper magnetization for magnetic particle testing purposes.

The ratio of the cross-sectional area of the part being magnetized, to the cross-sectional area of the coil is called "fill factor". When the coil is short low fill factors are necessary to produce a uniform field over the cross-section of the part when the part is placed in the center of the coil. The ratio of the part diameter to coil diameter in general should not be greater than one to ten. If the part is placed adjacent to the wall of the coil the fill factor becomes of less importance. This phase of magnetizing procedure will be discussed in more detail in Chapter 9.

10. EFFECT OF COIL LENGTH. When a solenoid is used for the magnetization of long parts, best results might appear to be obtained when the coil is as long as the part, thus forcing the field to traverse the part from end to end. When flexible cable is used to form the coil in place on the part over its entire length, this can readily be done, although the technique is not very practical. It is not much used in actual practice because the surface of the part is hidden by the coil, and it is therefore inaccessible for inspection, unless either the coil or the part is removed.

Coils for magnetizing purposes are usually short, especially those wound on fixed frames. But when a long part is being magnetized in such a coil, it will not be magnetized properly over its entire length unless either the coil or the part is moved. The lines of force

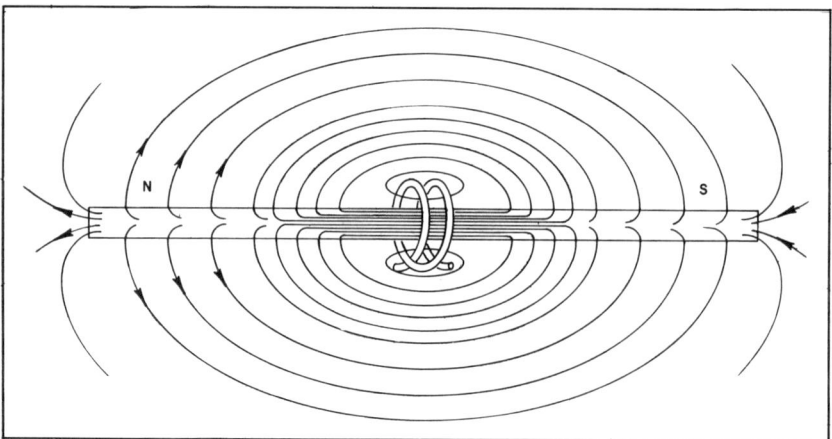

Fig. 76—Flux Path for a Long Bar in a Short Coil.

146

will leave their longitudinal direction at some distance on either side of the coil and return through the air on the path of least reluctance, consisting of optimum amounts of air and iron in the circuit. See Fig. 76. Long bars must be moved through the coil, or the coil moved along the bar, and several "shots" of current through the coil given along its length, with inspection following each shot. A single shot with the coil midway between the ends of a short bar may be sufficient.

11. CIRCULAR MAGNETIZATION. Longitudinal magnetization is used when the nature of the part and the orientation of the defects being sought are such that the latter lie in a direction transverse to the long axis of the part. But because of the polarity associated with longitudinal magnetization external leakage fields at the poles sometimes cause interfering background patterns. In circular magnetization there are seldom any external leakage fields except at a discontinuity. Therefore, all other things being equal, where a choice exists between longitudinal and circular magnetization, the latter is preferable.

When current is carried by a non-magnetic conductor, a circular field exists in and around the conductor; but if the conductor is of ferromagnetic material such as iron or steel, a very much stronger field is formed *inside* the conductor, whereas the field *surrounding* the magnetic conductor is the same as for the non-magnetic one. The strong field generated inside the ferromagnetic conductor is due to the high permeability of ferromagnetic material as compared with copper or other nonmagnetic metals.

When a bar is thus magnetized, the internal field will be zero at the exact center of the bar and increase to a maximum at the surface. If the conductor is a tube, there will be no field on the inner surface of the tube and maximum field strength at the outer surface. Circular magnetism can be better induced in a tube by passing the current through a *central conductor* threaded through the inside of the tube. In such a case, field strength will be a maximum at the *inside* surface of the tube, and somewhat less at the outside surface. A full discussion of field distribution in symmetrical objects will be found in Chapter 9.

12. EFFECT OF PLACEMENT OF CENTRAL CONDUCTOR. If a central conductor is so placed that it traverses the exact center of the tubular cross section the field in the tube will be symmetrical around

147

its cylindrical wall. If, however, the central conductor is placed adjacent to one point on the inner circumference of the tube, the field in the tube wall will be much stronger at this point, and weaker at the diametrically opposite point. Fig. 77 shows some magneto-

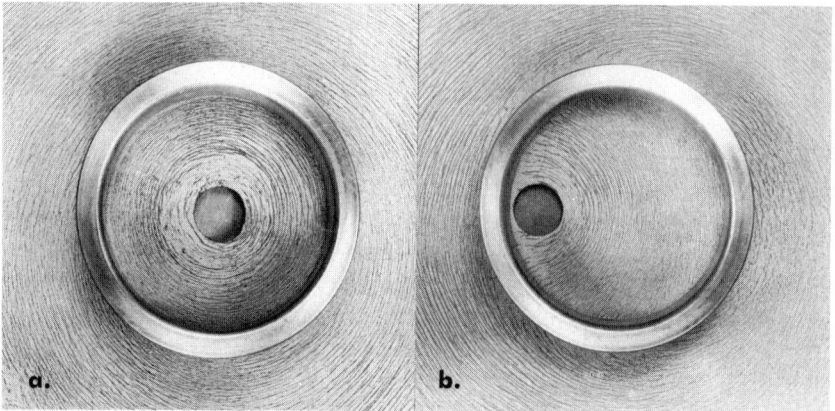

Fig. 77—Magnetographs of Field Showing Cross Section of a Tube Magnetized
a) With a Central Conductor Centrally Located.
b) With the Conductor Located Adjacent to the Wall of the Tube.

graphs illustrating this situation, and although only the flux lines of the *external* fields are shown, some idea of the unsymmetrical field is given by the illustration.

In the case of small tubes it is desirable that the conductor be centrally placed so that uniform fields for the detection of discontinuities will exist at all points on the tube's surface. In the case of larger diameter tubes, rings or pressure vessels, however, the current necessary in the centrally placed conductor to produce fields of adequate strength for proper inspection over the entire circumference become excessively large. By placing the conductor adjacent to the wall and leaving the current on, or giving a number of current "shots" as the tube is rotated about its axis, a field of sufficient strength is produced with much smaller currents at the points in the tube or ring wall adjacent to the conductor.

13. CIRCULAR FIELDS IN IRREGULAR-SHAPED PARTS. Both the direct contact method and the central conductor method are used to generate circular fields in all varieties of shapes and sizes of parts. For small parts having no openings through the interior, circular fields are produced by making contact directly to the part. This is done by means of clamping devices, generally on a bench-like unit which also incorporates the source of current. Fig. 78 illustrates such an arrangement.

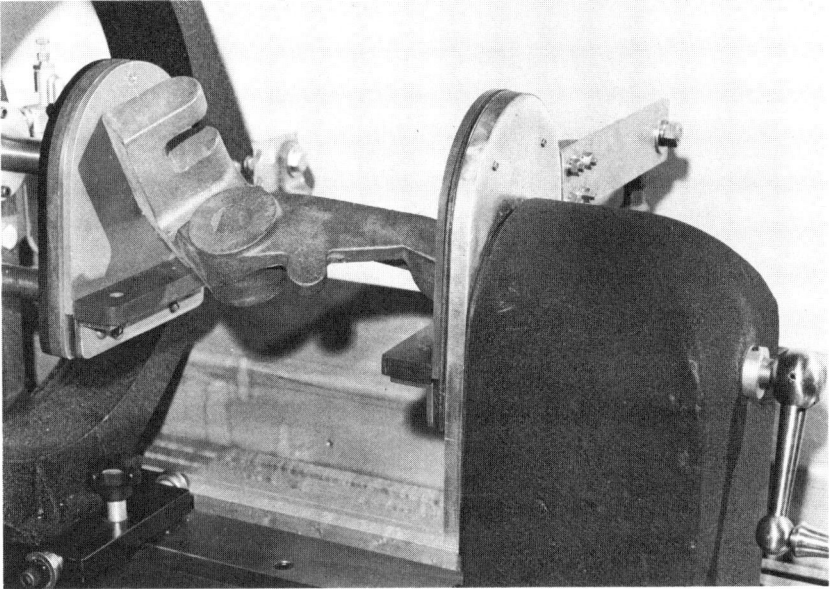

Fig. 78—Part Being Magnetized Circularly by Clamping It Between the Head Contacts of a Magnetizing Unit.

The contact heads must be so constructed that the surfaces of the part are not damaged, either physically by pressure, or structurally by heat produced by arcing or high resistance at the points of contact. Possible damage to finished parts from this cause is an important consideration and must constantly be borne in mind. It can be especially damaging to hardened surfaces such as bearing races.

Points of contact may have to be made at several places for complete inspection of an irregularly shaped part, in order to get fields in the proper directions at all points on the surface. This often necessitates several separate magnetizations and inspections. Modern techniques have greatly minimized the need for such multiple magnetizations by the use of the so-called "overall" magnetization method, by the use of multi-directional magnetization, or use of the induced current method of magnetizing. See Chapters 10, 19, and 24.

14. MAGNETIZATION WITH PROD CONTACTS. Magnetization by passing current directly through the part, or through a local portion of the part, is often resorted to for the inspection of large and massive articles too bulky to be put into a unit having clamping contact heads. Such local contacts do not always produce true circular fields, but are very convenient and practical for many purposes. Inspections of large castings and weldments are frequently made in this manner. Fig. 79 is a magnetograph showing the gen-

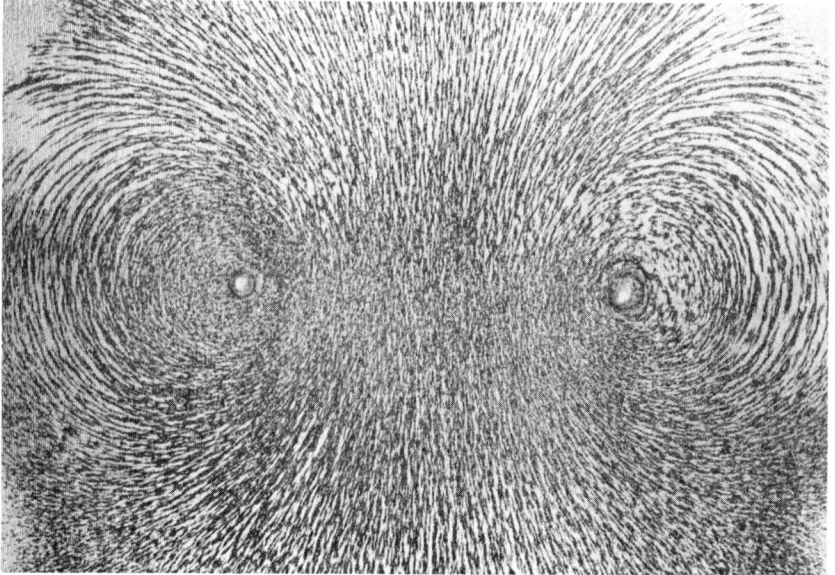

Fig. 79—Magnetograph of the Field Around and Between Prod Contacts.

eral direction of the field when contacts are made on the surface of a steel plate by means of two hand-held prods. The field here is somewhat similar to that produced by yoke magnets, except that in this case it is the current that passes between the contact points, and the field crosses the area between the contacts at 90° to the current. Cracks parallel to the line between the prods will be shown by this method of magnetization.

The method is widely used and has many advantages. Easy portability makes it the most convenient method for field use for the inspection of large tanks and welded structures. Sensitivity to defects lying wholly below the surface is greater with this method of magnetizing than with any other, especially when half wave current is used.

The use of prod contacts has some disadvantages of which the operator should be aware:

(1) It is necessary to scan the surface of the part being inspected in small sections, since suitable fields exist only between and near the prod contact. Since the spacings are seldom greater than 12 inches and usually much less, this means many separate contacts and a time consuming job. Modern techniques such as the "overall" method previously referred to, have in a great many cases replaced inspection with prods, and have shortened the time for such inspections by very large percentages, of 80% or more.

(2) Interference of the external field that exists between the prods sometimes makes observation of pertinent indications difficult. The strength of the current that can be used is limited by this effect.

(3) Great care must be used to avoid burning of the part under the contact points. Burning may be caused by dirty contacts, insufficient contact pressure or excessive currents. The chance of such damage is particularly great on steel of 30 to 40 points carbon or over. The heat under the contact points produces local spots of very hard material that can interfere with later operations, such as machining. Sometimes actual cracks are produced by this heating effect. Contact heating is not so likely to be damaging to low carbon steel such as has been used for structural purposes.

15. EFFECT OF TYPE OF MAGNETIZING CURRENT. There are basically two types of electric current in common use, and both are suitable for magnetizing purposes for magnetic particle testing. These are *Direct Current* (D.C.) and *Alternating Current* (A.C.). The strength, direction and distribution of the fields are greatly affected by the type of current employed for magnetization. Understanding the magnetizing characteristics of these types of current and the various modifications in their use, is of great importance for proper application of magnetic particle testing.

16. DIRECT CURRENT VS ALTERNATING CURRENT. Direct current as here being considered is a constant current flowing in one direction only. Alternating current is here considered to be commercial A.C., which is current reversing its direction completely at the rate of 50 or 60 cycles per second. Commercial A.C. is referred to as being 50 cycle or 60 cycle. One cycle consists of two complete reversals. See Fig. 81. There are also a few commercial services of 25 cycle A.C. 50 cycle A.C. is quite common in many countries, while in the United States the 60 cycle current is almost universally used.

The magnetic fields produced by direct and by alternating current differ in many characteristics. The difference which is of prime importance in magnetic particle testing is that fields produced by direct current generally penetrate the entire cross section of the part, whereas the fields produced by alternating current are confined to the metal at and near the surface of the part.

The phenomenon which causes A.C. to tend to flow along only the surface layers of metal of a conductor is known as "skin effect". In non-magnetic conductors the effect is not very noticeable until frequencies much higher than 60 cycles per second are reached. In magnetic materials, however, the effect is tremendously greater, so that at the commercial frequency of 60 cycles, alternating current passing through a ferromagnetic conductor will be pretty well confined to the surface layers of metal. Circular magnetic fields generated under these conditions will be similarly confined to surface or near-surface layers, as though the current were traveling through a tube sheathing the balance of the cross-section of the conductor.

Careful experiments have shown that in a steel shaft six inches in diameter, magnetized by passing heavy alternating current of 60 cycle frequency from end to end, there was little or no field detectable more than one and one half inches below the surface, and for all practical purposes the field was entirely confined in the outer half inch of metal. From this, it is obvious that when deep penetration of the field into a part is required—as, for instance, in the search for discontinuities lying wholly below the surface and possibly deep in the interior of the part—D.C. and not A.C. must be used as the source of magnetizing force.

17. SOURCES OF DIRECT CURRENT FOR MAGNETIZING PURPOSES. Commercial power is today very rarely furnished as direct current and need not be considered here as a practical source of D.C. Other possible sources are motor-generators, storage batteries, rectified A.C. and some very special types of D.C. power supplies providing a very short duration of current flow.

18. MOTOR-GENERATORS AND RECTIFIERS AS SOURCES OF D.C. Motor-generators for welding and electroplating commonly furnish direct current for these processes at voltages from around 6 or 12 volts for plating, up to around 75 volts for welding. Current outputs for welding are normally about 300 amperes, though some generators with larger outputs are available. Plating generators or rectifiers for large installations may deliver much higher amperages, depending on the size of the installation.

Neither of these sources of D.C. is very well adapted to furnishing magnetizing current for magnetic particle testing. The high voltage of the welding generators means a wasteful expenditure of power when high currents are being drawn. This same voltage means that any slipping of prod contacts can cause severe arcing and such flashes can be very damaging to the eyes of the operator as well as to the parts being tested. The sudden application of what constitutes practically a dead short on a welding generator may damage the generator windings in a short time. Furthermore, for many magnetic particle testing applications, the maximum current so delivered is entirely insufficient for the inspection of many parts or welds.

Plating generators or rectifiers delivering several hundred amperes D.C. at from six to twelve volts, are free from arcing troubles because of the low voltage; but their output of current is seldom large enough for most magnetizing purposes. *If* large enough, the installation would be very inflexible, and much more costly than the other sources of D.C. magnetizing currents available today.

19. STORAGE BATTERIES AS A SOURCE OF D.C. In the days before World War II, storage batteries were widely used as a source of direct current for magnetic particle testing. A group of six volt or 12 volt storage batteries connected in parallel will deliver many thousands of amperes momentarily when short-circuited through a coil or through a part, and in many ways provide a very satisfactory source of direct current for magnetic particle testing purposes. In order to magnetize parts with D.C., it is necessary for the current to flow for only a fraction of a second, and the drain on the batteries for each shot is therefore relatively small. The energy withdrawn by a shot may be at once replaced by a charger of sufficient capacity during the interval between shots.

Storage batteries, however, are practically never used today for magnetic particle testing because of a number of undesirable characteristics they possess. Storage batteries deteriorate if not carefully maintained, and when equipment is used only at irregular intervals, it is often found that the batteries are "down" when occasion arises to use the unit. Also the fumes from the charging of the batteries are corrosive, as is spilled battery acid, resulting in unsightly corrosion and damage to equipment. One further disadvantage is that the high currents can be drawn only momentarily, whereas it is often necessary to have the current flow for appreciable intervals of time.

20. D.C. FROM RECTIFIED A.C. By far the most satisfactory source of D.C. is the rectification of alternating current. Both single phase and three phase A.C. are furnished commercially. (See Section 4, Chapter 5.) By the use of rectifiers, the reversing A.C. can be converted to uni-directional current, and when three phase A.C. is so rectified, the delivered direct current is entirely the equivalent of straight D.C. for magnetic particle testing purposes. The only difference between the two is a slight ripple in the value of the rectified current, amounting to only about 3% of the maximum current value.

Advantages of such a source are obvious. Maintenance problems and corrosion, which characterize the use of batteries as a source of D.C., are eliminated. Equipment can be designed to deliver any desired current on any required work cycle. Variation of current value is relatively simple by the use of tapped transformers, saturable-core reactors (power-stats) or variable transformers, in the A.C. part of the circuit. There are many variations in the application of different types of rectified A.C. which are highly useful, and which could not be provided from any other source.

21. HALF WAVE RECTIFIED SINGLE PHASE A.C. When single phase alternating current is passed through a simple rectifier, current is permitted to flow in one direction only. The reverse half of each cycle is completely blocked out. The result is uni-directional current which pulsates—that is, it rises from zero to a maximum and drops back to zero. During the blocked-out reverse of the cycle no current flows. Then the half cycle forward pulse is repeated, and so on at the rate of 60 pulses per second.

When the half-wave current delivered by such a circuit is measured by a conventional D.C. meter, the meter reading will indicate the *average current of the entire cycle*. However, during the non-conductive half cycle, the current is very nearly zero. It is evident, then, that such a meter is actually indicating only one half of the average current that flows during the conductive half of each cycle. This is not a very realistic measurement of the effective magnetizing current level. Through the years it has been established that the average current *during the conductive half cycle* is a more representative measure of the effective magnetization. The average current for the conductive half cycle is double that read on the meter. A simple solution is to double the meter scale, and continue to use the conventional D.C. meter. "Magnetizing ampere" then replaces "ampere" as the unit of meter calibration for half wave current. This is still less than the peak value of current during the conductive cycle, which is 3.14 (π) times the average current. Since the meter scale has been doubled, the peak current of the conductive half cycle is 1.57 ($\pi/2$) times the "magnetizing ampere" reading.

There is a slight skin effect due to the pulsations of the current, but this is not pronounced enough to affect seriously the penetration of the field. The pulsation of the current is in fact useful, as it tends to impart some slight vibration to the magnetic particles, assisting them to arrange themselves to form indications.

The most predominant application for half-wave (H.W.) current is for weld inspection. In this application, half wave current is used with dry powder and prod magnetization, which combination provides the highest sensitivity for discontinuities which lie wholly below the surface. In many cases suitable amperages can be obtained from relatively small portable units. Half wave current is also widely used for casting inspection when sensitivity to sub-surface discontinuities is required. See Chapters 13 and 24.

22. FULL WAVE RECTIFIED SINGLE PHASE A.C. This source of pulsating uni-directional current is sometimes used for magnetic particle testing purposes for certain special-purpose applications. In general it possesses no advantage over half wave, and is not as satisfactory as three phase rectified current when straight D.C. is required, because of its extreme "ripple". Also it draws a higher current from the A.C. line than does half wave for a given magnetizing effect, which is a distinct *disadvantage*.

23. THE SURGE METHOD OF MAGNETIZATION. There is another way to get stronger fields in a part when the current must be left on for more than a fraction of a second. A rectifier unit capable of delivering 400 amperes continuously can put out much higher currents for short intervals. It is therefore possible by suitable current control and switching devices, to pass a very high current for a short period—less than a second—and then reduce the current, without interrupting it, to a much lower steady value. Figure 80 illustrates how this effect is produced. In the figure, the current value for both initial and steady currents are projected onto the hysteresis loop, and from it the values of the fields that result are indicated. The method is called the "surge" method of magnetization. The "surge" current carries the magnetization of the unmagnetized part high on the hysteresis loop to point "A". When the current drops back to a steady value, magnetization drops along the hysteresis curve to point "M", resulting in a stronger field than the same steady current, without the initial surge, can produce. In the latter case the magnetization follows the "virgin" curve, and the field strength rises only to point "N".

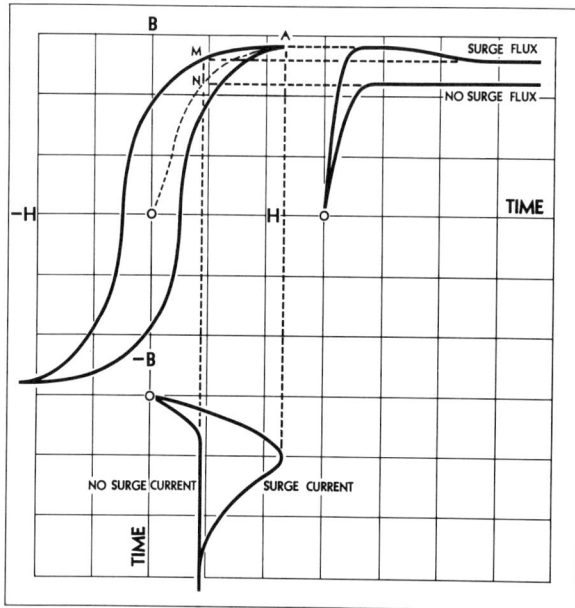

Fig. 80—Effect of a Short Surge of High Current Followed by a
Drop to a Steady Current of Lower Value.

The method was formerly widely and usefully employed for weld inspection, although today the half-wave system has almost universally replaced it.

24. THREE PHASE RECTIFIED A.C. By far the most useful and most widely used source of direct current for magnetic particle testing is rectified three phase A.C. Three phase current is generally used for power equipment in most plants and is preferred over single phase current because of more favorable power transmission and line load characteristics. From the magnetic particle testing point of view it is also preferred because it delivers, when rectified, current which for all practical purposes *is* direct current, and produces all the effects which are required when D.C. magnetization is indicated. (See Fig. 59, Chapter 5.)

From both design and operational viewpoints this system has many advantages. Switching and control devices are operated at A.C. line voltage, which may be 230 or 460 volts or higher, *before* the voltage is stepped down to the low values at which the rectified current must be delivered. By simply increasing the size of transformers and rectifiers, units delivering 20,000 amperes D.C. at low voltage have been built. Much of the equipment that will be described and discussed later in this book will be of this type.

25. SOURCES OF ALTERNATING CURRENT. Alternating current, which must be single phase when used *directly* for magnetizing purposes, is of course taken from commercial power lines, and is usually 50 or 60 cycles per second in frequency. When used for magnetizing purposes, the line voltage, which may be 115, 230 or 460 volts, is stepped down by means of transformers to the low voltages required. At these low voltages, magnetizing currents up to several thousand amperes are often used. Switching and current control devices, as in the case of D.C. rectifier units, are operated at line voltages before step-down.

26. PERMANENT MAGNETIZATION WITH A.C. One of the problems in the use of A.C. is the fact that, normally, the part is not necessarily left with the level of residual magnetism which the peak current of the A.C. cycle generates. When the current is broken the remanent magnetism will depend on the part of the cycle at which the current is when interrupted.

Fig. 81 is a drawing of the sine wave form of a single phase alternating current. The field generated in a part by such a current will rise during the first quarter of the cycle till the current reaches the peak at point "b". As the current drops to zero at point "c", the field in the part drops to the level of residual magnetism which it can retain. As soon as the current reverses, however, the magnetizing cycle also reverses and the field diminishes to zero, and then increases to a negative maximum when the current reaches point "d" on the reverse cycle. The problem then, if it is desired to leave the part permanently magnetized, is to break the current on

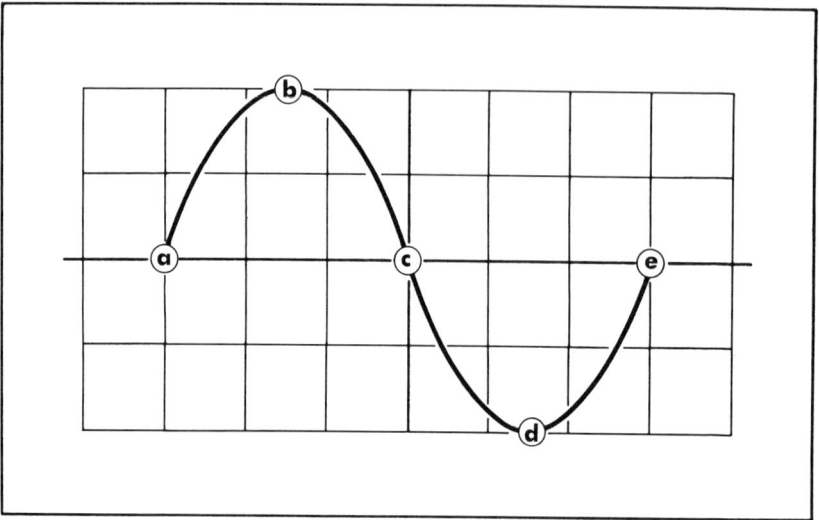

Fig. 81—Sine Wave Form of Single Phase Alternating Current.

that portion of the cycle between points "b" and "c" (or between "d" and "a" on the reverse cycle). This may be accomplished by the arc which is drawn between the switch points when the current is broken. As the arc becomes long it is quenched at the point of zero voltage on the cycle. Fig. 82 is an oscillograph of such a single

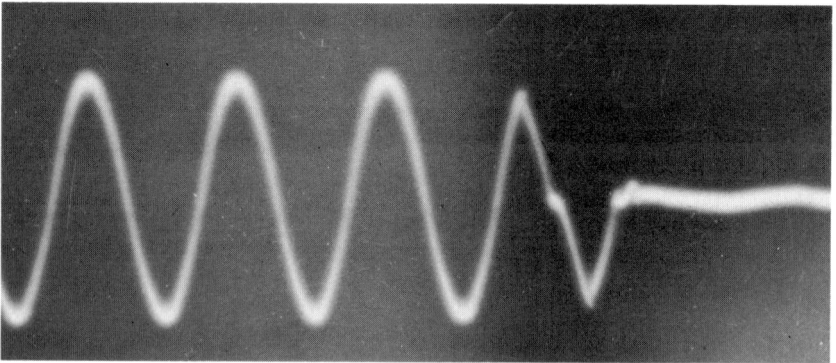

Fig. 82—Oscillograph of the Voltage Across the Switch Points When a Single Phase A.C. Circuit Is Broken, Showing the Arc Quenched at the Point of Zero Voltage.

phase A.C. break. Since the current lags behind the voltage by one-quarter of a cycle, if the voltage is interrupted at the zero point of its cycle, the current will be broken at the peak (point "b" or "d"), thus leaving the maximum residual magnetism in the part. Complete consistency in breaking the cycle at zero voltage is not obtained with ordinary switch gear. An arc must be drawn between breaker points to secure this result. Switch gear is usually designed to suppress such arcing rather than to allow the arc to be drawn. Further, the character of the electric circuit affects the point at which the arc is quenched. As a result of these uncertainties, when the residual method is being used, full-wave rectified single phase A.C. is employed. This form of rectified A.C. is known to give 100% assurance of the maximum residual field in the part.

27. SKIN EFFECT. The skin effect when magnetizing with A.C. has already been described (Section 16). For magnetic particle testing purposes this can be turned to advantage if only surface cracks are being sought. Because the power used and the heating effect on both part and equipment are determined by the "R.M.S."* value of the voltage and current, equipment can be lighter and use less power than a D.C. unit of comparable magnetizing capacity. In many cases surface cracks are the only interest, as for instance in overhaul or maintenance inspection where location of fatigue cracks is the objective. Therefore fields confined to the surface layers of metal are entirely satisfactory and desirable for this purpose.

28. MAGNETIZING WITH TRANSIENT CURRENTS. When a direct current in a circuit is suddenly cut off, the field surrounding the conductors collapses, or falls rapidly to zero. The rapid change of field tends to generate a voltage (and current) which is opposite in direction to that which had been established in the circuit. When ferromagnetic material is under the influence of such a collapsing field the effect is much increased. Under certain conditions the rapid collapse of the field can generate very high currents inside the ferromagnetic material, and the phenomenon can be made useful in some magnetizing problems.

For example, when a bar is magnetized longitudinally in a solenoid, the two ends become poles and the flux lines leave and enter the bar at 90° to the surface. The near-end portions of the bar are therefore not truly longitudinally magnetized, and transverse cracks in this zone will probably not give readable indications.

---

*Alternating current meters are read in "R.M.S." values. This means "root of mean squares", and represents the true measure of the power consumption in the A.C. circuit.

When the current in the solenoid is interrupted with sufficient rapidity, however, a transient current is generated which flows circularly inside the bar, and generates a sort of doughnut-shaped field close to the surface, which is truly longitudinal clear to the ends of the bar. The effect has been called "quick break" or "fast break", and is built into quality D.C. magnetizing units today. Figure 83 illustrates this effect diagramatically.

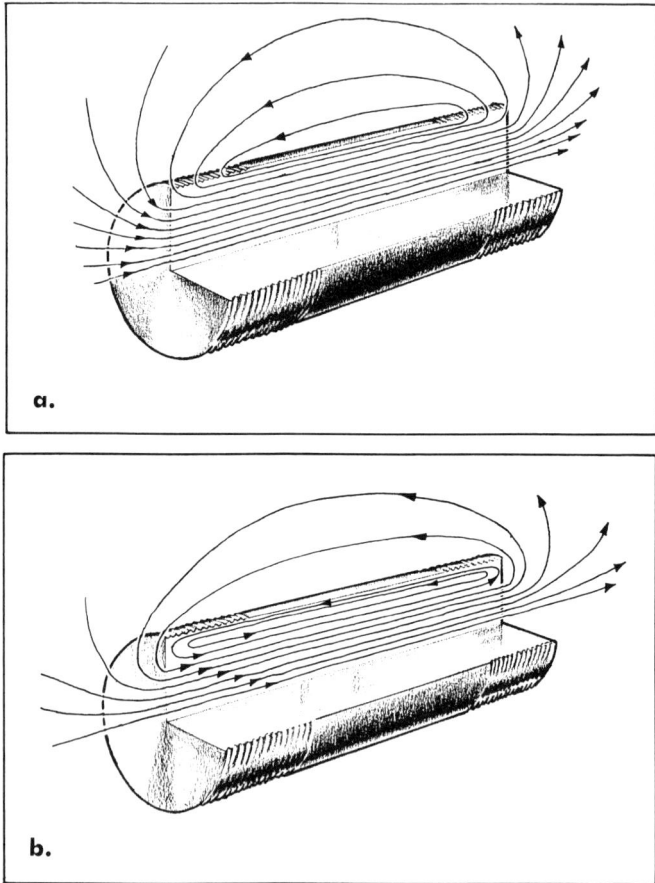

Fig. 83—Residual Field Inside a Bar Generated by
a) Slow Break.   b) "Quick Break" Transient Current.

29. INDUCED CURRENT MAGNETIZATION. An extremely useful application of this collapsing field method of magnetization has been developed for the magnetizing of ring-shaped parts such as bearing races, without the need to make direct contact with the sur-

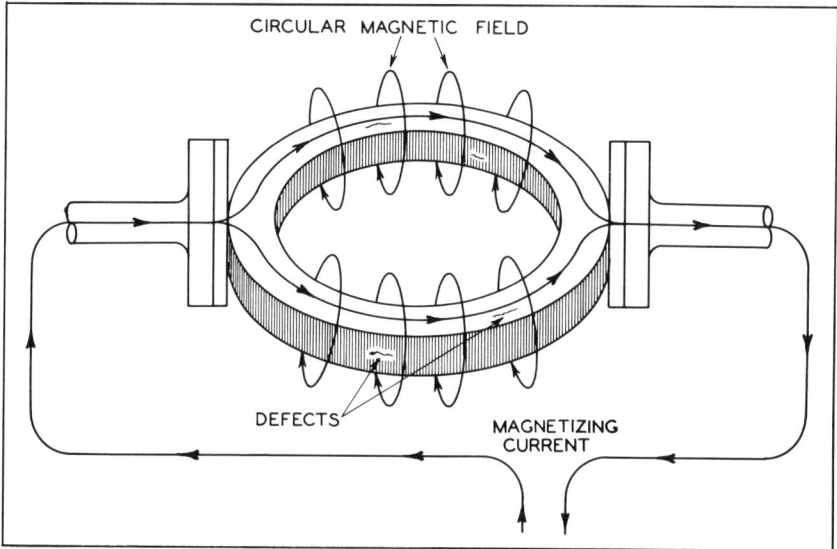

Fig. 84—Direct Contact Method of Magnetizing Ring-Shaped Parts
to Locate Circumferential Defects.

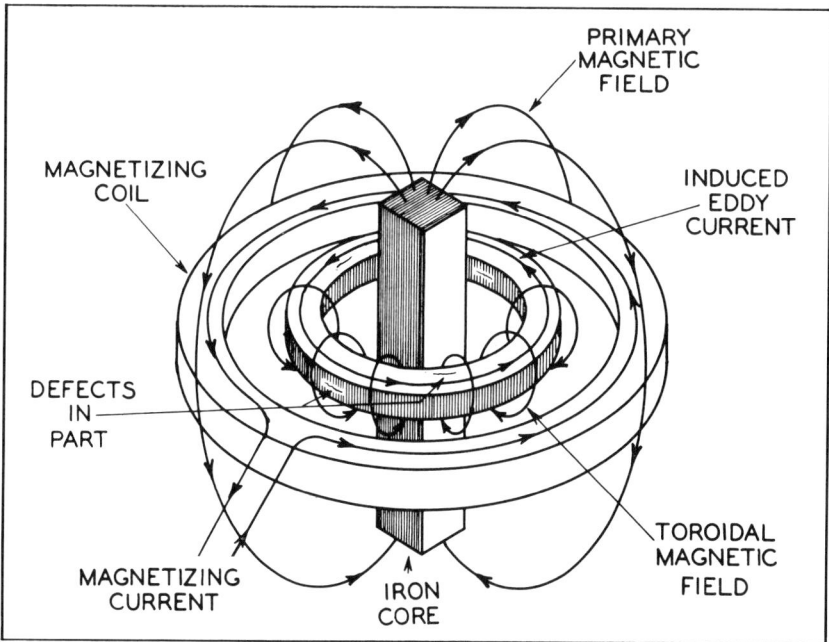

Fig. 85—Induced Current Method of Magnetizing Ring-Shaped Parts
to Locate Circumferential Defects.

face of the part. Figure 84 shows the direct contact method for producing circular fields in a ring to indicate circumferential cracks. To obtain reliable testing of the entire cylindrical surface two magnetizations are required, because the areas near the points of contact where the current enters and leaves the ring are not properly magnetized for this purpose. The ring must be turned 90° and then retested.

By making the ring the one-turn, short circuited secondary of a "D.C. transformer" a large current flowing circumferentially around the ring can be induced. Figure 85 illustrates this effect. To accomplish this a standard magnetizing coil can be used. The ring should be placed inside the coil with its axis parallel to that of the coil. When the coil is energized this arrangement constitutes an air-core transformer, the magnetizing coil being the primary and the ring-shaped part a single-turn secondary. The total current induced in the ring is greatly increased by inserting a laminated core of ferromagnetic material through the ring.

The choice of type of magnetizing current for this method is determined by the magnetic properties of the material to be tested. For parts which have high retentivity, D.C. with "quick-break" is usually used and the parts tested by the residual method. When the D.C. field is caused to collapse suddenly by the abrupt interruption of the magnetizing current, the circular field generated by the transient current so produced leaves the part with a strong residual field. A bearing race is a good example of the type of part that can be tested advantageously by this method.

For parts that are soft, with little or no retentivity, the continuous method must be used. The collapsing D.C. field method is not applicable in this case, since the magnetizing current must flow for an appreciable length of time. By using A.C. or half-wave in the magnetizing coil, however, the current may be left on, and an A.C. or half-wave current is induced in the ring, of the same frequency as the magnetizing current. This current can be allowed to flow long enough to produce indications by the continuous method.

Regardless of whether D.C. and the residual method, or A.C. or half-wave and the continuous method is employed, the induced current method is usually faster and much more satisfactory than the contact method. Only one operation is required and the possibility of damaging the part due to arcing is completely eliminated, since no external contacts are made to the part.

30. FLASH MAGNETIZATION. There are several ways of generating high values of current for very short intervals that have some special applications. These methods have some interesting possibilities, although, to date, they have not been used very often in the field of magnetic particle testing. One example is the use of A.C. (or D.C.) at line voltage, passed through a part, or a solenoid around the part. In the circuit is an ordinary fuse and a switch. When the switch is closed the line is shorted through the part or the coil and current rises to a high value but is quickly cut off again when the fuse burns out. Much higher currents can thus be obtained from line voltages without the use of transformers than would otherwise be possible. Parts are left with high residual magnetism, since the arc produced when the fuse burns out is one that causes the A.C. to be stopped at the peak of the cycle. The method is not very practical, especially at high line voltages, and this magnetizing technique is not generally recommended.

Another example is the use of the discharge current from a large condenser. Such discharges may rise to several thousand amperes for durations of from one ten-thousandth to one one-hundred-thousandth of a second. The condenser is charged to a high voltage by means of a rectified A.C. circuit, and discharged when the voltage rises to a pre-determined point.

31. SUITABLE FIELD STRENGTHS FOR MAGNETIC PARTICLE TESTING. Over the years some rule of thumb values for suitable current strengths for magnetic particle testing have been in use. These have been based on experience rather than on any measured determination of what the requirements actually are. For circular magnetization it has been customary to use the figure of 1000 amperes per inch of diameter of the part being magnetized. If literally followed, this figure quickly gets one into trouble, and must be radically modified in most applications. It is apt to call for currents too high for small diameter parts, and for large diameters is usually unattainable and unnecessary. However, it is pretty safe to say that *if* this amount of current is used, satisfactory field strengths are most likely to result.

No comparable figure for coil magnetization had been determined until quite recently. 200 ampere-turns per inch of coil diameter has been suggested and used, and in many cases has been a fairly good guide.

The trouble with these rules of thumb is mainly that they take no account of varying shapes and magnetic properties of parts which are to be magnetized. Permeability of steels varies widely, and a soft steel of high permeability requires a far lower magnetizing force to produce a suitable field than does a high carbon or alloy steel which may have a very low permeability.

In recent years much work has been done in the laboratory to establish the current requirements more realistically, based on actual conditions met when steel parts of varying composition, sizes and shapes are to be magnetized. The effect of the varying properties of steels, as well as the sizes and shapes of parts, on the strength and distribution of magnetic fields will be discussed in Chapter 9. The methods of arriving at suitable current values to produce fields for satisfactory magnetic particle testing under these varying conditions will also be given in that chapter.

CHAPTER 8

DETERMINATION OF FIELD STRENGTH
AND DISTRIBUTION

1. IMPORTANCE OF KNOWING FIELD STRENGTH AND DISTRIBUTION. It has been emphasized in the two foregoing chapters that field strengths must be adequate and field direction favorable, in relation to the size and direction of the discontinuity, in order that a good indication be produced with magnetic particles. It is, therefore, obviously important that the operator applying the magnetic particle method know how strong the field is inside the part being tested; and also the direction which it has taken at any point where a defect is being looked for. In other words, he should know the field distribution and strength, inside the part.

The *need* to know exists and is important; but the *means for determining* field strength and distribution inside a part with any degree of accuracy, are less than satisfactory.

2. MEASUREMENT OF FIELD INSIDE A PART. There is no generally applicable method known today which permits the exact measurement of field intensity at a given point *within* any given piece of magnetized iron or steel. In order to measure field strength it is necessary to intercept or cut the flux lines, and this it is impossible to do without cutting the steel itself. Cutting the steel would at once change the value of the field density inside the part.

Explorations of flux density inside a piece of magnetized steel have been carried out by drilling a small hole into or through the piece, and inserting a small rotatable coil or other measuring device. The field strength actually being measured by this procedure is, however, *not* the field inside the metal, but only that which jumps the air gap of the drilled hole. Such measurements are useful in indicating the gradients of field intensity, as for instance from surface to center of a part; but the actual densities obtained are comparative only.

3. EXPERIMENTAL FIELD MEASURING TECHNIQUES. There are of course methods for measuring magnetic fields and these will be described. However, only one of these—the flux meter—measures the flux density *inside* a part; and this measurement must be made on test specimens of suitable special shape. The procedure is not useable for a point to point measurement of field strength inside a part already magnetized. It has some useful applications but is not an answer to the general problem.

4. FLUX METERS. The flux meter is an instrument that measures the *total change of flux* through a coil, independent of the *rate* of change. It consists of a coil connected to a ballistic galvanometer through suitable long-time-constant circuitry. When the magnetic flux through the coil changes, either by moving the coil or the part, or otherwise varying the field in the part by some means, the needle of the galvanometer swings and indicates the amount of change. The change in flux may be in either the positive or the negative direction.

If the flux is zero to begin with and is increased to a value "A" the total change will be the flux at the value "A". If the area of the coil is known, the average flux density within the coil—that is, the average flux per unit of area—can be readily calculated.

This device measures the *average* flux, and consequently the flux density, through the coil, provided the flux through the coil is zero either at the start or at the end of the measurement.

If an unmagnetized piece of ferromagnetic material is magnetized while in the coil, the total change of flux is indicated by the meter. Or a piece of steel, already magnetized, may be inserted from outside the coil. The total change of flux, from zero to the flux in the part will be indicated. Or, the magnetized part may be at rest inside the coil and then be withdrawn, reducing the flux through the coil to zero. Again, the meter registers the change, which is the total flux in the part.

The coil is separate from the meter, and is wound by the user to suit the particular problem at hand. It is connected to the flux meter by a pair of electrical leads. The flux measured is that parallel to the axis of, and within the coil. Therefore the *direction* of the field must also be known before a measurement can be made. In an irregular shaped part the direction of the field is not easily determined. But with a symmetrical part, such as a bearing race, which has uniform cross-section around its circumference, the direction of the field when magnetized with a central conductor is circumferential. In such a case the measurement of the flux is very accurate. The coil is wound around the race so as to be perpendicular to the cross-section. Figure 86 illustrates this arrangement. Generally the coil should be made to fit snugly; however, little error is introduced by coil fit unless the fit is very poor, since, due to the high permeability of the steel compared to air, nearly all the flux will be in the steel ring.

The change in flux is obtained by first setting the meter to zero with the magnetizing current off, and then noting the reading when the current is turned on. The value of the flux is obtained by the formula

$$\text{Flux} = \frac{\text{K} \times \text{Deflection}}{\text{Number of Turns}}$$

$$= \frac{10^4 \times \text{D}}{\text{N}}$$

K is a constant for the meter used, which in the example is 10^4
D is the deflection read on the galvanometer
N is the number of turns in the coil

The flux density is obtained by dividing the total flux by the cross-section of the coil

$$\text{Flux Density} = \text{B} = \frac{\text{Flux}}{\text{Coil Area}}$$

$$= \frac{10^4 \times \text{D}}{\text{NA}}$$

A is the area of the coil. The result is in terms of lines per unit of area. If the cross-section is expressed in square centimeters, the resulting value of flux density will be in Gausses.

The instrument may be used to measure the flux density in a steel plate by drilling small holes through the plate so that a flux meter coil can be wound through the holes. The meter will measure the flux at right angles to a line between the holes and within the coil when the plate is magnetized.

5. MAGNETIC FIELD METERS. Many field meters are available which measure the magnetic field in air. As used in determining field distribution in magnetic particle testing these meters almost

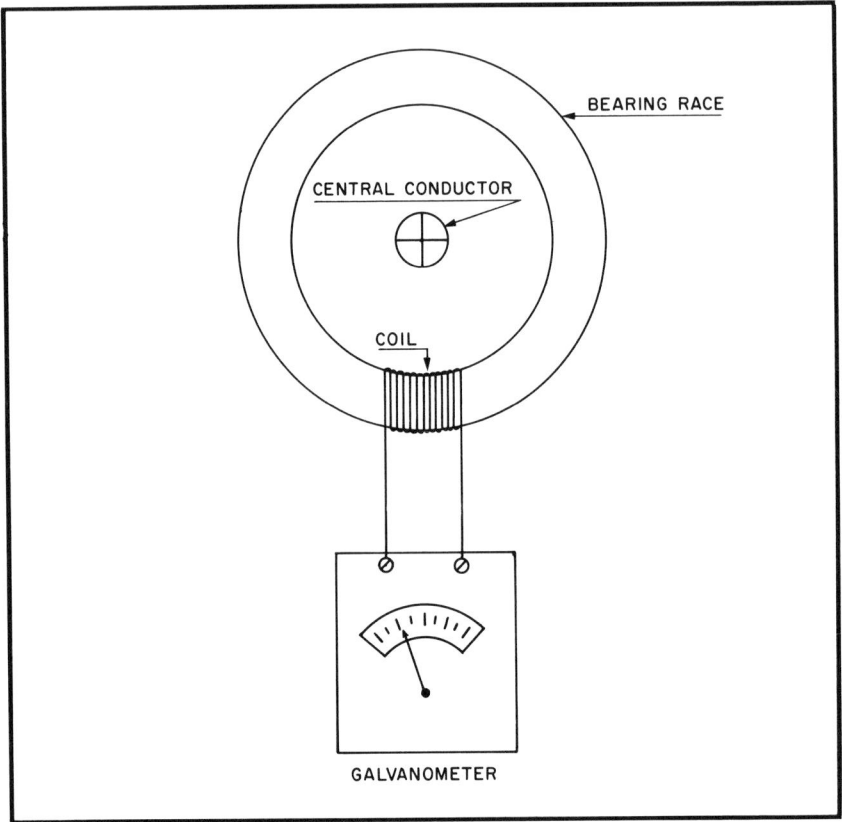

Fig. 86—Flux Meter Arranged to Measure the Flux in a Bearing Race.

always measure "H" (magnetic field strength) rather than "B" (flux density), even though the meter may be calibrated in Gausses which is the unit of flux density. See the definitions of H and B in Chapter 5, Sections 3(6), 3(9) and 3(15). If the instrument is direction sensitive — for example, based on the Hall effect (Chapter 5, Section 3(7)), then it may be used to measure the tangential component of H *at the surface of the part*. The tangential component of H is that component of the field having direction parallel to the surface of the part. Since the value of the tangential component of H is the same on either side of the boundary between the steel and the air, the value measured in air is also a measure of the field strength close to the surface inside the part. If the permeability of the steel is known the flux density can be calculated.

Instrumentation of this type is useful whether or not the permeability is known, because it does give information on the direction of the field, and because it can be used as a comparative measuring device. Thus, if the level of magnetization required for the satisfactory testing of a part has been determined by trial, the tangential component of H can be measured for that field at a given point on the surface of the part. By requiring that similar pieces when magnetized give the same reading on the meter at that same point, assurance can be had that all pieces of that type will be tested at the same level of magnetism, even when the test is made by different operators and on different equipment.

Meters which measure H, and are sensitive to its direction, can be based on a variety of principles, of which the Hall effect is only one. One of the simplest meters is based on the force exerted on a permanent magnet placed in the magnetic field. The magnet will tend to align itself in the direction of the field—as, for example, the magnetic compass—and, if restrained by a spring, can yield quantitative information regarding the strength and direction of the field. Such magnets can be made very small, can be attached to a rigid shaft and located some distance from the meter movement. Such devices have been used to explore the field *distribution* inside large parts, by drilling holes into the part and slipping the magnet down the hole, taking readings as a function of depth.

It should be understood, however, that this technique of drilling holes into a sample and then inserting a field-sensitive device into the hole, does not yield the true value of the field *inside the part,* since the removal of the metal in the drilling of the hole causes a redistribution of the field in the same manner as when caused by a (sought-for) discontinuity. The measurement is still made "outside" the part.

Another version of this kind of field measuring device utilizes a soft iron vane which will, when placed in a magnetic field, tend to align itself with the direction of the field. This tendency can be made quantitative by restraining the motion of the vane with a spring or with the field of a permanent magnet, and transmitting the movement of the vane to a pointer and scale. Figure 87 is a sketch of such an instrument.

6. MAGNETOGRAPHS. Magnetographs are made by placing a paper over a magnetized piece of iron or steel, or around a conductor or coil carrying current, and sprinkling fine iron particles over the

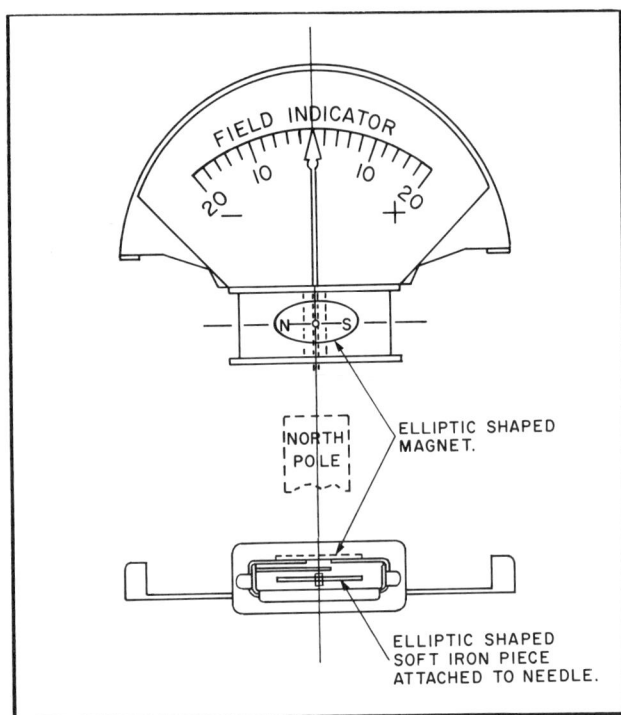

Fig. 87—Diagram of Soft Iron Vane Type of Magnetic Field Meter.

paper. Such magnetographs can be used to give some idea of how fields are distributed inside and around the part or conductor. The patterns of the magnetographs, however, are merely pictures of the field distribution *outside* the part, and give only indirect information as to how the field may be distributed *inside*. However, even though this is true, some inferences can be drawn from the magnetograph pattern, as illustrated in Fig. 88. Here a piece of soft iron has been placed adjacent to a conductor carrying direct current. The pattern shows lines entering and leaving the iron, and the path can be pretty well inferred.

7. FLUX SHUNTING DEVICES. A variety of devices have been developed for the intended purpose of insuring that the field distribution in a particular part being tested is of proper magnitude and direction. These devices operate by attempting to "shunt" some of the field out through the surface into an external test piece and back through the surface again. One of these devices is known as the Berthold Field Gauge. Another, developed in Japan, is called

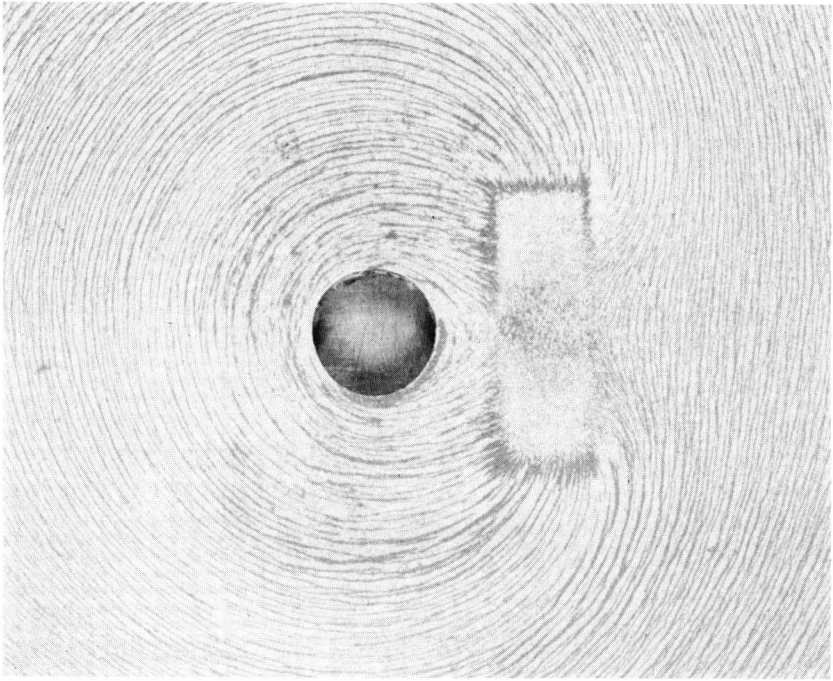

Fig. 88—Magnetograph of the Field Around an Iron Piece Placed Adjacent to a Conductor Carrying Direct Current.

the "Magnetization Indicator", and is reported in the proceedings of the Third International Conference on Nondestructive Testing which was held in Tokio, Japan, in March of 1960. These indicators consist of soft iron pieces into each of which have been machined an "artificial defect" in the form of a straight slot. In use, the indicator is placed on the part to be inspected so that the artificial defect is in the direction of the cracks that are expected to occur in the part. The part is then magnetized and the magnetic particles applied in the usual manner. If the artificial defect in the indicator is shown, then magnetization is considered to be proper. The proper level of sensitivity for various sizes of defects is achieved by varying the width and depth of the artificial defect.

If properly used these devices can be very valuable. It should be born in mind however, that if the continuous method is being used, then the indicator is more truly indicating H, the magnetizing force through the indicator, than it is the field strength, B, through the surface layers of the magnetized part. On a symmetrical part the indication would be the same on copper, aluminum or other non-magnetic material as it would be on a magnetic part.

As long as the operator understands that it is H that is being measured and that he is not actually measuring the field inside the part, the device can be useful for assuring that the magnetization of the part is of the order of strength required.

8. CALCULATION OF FIELD DISTRIBUTION. It would be convenient if the field distribution within a part could be calculated from dimensional and magnetizing current data. However, no method is available for doing this on any broadly useful basis. In some cases it *is* possible to obtain a magnetic field distribution plot by calculation from theoretical considerations. In general, however, calculation of field distribution is limited to a few relatively simple shapes. As a practical means of arriving at field distribution inside a part for magnetic particle testing purposes, calculations of this sort are of very little value. Still, a brief review of mathematical approaches may be helpful in understanding what the problem really is.

The mathematical solutions of field distribution problems usually involves the use of LaPlace's equation. This mathematical procedure is difficult and cumbersome, and not to be undertaken as an exercise by any but a skilled mathematician. For such, the calculations necessary to arrive at the field distributions for example, for a magnetic sphere and a magnetic elipsoid in a uniform magnetic field, as plotted in Fig. 89, would still be laborious but perhaps not too difficult. Figure 89 shows the calculated field distribution both inside and outside the sphere and the elipsoid. In both cases the field inside is uniform. The set of conditions assumed for this case apply for coil or longitudinal magnetization. Details of the mathematical calculations will be found in reference 8, cited at the end of this chapter.

Mathematical methods are also applicable where parts are magnetized by the direct contact or central conductor method (that is, circular magnetization) provided the geometrical shape of the part is extremely simple. Figure 90 shows plots of the calculated field distribution in and around two magnetic tubes carrying direct current, having two different degrees of eccentricity.

9. TRANSFORMATION METHODS. The transformation method is frequently called the conformal mapping method. It is based on the fact that there exists a class of complex mathematical functions, known as analytical functions, whose real and imaginary parts

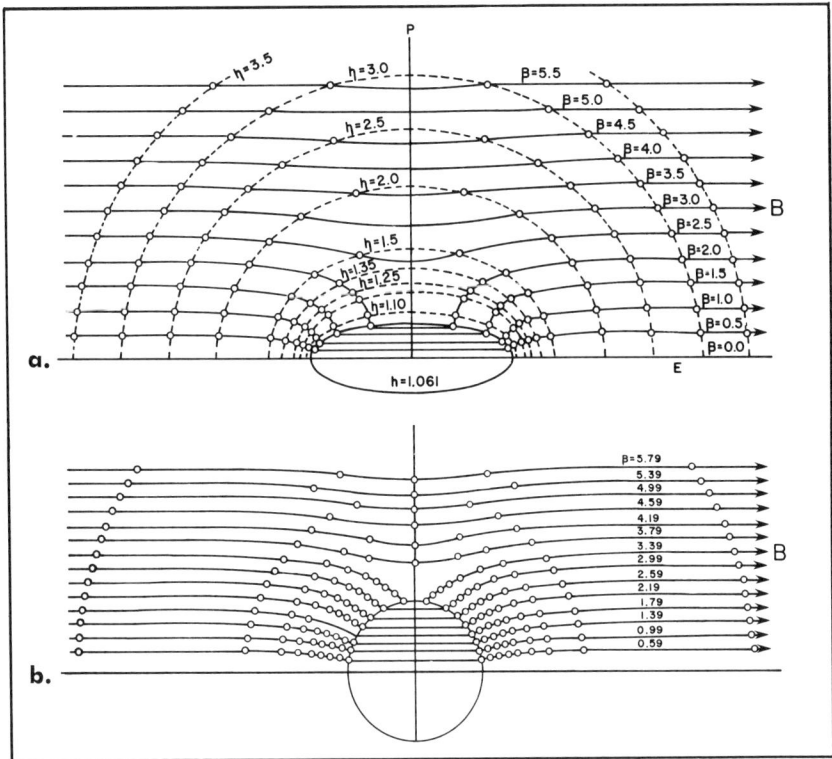

Fig. 89—Calculated Field Distribution In and Around a Magnetic Sphere
and an Elipsoid in a Uniform Magnetic Field.

are separate solutions of LaPlace's equations, and that these functions can be transformed (by changing the variable) while still retaining their analytical properties. This means that a solution of
LaPlace's equation which satisfies a particular set of conditions, can
be transformed into a solution which fits other sets of conditions,
if only the transformation can be found which changes the conditions of the original problem into the conditions of the problem the
solution of which is desired. A simple example is shown in Fig. 91,
where the proper transformation changes the plot of a simple
parallel field distribution just under the flat surface of a piece of
steel (a) into the flux distribution at a right angle corner in a large
piece of steel (b). Many elaborate transformation functions have
been worked out and are compiled in "dictionaries". The method
applies in many other areas of engineering as well as magnetic field
analysis.

173

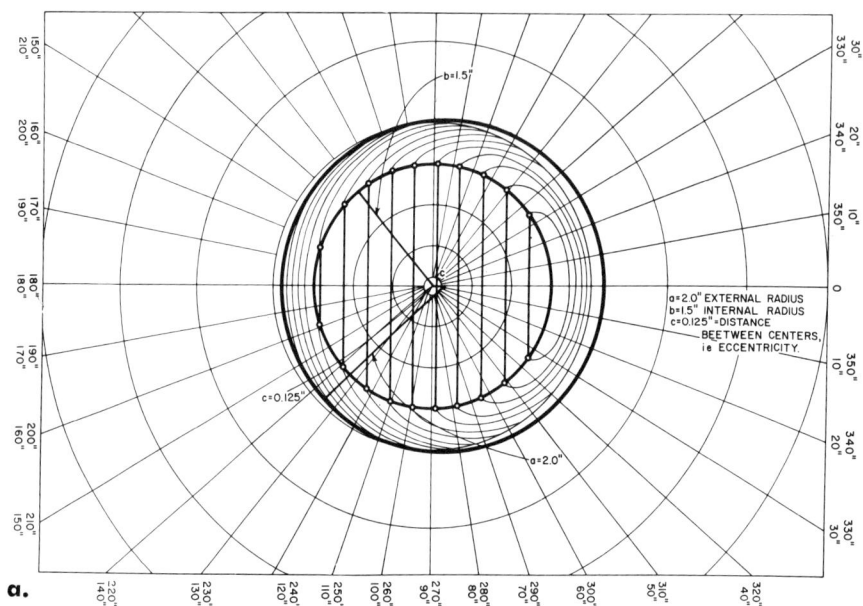

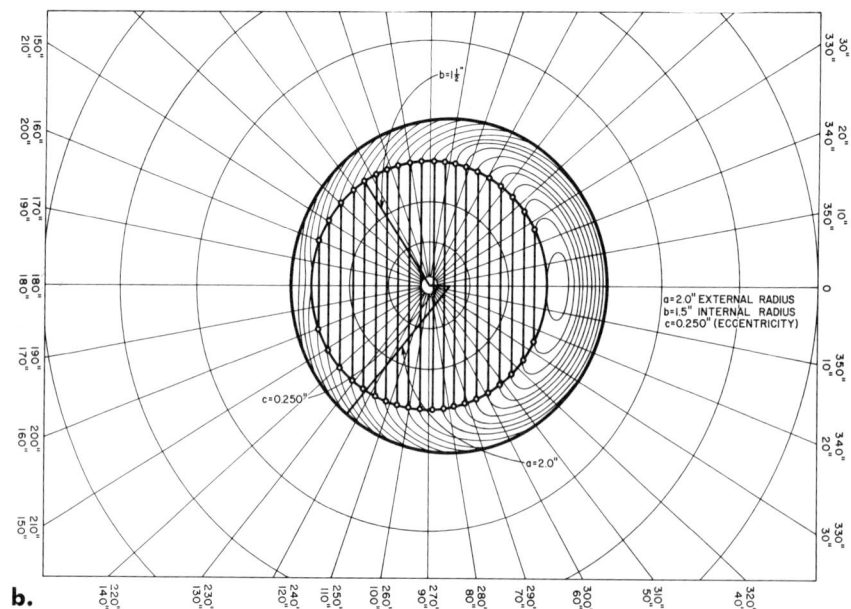

Fig. 90—a) Calculated Field Distribution In and Around a Ferromagnetic Tube of
Moderate Eccentricity, Carrying Direct Current.
b) Calculated Field Distribution In and Around a Ferromagnetic Tube of Extreme
Eccentricity, Carrying Direct Current.

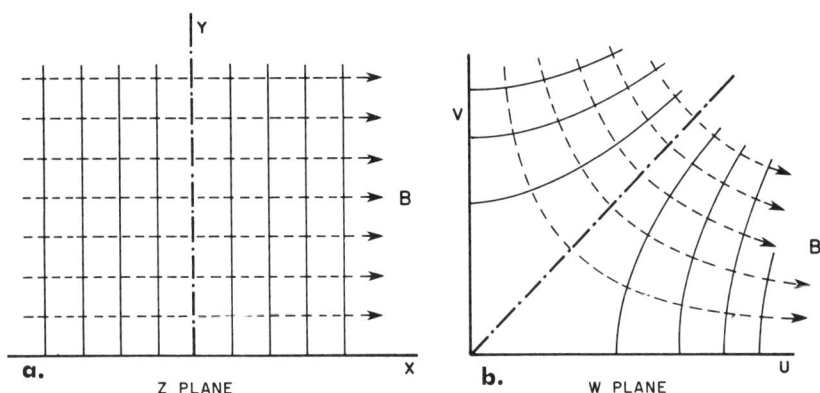

Fig. 91—Example of Conformal Mapping Technique for
Analyzing Field Distribution.

10. ANALOG METHODS. The fundamental differential equations governing the field distributions in electrostatics, magnetostatics, stationary electric currents, stationary temperatures, fluid dynamics and gravitational dynamics, are the same. Thus, solutions to problems in one field can be transferred directly for use on similar problems in another field. This means that it is possible to obtain magnetic field distributions measuring the electric flux density, or electric current density, or the heat power flow density, etc., in models which are constructed to be similar to the magnetic problem whose solution is desired. This technique has been used widely to determine the electric field distributions in vacuum tubes. It is not often used for magnetic measurements because a direct analog of permeability is almost impossible to obtain.

11. FIELD PLOTTING. It is possible, with a little practice, to sketch fields with considerable accuracy without recourse to mathematical methods. It is known, for instance, that the flux lines and the lines of equal magnetic potential are everywhere at right angles to each other. Following is a list of the facts and the properties of magnetic fields which are made use of in such field distribution sketching;

1. The flux and potential lines form families of curves, intersecting each other at right angles.

2. Flux lines generally enter or leave a magnetic surface at right angles to that surface.

3. Flux and potential patterns should be drawn as curvilinear squares—that is, four-sided figures with right angles at the

175

corners and with equal average lengths of opposite sides. These squares are drawn such that
(a) The same potential difference exists across each square,
(b) The same flux passes through each square.
Examples are shown in Fig. 92, (a) and (b). Fig. 92-a gives the sketch for an iron bar having a change of cross-section, and Fig. 92-b gives the partially finished sketch for a rectangular bar having a notch.

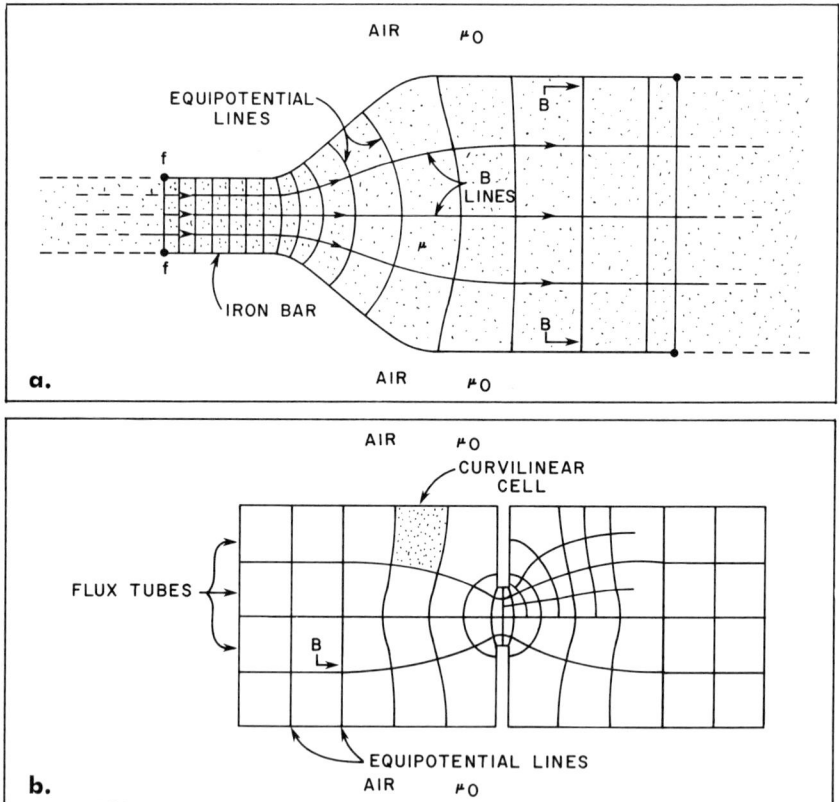

Fig. 92—Sketches of Fields for
a) An Iron Bar Having a Change in Cross-Section.
b) A Rectangular Bar Having a Notch (Sketch Only Partially Complete).

BIBLIOGRAPHY FOR CHAPTER 8
1. NEHARI, Z., *"Conformal Mapping."* McGraw-Hill (1952).
2. CHURCHILL, R. V., *"Introduction to Complex Variables."* McGraw-Hill (1948).

3. KOBER, H., *"Dictionary of Conformal Representations."* Dover (1952).

4. KRAUS, JOHN D., *"Eloctromagnetics."* McGraw-Hill (1953).

5. PAGE, L. AND ADAMS, Y. I., *"Principles of Electricity."* Van Nostrand (1949).

6. WEBER, ERNST, *"Electromagnetic Fields, Vol. I, Mapping of Fields."* John Wiley and Sons (1950).

7. SMYTHE, W. R., *"Static and Dynamic Electricity."* McGraw-Hill (1950).

8. McCLURG, G. O., *"Magnetic Field Distribution for a Sphere and an Elipsoid."* Amer. Jour. of Physics, Vol. 24, No. 7, pp 496-499. October, 1956.
and *Ibid*, Vol. 25, No. 4, P 266. April, 1957.

9. McCLURG, G. O., *"Theory and Application of Coil Magnetization."* Journal Soc. for Nondestructive Testing. Jan.-Feb., 1955, pp 23-25.

CHAPTER 9

FIELD STRENGTH AND DISTRIBUTION
IN SYMMETRICAL OBJECTS

1. INTRODUCTION. In the previous chapter we have considered available methods for measuring the strength, direction and distribution of magnetic fields in and around magnetized ferromagnetic materials without, however, taking into account how these fields were produced. We have also surveyed the possible methods of arriving at field distribution by calculation or other theoretical procedures. It is apparent that there are no quick and easy methods for accomplishing either objective for the purposes of magnetic particle testing. An approach *is* possible by inference and extrapolation from the information provided by the use of various field-measuring devices which give the strength and direction of the field in the air outside the magnetized part.

Fortunately, the requirements for field strength for magnetic particle testing are not particularly critical, and in most cases a considerable range is permissible. Experience has built up a large amount of "know-how" for the proper magnetization of various types of material, and of sizes and shapes of parts, and for the use of available magnetizing means.

2. ELECTRO-MAGNETIC FIELDS. Magnetization of parts for magnetic particle testing is accomplished, with very few exceptions, by the use of electric current. The exceptions are those few cases where permanent magnets are used—usually when electric current is not available, as in some inspections in the field, or where explosion hazards exist in a plant and use of electricity is excluded.

Magnetic fields in and around conductors carrying current follow certain laws which are well understood. These laws are very helpful in arriving at approximate values for field strengths under various conditions, including the directions and distributions of these fields. In the following sections we will discuss the behavior of fields in and around conductors carrying currents, and how this behavior affects proper application of magnetization for magnetic particle testing purposes. The discussion will be confined to magnetic and non-magnetic conductors which are relatively small in size and symmetrical in shape.

3. PERMEABILITY OF MAGNETIC MATERIALS. The term "permeability" is constantly used in magnetization discussions in connection with magnetic particle testing. The concept of permeability is, however, very often misunderstood. The question, "What is the permeability of a certain kind of steel?" cannot be answered in a simple statement. There is no such thing as *the* permeability of a piece of steel. Permeability is a ratio—B/H—and therefore varies as H is varied. Quite a few permeabilities have been defined, such as material permeability, maximum permeability, initial permeability, and apparent permeability among others.

4. MATERIAL PERMEABILITY. In magnetic particle inspection with *circular magnetization* it is *material permeability* which is of interest. By material permeability we mean the ratio of the flux density, B, to the magnetizing force, H, where the flux density and magnetizing force are measured when the flux path is entirely within the material. In such experiments one measures H, the magnetizing force, and B, the flux density produced by that magnetizing force, point by point for the entire magnetization curve, using a flux meter and a prepared specimen of the material. Figure 93-a shows such a curve, plotted for a hypothetical steel having a relatively low material permeability. On this figure the heavy curve is called the normal magnetization curve. Figure 93-b shows a point by point plot of the ratio of B to H. This curve shows how the material permeability, μ, varies as a function of H. Notice that μ varies from some initial value to a maximum and then tapers off to an eventual value of one at saturation.

For magnetic particle testing, the level of magnetization is generally chosen to be just below the knee of the magnetization curve. As can be seen in Figures 93, -a and -b, the maximum permeability occurs near this point. Consequently, for circular magnetization we are interested in the *maximum material permeability*. For most engineering steels the maximum material permeability is in the range from 2000 to 5000 Gauss per Oersted.

In order to emphasize the fact that the material permeability is not constant for a given steel, the B vs H curves have been plotted in Fig. 93-a, which would result if μ *did* remain constant and had the values, respectively, of 1, 10, 100 and 1000. These are the dotted lines on the chart.

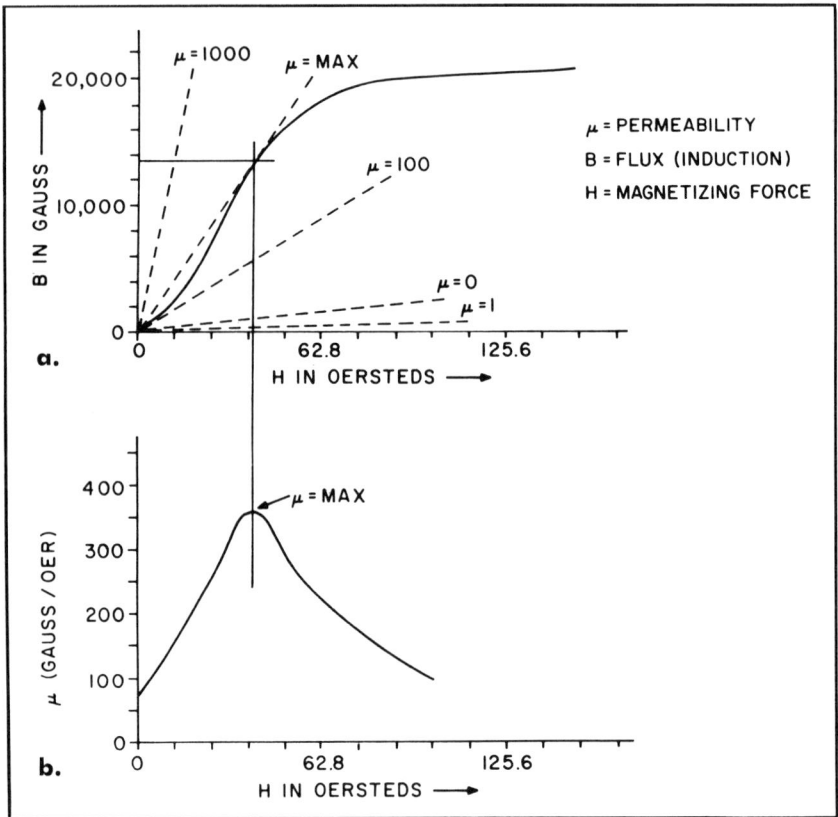

Fig. 93—a) Magnetization Curve for a Steel Having Relatively
Low Material Permeability.
b) Plot of the Material Permeability of the Steel Relative to H.

5. EFFECTIVE PERMEABILITY. In the case of coil or longitudinal magnetization the permeability of prime importance and interest is quite different from the material permeability. The name for this permeability in common useage is "apparent permeability", but for magnetic particle testing purposes the term "effective permeability" is preferred as being more descriptive. The effective (apparent) permeability is defined as the ratio of B in the part to H, when H is measured at the same point in absence of the part. The effective permeability is not solely a property of the material, but is largely governed by the shape of the piece.

6. INITIAL PERMEABILITY. Initial permeability is the permeability of an un-magnetized material as the magnetizing force, H, is applied for the first time. With increasing H the field in the part increases along the "virgin" curve of the Hysteresis loop.

It should be noted that when the term permeability is used without qualification, it is the *maximum material permeability* that is being referred to.

7. SELF-DEMAGNETIZING EFFECT. When a part is placed in the magnetic field of a coil, magnetic poles appear near the areas at which the field enters and leaves the part. These poles produce in the part another magnetic field which is opposite in direction to the applied field of the coil, and consequently weakens the field in the part. This weakening is called "self-demagnetization", and the amount of it depends on the distance between the poles and on their concentration. Thus the self-demagnetization depends on the geometric shape of the part. Obviously the effect also depends on the orientation of the part with respect to the applied field, since the distance between the induced poles will depend on whether the long or short axis of the part is held parallel to the applied field. In magnetic particle testing most parts magnetized in a coil are placed with their long axes parallel to the applied field.

The self-demagnetizing effect is responsible for the longitudinal field in the part being much less than might be expected from the strength of the magnetizing force applied. This weakening of the field in the part is taken into account by the effective permeability. The effect is discussed more fully in the paper entitled "Theory and Application of Coil Magnetization" which was published in the Jan.-Feb., 1955 issue of the Journal of the Society for Nondestructive Testing—Vol. 13, No. 1, pp 23-25. In that paper it is shown that the effective permeability can be given by the equation

$$\mu_{eff} = 6 \frac{L}{D} - 5, \text{ where } \frac{L}{D} \text{ is the length-to-diameter ratio of the}$$

part to be magnetized. This equation is valid providing the $\frac{L}{D}$ ratio is not greater than 15, and the material permeability is greater than or equal to 500. Neither of these conditions is very restrictive, since the material permeability of most engineering steels is considerably greater than 500, and few parts magnetized in a coil have L/D ratios greater than 15.

8. RULE FOR DETERMINING AMPERE TURNS FOR LONGITUDINAL MAGNETIZING. Since the effective permeability is independent of the material permeability (if the material permeability is greater than 500) a coil magnetization thumb rule can be developed which fits all steels and which specifies the number of ampere turns required to produce sufficient magnetizing force to magnetize the part adequately for inspection. This gives a far more dependable way than earlier attempts to arrive at the proper number of amperes which must pass through a coil of a known number of turns, to magnetize satisfactorily any given part for longitudinal inspection.

The paper on theory and application of coil magnetization referred to above shows that this thumb rule is

$$NI = \frac{45,000}{L/D}.$$ In the equation, N represents the number of turns in the coil, I is the current in amperes, and L/D is the length-to-diameter ratio previously defined. When the part is magnetized by placing it on the *bottom* of the circular magnetizing coil, adjacent to the coil winding, this thumb rule will produce a flux density of about 70,000 lines per square inch. Extensive experimental work has shown that a field of 70,000 lines per square inch is a satisfactory field strength for almost all cases of coil magnetization.

Note that the rule does not require information regarding the diameter of the coil. This independence of coil size holds, provided the cross-sectional area of the part to be magnetized is not greater than 1/10 of the cross-sectional area of the coil.

For those cases when it is desirable to magnetize the part by centering it in the coil, a somewhat different rule applies. The equation for this case becomes

$$NI = \frac{43,000 \; R}{\mu_{eff}}$$ where R is the radius of the

coil in inches and $\mu_{eff} = 6\frac{L}{D} - 5$. Note that when the part is centered in the coil, one must take into account the size of the coil since larger coils require more ampere turns to produce the same flux density in the piece. This equation applies only to parts which have a cross section area not greater than 1/10 that of the coil.

9. MINIMUM PERMEABILITY REQUIRED FOR MAGNETIC PARTICLE TESTING. The increasing use of high strength magnetic stainless steels has made it necessary to know the answer to the question

"What is the minimum permeability of a material that will permit satisfactory testing with magnetic particles?" In answering this question, circular magnetization and coil magnetization must be considered separately.

10. MINIMUM PERMEABILITY FOR COIL MAGNETIZATION. As has already been said, for longitudinal magnetization (coil magnetization) the permeability of interest is the effective permeability μ_{eff}. The effective permeability is very nearly independent of the material permeability as long as the maximum material permeability is greater than 500. Therefore, for coil magnetization, the question regarding minimum permeability for satisfactory testing with magnetic particles can be answered by saying that the inspection will be satisfactory if the steel has a maximum material permeability greater than 500. Under these circumstances the currents required for coil magnetization should be the same as those required for steels whose maximum material permeabilities are in the usual higher range.

11. MINIMUM PERMEABILITY FOR CIRCULAR MAGNETIZATION. For circular magnetization the permeability to be considered is the *maximum material permeability* (see Section 4). Experience has shown that nickel tubing can be successfully tested with magnetic particles, using circular magnetization. The maximum material permeability of nickel is approximately 500. Since it has been shown that coil magnetization works satisfactorily on steel whose maximum material permeability is as low as 500, and in view of the successful results on nickel at this level, it is safe to conclude that defect detection by circular magnetization will also be satisfactory at a maximum material permeability of 500. As a matter of fact, this is probably a conservative figure, and satisfactory performance might well be expected from considerably lower values for material permeability, at least for the detection of surface discontinuities.

12. CURRENTS REQUIRED. The values of currents required for successful circular magnetization will probably be higher for a steel of 500 maximum material permeability than for a similar piece made of material whose maximum permeability is 2000. Currents would be still higher if the maximum permeability were as low as, say, 100. Therefore, although theory indicates that magnetic particle testing could probably be carried out successfully on materials whose maximum material permeability is as low as even 100, it might be impractical to carry out the test (using circular magnetization) because of the heavy current requirements. The same reasoning applies equally to longitudinal or coil magnetization. Because of the self-magnetizing effect (see Section 7) the magnetizing force to produce a field of 70,000 lines per square inch would be even greater than for circular magnetization. Ampere turns required of a coil under these circumstances might again be beyond the limits of readily available equipment.

13. EFFECT OF SHAPE ON FIELD DIRECTION. As long as the piece to be magnetized is of regular shape and moderate size, the direction a field will take under the application of various magnetizing forces is quite readily predictable. Bars, square or round, or tubes or rings are examples of such simple shapes. And in a qualitative way at least, the relative intensity of the field may also be predicted, given some knowledge of the hardness and the composition (and therefore the permeability) of the metal.

14. LONGITUDINAL MAGNETIZATION. When a length of bar or tube is magnetized by means of a coil we have what seems to be a simple problem of distribution. Magnetographs of bars so magnetized indicate a concentration of flux lines at the ends, suggesting that the field traverses the bar from end to end, with some flux loss along the sides as indicated in Fig. 94. If the bar is relatively

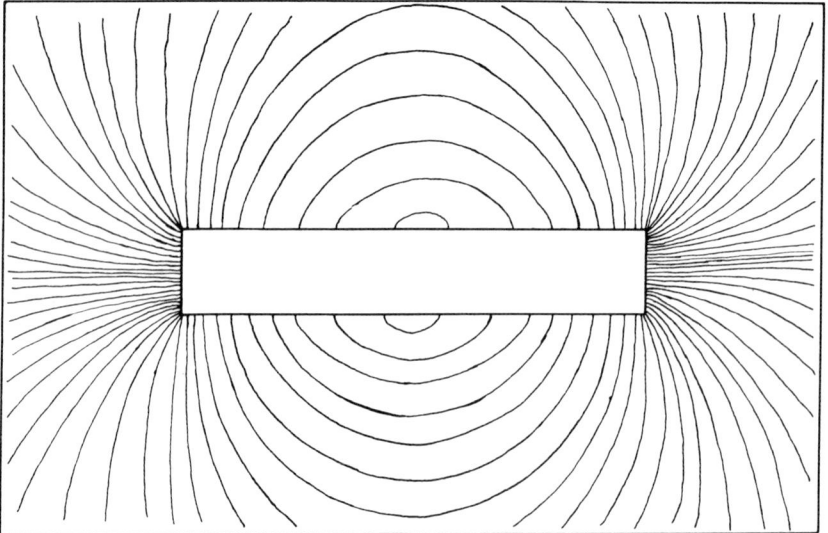

Fig. 94—Diagram of the Field Around a Bar Magnet.

...

uniform as to permeability, across its section and along its length, it is a pretty safe assumption that the flux density is approximately uniform over the cross-section at any point except at each end of the bar.

15. DISTORTION OF FIELD DUE TO SHAPE. If, however there is an upset section along the length of the bar, as shown in Fig. 95, the field tends to flow out into the upset portion, but does not do so uniformly. The larger the relative diameter of the upset portion, the farther will the field in this section depart from a strictly longitudinal direction. External poles will tend to form at "C" and "X", and the field direction will tend to become radial along the surface "BC". Such a field distribution is favorable for finding circumferential cracks in the fillet at "B", and in the surface "BC" —but may not be favorable for locating circumferential cracks on the surface "CX". This would be especially true if the diameter of the upset portion is large with respect to the distance "CX". This distribution of field can be verified by the indication of poles at "C" and "X" by means of a small compass or field meter.

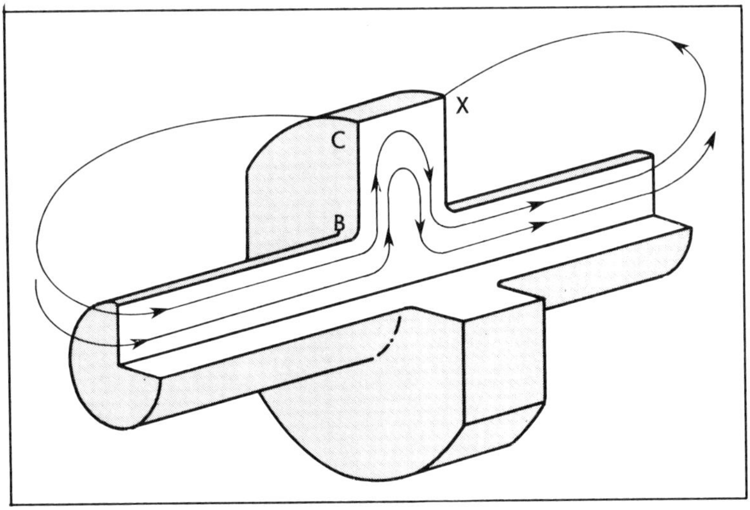

Fig. 95—Behavior of a Longitudinal Field in the Upset Portion of a Magnetized Bar.

When attempting to magnetize a part of irregular shape for the first time, some such analysis of the probable path of the field should be made. As the shape of the part becomes more complex the problem becomes correspondingly more difficult. Sometimes separate coil magnetization must be applied to various projections of the part to insure proper field direction at all locations. A satisfactory procedure can usually be worked out for the most complicated shapes after some experimentation.

In predicting the direction the field will take when magnetizing with a coil, it is well to remember that flux lines always must close upon themselves to form a complete circuit, and that they tend to follow the path of lowest total reluctance. Further, because of the very low reluctance of iron as compared to air, they will follow the iron path as far as possible.

16. CIRCULAR FIELDS. In predicting or determining the distribution and intensity of fields produced by passing current through the part or through a central conductor threaded through an opening in the part, a different set of rules applies. Some of these rules are simple, and when applied to relatively simple shapes, enable the operator to carry out the inspections with the assurance that he has the correct fields. But often in complicated cross-sections, cut-and-try methods must still be resorted to.

17. FIELD AROUND A CONDUCTOR. The basic rules regarding fields around a circular conductor carrying direct current are;

(1) The field outside a conductor of uniform cross-section is uniform along the length of the conductor.

(2) The field is at right angles to the path of the current through the conductor.

(3) The field strength outside the conductor varies inversely with the radial distance from the center of the conductor.

In passing current through a part to be magnetized we therefore know that in general the field set up will be 90° to the direction of the current path—*a most useful rule to remember*. This means that the current should always be passed in a direction parallel (as nearly as possible) to the direction of expected discontinuities, since the resulting field, being at right angles to the current, will then also be at right angles to the discontinuities.

The *length* of a part being magnetized by passing current from one end to the other, does not affect the strength of circular field produced. The *diameter*, however, is very important, since the field strength at the surface decreases with an increase in diameter. A two inch bar carrying 1000 amperes will have a field at its surface twice as strong as would be present in a bar four inches in diameter carrying the same current—or half as strong as would be the case in a one inch bar.

18. FIELD IN AND AROUND A SOLID NON-MAGNETIC CONDUCTOR, CARRYING D.C. The distribution of the field *inside* a non-magnetic conductor, such as a copper bar, when carrying direct current, is different from the field distribution *external* to the bar. At any point inside the bar the field is due only to that portion of the current which is flowing in the metal between the point and the center of the bar—that is, through a cylinder whose radius is the distance from the center of the bar to the point in question. Thus the field increases from zero at the center along a straight line, to a maximum at the surface. Outside the bar the field decreases along a curve as shown in Fig. 96. In calculating field strengths *outside* the bar the current

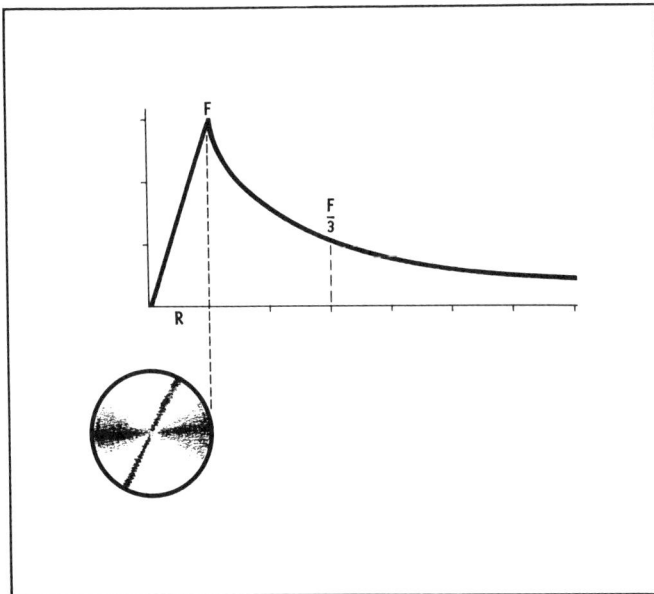

Fig. 96—Distribution of the Field In and Around a Solid Non-Magnetic Conductor Carrying Direct Current.

must be considered as being concentrated at the *center of the bar*. If the radius of the bar is R, and the field at the surface is F, then the field at a distance 2R from the center will be $\dfrac{F}{2}$; at 3R, $\dfrac{F}{3}$, etc.

19. FIELD IN AND AROUND A HOLLOW NON-MAGNETIC CONDUCTOR CARRYING D.C. In the case of a hollow non-magnetic conductor carrying direct current, somewhat different conditions exist. Obviously, there is no current flowing between a point on the inside surface and the center of the tube. Since the field at a given point in a conductor is due only to the current flowing between the point and the center of the conductor, it follows that there is no field at the inside surface of a hollow conductor. At the outside surface, however, the field will be the same as in the case of a solid conductor of the same diameter carrying the same amount of current. Here, in calculating external field strength, distances must be taken from the center of the tube, not from its inside surface.

The gradient of field strength from inside to outside surface of the tubular conductor is steeper (depending on the wall thickness) than from the center to the outer surface of a solid bar of the same outside diameter.

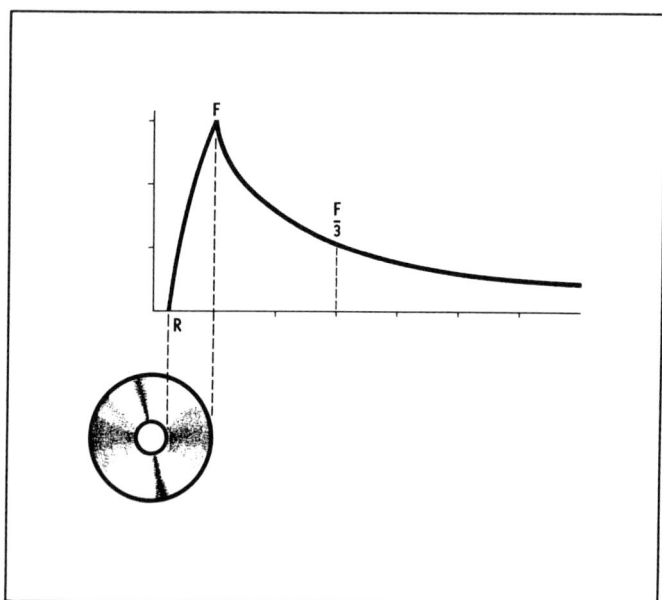

Fig. 97—Distribution of the Field In and Around a Hollow Non-Magnetic Conductor Carrying Direct Current.

The field *external* to the hollow conductor will be exactly the same as for the solid conductor, and will decrease along the same curve as in the case of the solid bar. This condition is shown in Fig. 97.

20. THE CASE OF A SOLID MAGNETIC CONDUCTOR CARRYING

DIRECT CURRENT. If the conductor carrying direct current is a solid bar of steel or other magnetic material, the same *distribution* of field will exist as would be the case in a similar non-magnetic conductor, but the *strength* of the field will be much greater.

Consider a conductor of the same diameter as that shown in Fig. 96. The field at the center would again be zero, but the field at the surface would be $\mu \times$ F, where μ is the material permeability of the magnetic material. The actual field, therefore, may be 1000 or 2000 times the field in the non-magnetic bar. Just *outside* the surface, however, the field strength drops to *exactly* the same value as for the non-magnetic conductor, and the falling off of field strength with increasing distance follows the same curve. See Figure 98.

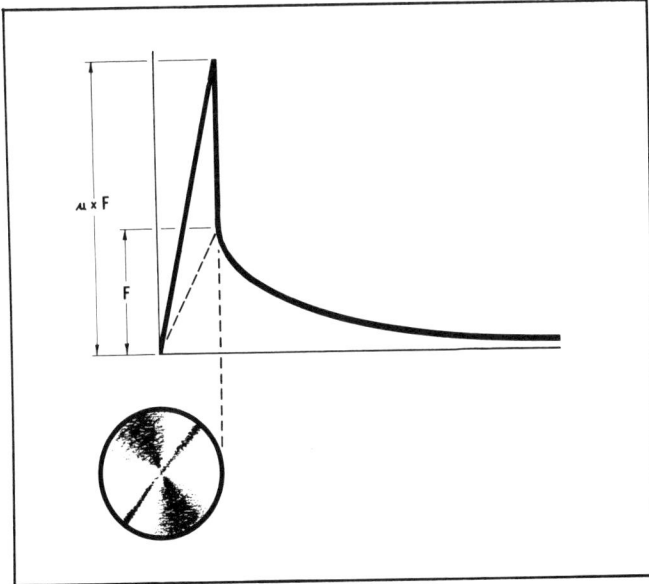

Fig. 98—Distribution of the Field In and Around a Solid Conductor of Magnetic Material Carrying Direct Current.

21. THE CASE OF A HOLLOW MAGNETIC CONDUCTOR CARRYING DIRECT CURRENT. A similar relationship exists in the case of the hollow conductor made of magnetic material as compared to the hollow non-magnetic conductor. The field again is zero at the inside surface, and increases along an approximately straight line to a maximum at the outside surface, but is again μ times the field strength in the non-magnetic tube.

This distribution indicates an unfavorable field for the detection of defects existing on the inside surface of the tube, and this method of magnetizing should preferably not be used for the inspection of tubes for inside surface defects where maximum sensitivity is required.

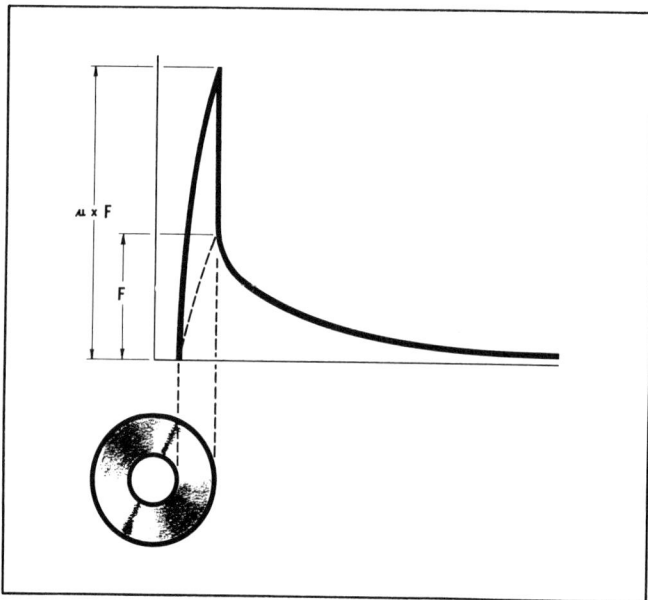

Fig. 99—Distribution of the Field In and Around a Hollow Conductor
of Magnetic Material Carrying Direct Current.

Cracks, however, have depth, and as can be seen from Fig. 99 the field strength curve rises steeply, progressing from the inside to the outside surface. Thus, defects several thousandths of an inch in depth will intercept some field, and may be reliably detected. Before using this method for the detection of I.D. defects, experimental proof of its effectiveness should be obtained.

Outside the tube, the field strength drops off with distance from the surface along the same curve as with non-magnetic conductors. This situation is shown in Fig. 99.

22. FIELD INSIDE A CONDUCTOR—THE GENERAL CASE. Figure 100 is a plot of the field produced by direct current inside a conductor, and is applicable to both magnetic and non-magnetic conductors, either solid or tubular. In the plots, the field strength is given as a ratio, and the dimensions are in "units" to which any numerical value can be assigned.

The conductor is defined as having a radius to the outside surface of "a", and a radius to the inside surface (if tubular) of "b". The case of b = zero is the case of the solid conductor. The point "R" is the point inside the conductor for which information regarding field strength is desired. The field strength in the plot is given as a ratio of the field at point R to the field at the outer surface—that is $\frac{H_R}{H_a}$. In the curve, the vertical axis could be labelled B_R/B_a, since

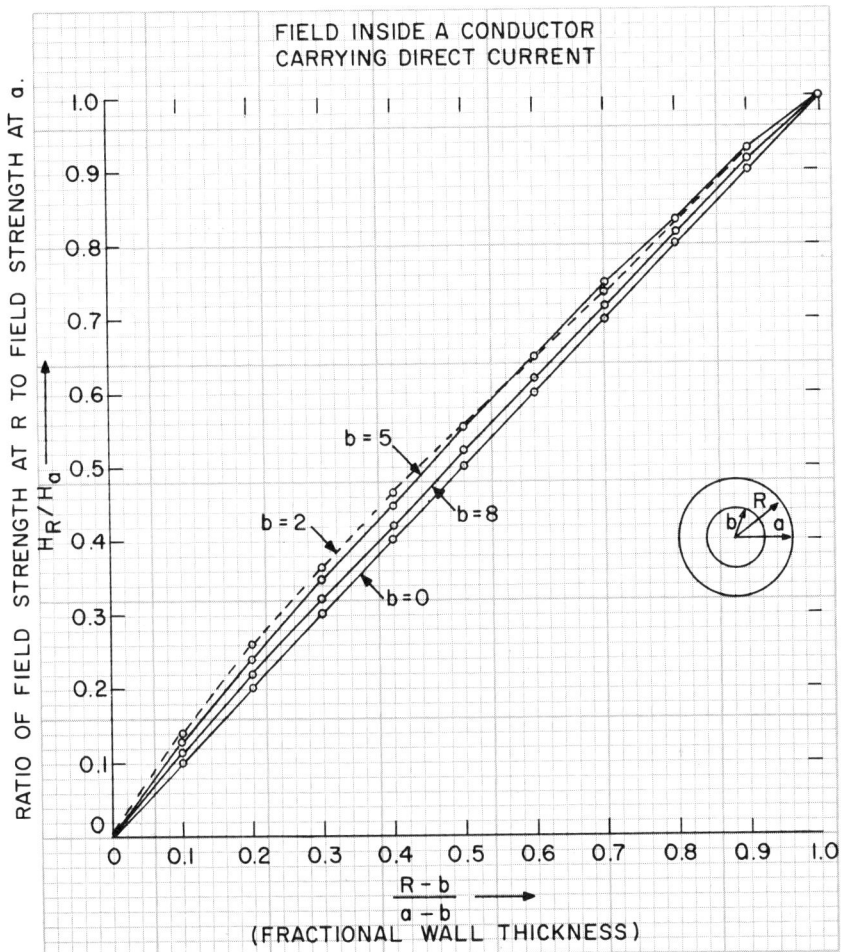

Fig. 100—Graph Showing the Variation of the Field Inside Any Conductor Carrying Direct Current.

$B_R = \mu H_R$ and $B_a = \mu H_a$. The ratio then, is independent of the permeability, and so can be applied to either non-magnetic or magnetic materials by substituting the appropriate values.

Similarly, since the dimensions are in "units", plotted for "a" (outside radius) = 10 units, any dimensions desired can be assigned to "a" and "b" and "R" by choosing the right units. For example, for a cylinder having outside and inside radii of respectively two inches and one inch, the "unit" is 0.2 inch. Since on the plot "a" = 10 units, the outside radius would be $0.2 \times 10 = 2$ inches, and the inside radius "b" would be 5 units, or $0.2 \times 5 = 1$ inch. The horizontal axis of the plot is in terms of the position of the point R.

This plot is useful in that it shows on a single figure how the field strength varies inside any conductor carrying direct current.

23. THE CASE OF A CYLINDER OF MAGNETIC MATERIAL WITH DIRECT CURRENT FLOWING THROUGH A CENTRAL CONDUCTOR. A much better way to magnetize a tube when defects on the inside surface are sought is to pass direct current through a conductor threaded through the interior of the tube. This case is shown in Fig. 101. Here the field due to the current in the central conductor is a maximum at the surface of the conductor, and decreases along the same curve (for air outside the conductor) as before, through

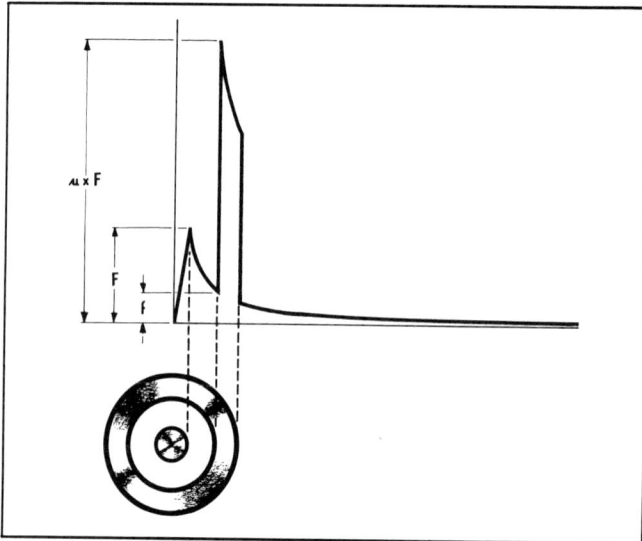

Fig. 101—Distribution of the Field In and Around a Hollow Magnetic Cylinder with Direct Current Flowing Through a Central Conductor.

the space between the conductor and the inside surface of the tube. However, at this surface the field is immediately increased by the permeability factor μ, and then *decreases* to the outer surface. Here the field again drops to the same decreasing curve it was following in the air space inside the tube.

This method, then, produces a maximum field at the inside surface, and thus gives strong indications of defects on the inside surface. Sometimes these indications may even show on the *outside* surface of the tube.

It should be noted that, as it affects the strength of the field in the tube, it makes no difference whether the central conductor is of magnetic or non-magnetic material, since it is the field *external* to the conductor itself that constitutes the magnetizing force for the cylinder.

24. MAGNETIZING WITH ALTERNATING CURRENT. In the foregoing discussion, the magnetizing current has in all cases been considered to be D.C. Most of these rules do not hold when the magnetization is done with alternating current.

It is a well known electrical fact that alternating current tends to flow only along the surface of a conductor. This tendency is in part a function of the frequency of the current, and is extremely

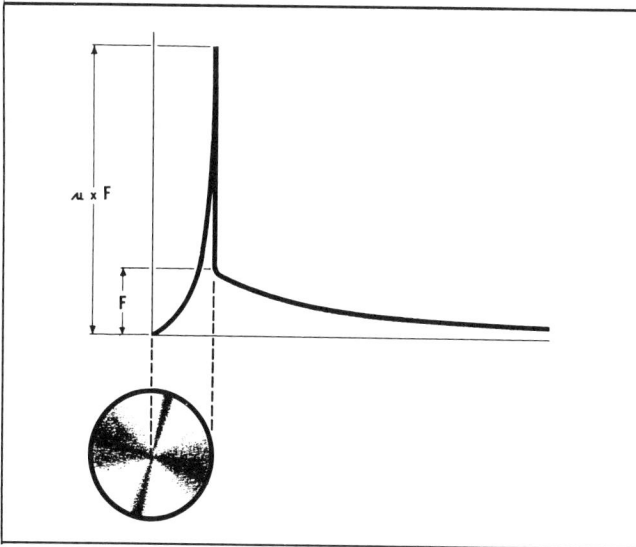

Fig. 102—Distribution of the Field In and Around a Solid Conductor of Magnetic Material Carrying Alternating Current.

pronounced at very high frequencies. Even at commercial frequencies (60 cycles) the tendency is appreciable, especially in magnetic materials. The phenomenon is referred to as "skin effect".

From the principles already set forth in discussing field distribution when D.C. is used for magnetization, it is obvious that if the current density is greater in the outer layers of the conductor, the field density will be correspondingly greater at such locations. Experiments have indicated that in a steel shaft six inches in diameter, magnetized by passing heavy alternating current through it from end to end, there was little if any field detectable more than one and one-half inches below the surface, and that the field was almost entirely concentrated in the outer half inch of metal.

25. THE CASE OF A SOLID CONDUCTOR MADE OF MAGNETIC MATERIAL, CARRYING ALTERNATING CURRENT. The field distribution in a solid magnetic conductor carrying alternating current is shown in Fig. 102. *Outside* the conductor the field strength at any point is decreasing in exactly the same way as when direct current is the magnetizing force, but it must be remembered that while the A.C. is flowing, the field is constantly varying, both in strength and

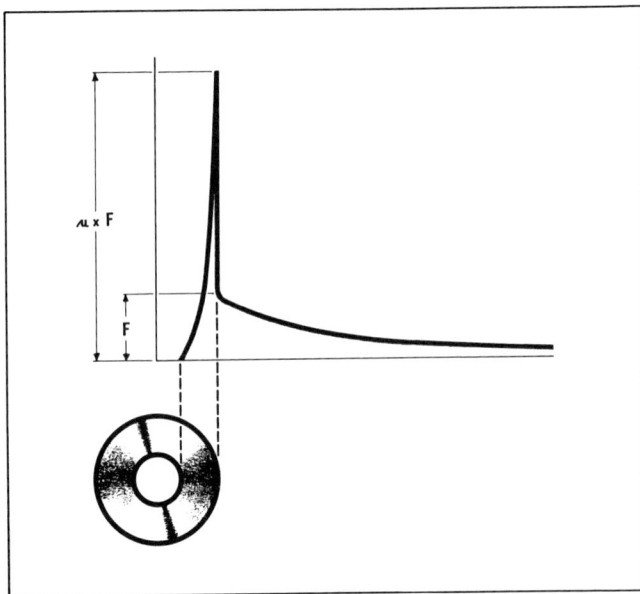

Fig. 103—Distribution of the Field In and Around a Hollow Conductor of Magnetic Material·Carrying Alternating Current.

direction. *Inside* the conductor the field is zero at the exact center, and increases toward the outside surface, slowly at first, then with increasing rapidity to reach a high maximum at the surface.

26. THE CASE OF A HOLLOW CONDUCTOR MADE OF MAGNETIC MATERIAL, CARRYING ALTERNATING CURRENT. Similar differences in field distribution occur in the case of the hollow conductor of magnetic material when A.C. is used for magnetization. Again there is no field at the inside surface of the tube. Between the inside and the outside surface, the field increases at an accelerating rate

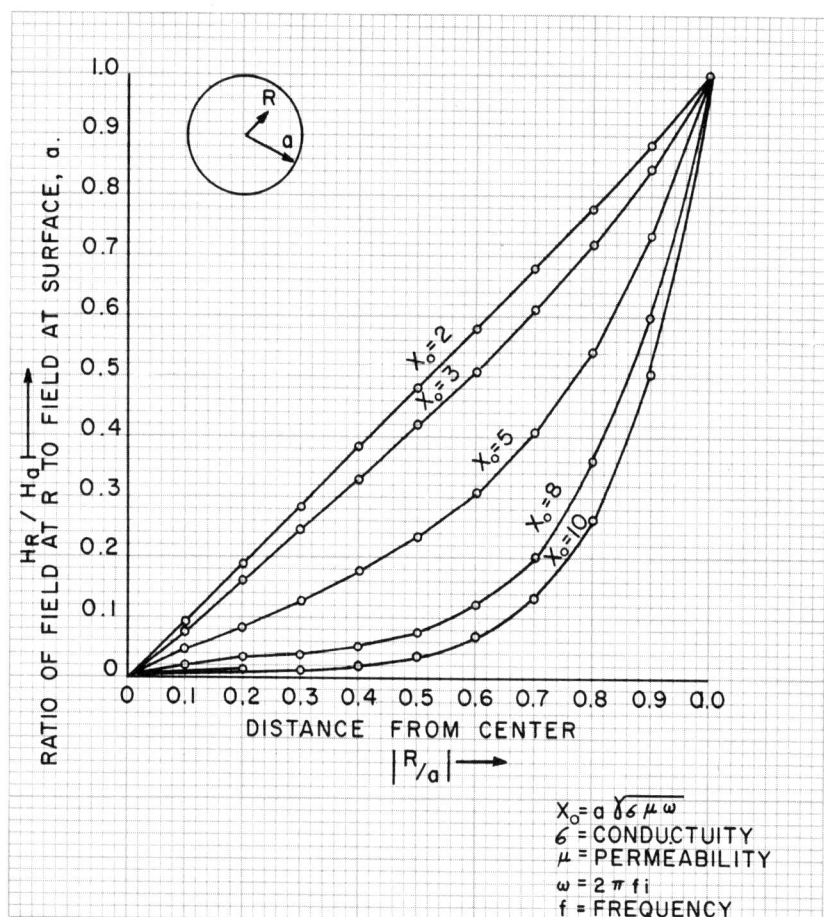

Fig. 104—Graph Showing the Effect of Conductivity, Permeability and Frequency on the Skin Effect of Alternating Current.

from zero on the inside surface to reach a high value near the out-side surface with a maximum field at that surface.

27. FIELD INSIDE A SOLID CONDUCTOR CARRYING ALTERNATING CURRENT—THE GENERAL CASE.

Figure 104 is a graph on which is plotted the field distribution at several points between center and surface of a solid conductor (magnetic or nonmagnetic) carrying alternating current. Dimensions are again in "units" so that any values can be assigned to them, as in the case of the plot for direct current (Fig. 100). The magnetizing force is again expressed as a ratio, $\dfrac{H_R}{H_a}$, and as before is

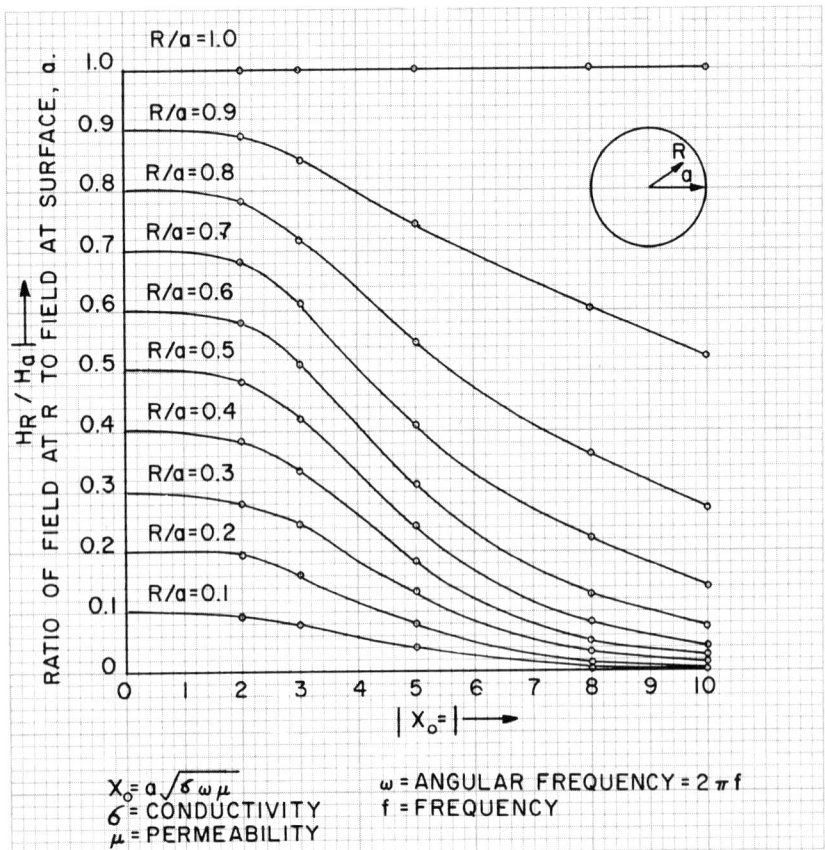

Fig. 105—Graph Illustrating the Skin Effect of Alternating Current.

independent of the permeability of the material composing the conductor. The horizontal axis represents the distance of the point R from the center in terms of fractions of the radius of the bar.

The several curves are plots of the field strength for various values of the term X_o, which is a function of the conductivity and permeability of the conductor, and of the frequency of the alternating magnetizing current. The value of X_o increases with an increase in any or all of the values for conductivity, permeability and frequency. The plots show at a glance the falling off of the field strength from the outside surface to the center of the conductor. The rate of this falling off is more rapid if the conductivity is high, or if the permeability is high, or if the frequency is high. The graph is a clear illustration of the skin effect of alternating current.

Figure 105 also applies to the case of a solid conductor carrying alternating current, and is merely a re-plot of the data used for Fig. 104. In this case the graph of the ratio of the field intensities $\dfrac{H_R}{H_a}$ is plotted against the function X_o, for various points along the radius from center to surface. This plot shows particularly well how rapidly the field strength falls off as a function of depth for the case of alternating current in a conductor.

CHAPTER 10

FIELD DISTRIBUTION IN LARGE OR IRREGULAR SHAPED BODIES

1. INTRODUCTION. The rules discussed in the foregoing two chapters for arriving at suitable field strengths for magnetic particle testing, both for longitudinal and circular magnetization, are valid regardless of the size of the object being magnetized. However, their application becomes impractical as the size of the object becomes larger. For example, although the rules regarding coil magnetization in terms of the L/D ratio will remain valid, coil diameter will become very large, since one of the restrictions of the rule of thumb for this case is that the cross-sectional area or the part be not greater than one tenth the cross-sectional area of the coil. For a six inch shaft this would require a 19 inch coil, and for a part 12 inches in diameter a 38 inch coil.

Similarly, the one-thousand-amperes-per-inch rule for circular magnetization becomes impractical in most cases, since even a 12 inch part would require 12,000 amperes. This amount of current is generally not available unless equipment for the purpose has been especially built. Further, the rules for field variation described in Chapter 9 apply, strictly speaking, only to uniform objects of cylindrical shape. When we pass to irregular shapes and cross-sections it becomes more difficult to predict current and field distribution. The difficulty becomes even greater as these irregular shapes become quite large.

Two points, however, can be emphasized. First, if the rules *are* followed the fields produced *will be adequate* for magnetic particle testing, and in most cases the fields so produced in large objects will be *larger* than required.

Second, when high permeability steels are being tested, current values much lower than indicated by the rules are often adequate. When they are not, local magnetization with prods or yokes is a means of securing higher strength fields without the use of extremely high currents. Overall magnetizing of parts is often not necessary.

2. THE CASE OF A SQUARE BAR, CIRCULARLY MAGNETIZED WITH D.C. The 1000 ampere-per-inch-of-diameter rule would require 2000 amperes to magnetize a round bar 2 inches in diameter. If we take a square bar, 2 inches on a side, the longest diametrical dimension is the diagonal, or 2.82 inches. In this case the current should be stepped up to 2820 amperes, according to the rule.

Measurements made with the Hall Effect probe, of the tangential component of H (See Chapter 8, Section 5) on such a bar, indicate that the field strength at the corners of the bar is about half that at the center of a face. In other words, the field strength is not uniform over a square cross-section. Furthermore it does not follow the simple rule of decrease with distance from the center, as in the case of a round bar. A round bar, having a diameter of 2.82 inches, carrying 2000 amperes, would have a field strength at the surface of $\dfrac{2}{2.82}$, or 0.71 of the surface field of a 2 inch diameter round bar carrying 2000 amperes. In the case of the square bar, however, the field falls off more rapidly at the corners, where "diameter" or diagonal is 2.82 inches, and has a strength of only half, or 0.50 that of the field at the center of a face (where "diameter" is 2 inches). Figure 106 illustrates this case.

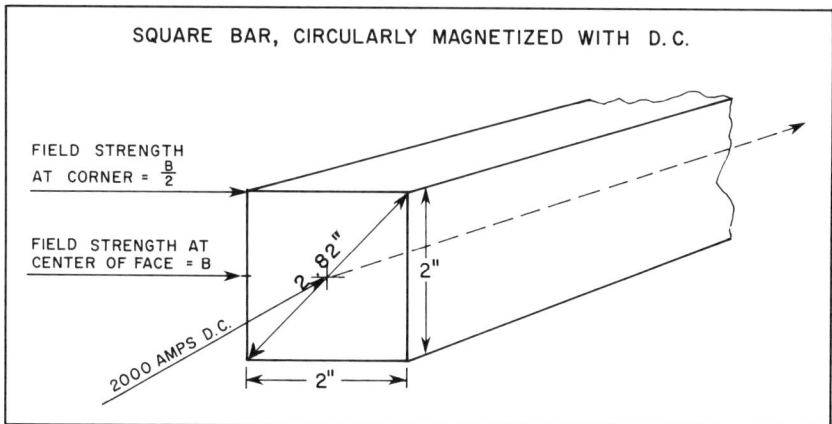

Fig. 106—Field Distribution in a Square Bar, Circularly Magnetized with Direct Current.

3. THE CASE OF A RECTANGULAR BAR, CIRCULARLY MAGNETIZED WITH D.C. Suppose we assume a rectangular bar, 2 inches by 6

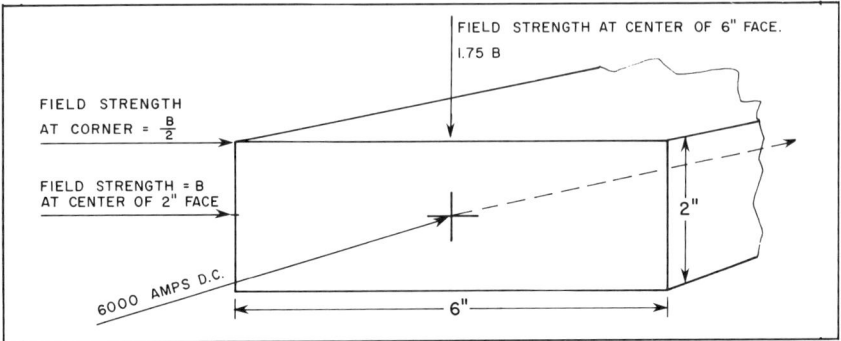

Fig. 107—Field Distribution in a Rectangular Bar, Circularly
Magnetized with Direct Current.

inches in cross-section, and that we wish to produce in it a field of
the same strength as we had in the 2 inch square bar, again using
direct current. This case is illustrated in Fig. 107. If we interpret
the 1000 ampere-per-inch rule as applying to the long dimension of
the cross-section of such a bar, a current of 6000 amperes is
indicated.

When such a bar was magnetized in the laboratory with 6000
amperes D.C., and by means of the Hall probe the tangential component of H measured and compared with the results obtained on
the 2 inch square bar, the following values for field strength at
various points were found;

(a) The field strengths at the corners of the 2 by 6 inch bar and
the 2 by 2 inch bar were approximately the same.

(b) The field at the center of the 6 inch face of the rectangular
bar was 1.75 times the field at the center of the 2 inch face
of the 2 inch square bar.

In this case it can be assumed that the distribution of *current* is
uniform over the 2 by 6 inch cross-section, but the *field distribution*
is not uniform. However, this non-uniformity tends toward producing stronger fields than needed.

4. NON-UNIFORM CROSS-SECTIONS, CIRCULARLY MAGNETIZED
WITH D.C. When the cross-section departs radically from a substantially uniform distribution of material, the above method of
estimating probable field strength and distribution breaks down.
Fundamental rules are:

(a) Over a non-uniform cross-section *current* will distribute itself evenly over the cross-section, so that the current flowing

in each element of the section is the same, provided contact is made evenly over the entire cross-section. If the piece is long enough the distribution of the current will be uniform even if contact is over a limited portion of the cross-section. Experiments have shown that the current spreads out very quickly over the entire cross-section.

(b) If the cross-section varies along the length of the part, the current will spread out (or condense) to occupy the changed section, and field strength will vary with current distribution, up or down. Field strength will be greater at smaller diameters and weaker at larger.

5. THE CASE OF AN I-SHAPED CROSS-SECTION, CIRCULARLY MAGNETIZED WITH D.C. Figure 108 shows the cross-section of an I-shaped bar. The two sections at the edges are two inches square, and they are joined by a one half inch by two inch web. This can be considered to be the two by six inch rectangular bar of Fig. 107 with the middle two inches machined away to leave a one half inch thick web. Suppose that it is desired to magnetize such a bar circularly, so that the field over the cross-section will be that called for by the 1000 ampere-per-inch rule. If we break the section down

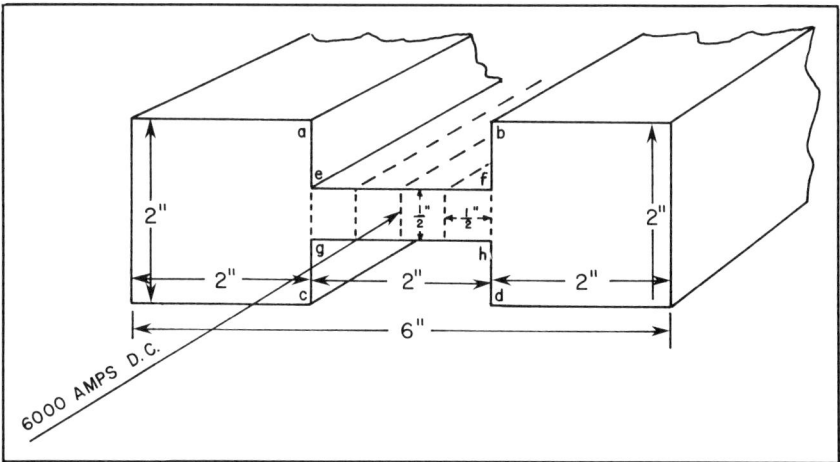

Fig. 108—Field Distribution in an I-Shaped Bar, Circularly
Magnetized with Direct Current.

into three parts—two 2 by 2 inch blocks at the outer edges, and a two by one half inch web between, we can say that each of the two by two inch edge areas would require 2000 amperes, or 500 amperes per square inch. Since the current distributes itself uniformly per unit of area, the center section, having an area of one square inch, would take only 500 amperes for the same current density as the end sections—a total of 4500 amperes for the entire section. But if the central web area is broken down into four one half inch square elements, each of these would carry only 125 amperes—one fourth of the 500 ampere per square inch current density of the end sections. But by the 1000 ampere-per-inch rule each one half inch diameter element of the web would require 500 amperes, four times what it would carry using the 4500 total amperes calculated above. And if the current is stepped up to give each one half inch square element 500 amperes, the web would take 2000 amperes total. The two-inch square edge sections would take the same current density, or 8000 amperes each. With the 2000 amperes total through the web, 18,000 amperes altogether would be needed.

Obviously this is an absurdly high value and experience has shown that it need be nowhere near as large as this. On the basis of the six inch longest dimension of the I section, only 6000 amperes would be called for by the 1000 ampere rule, which on the rectangular bar of Fig. 107 gave nearly double the field in the center sections as was found at the edges. We can conclude therefore, that at 6000 amperes for the I-shaped section, the field in the edge areas would be adequate, whereas the field at the center of the web would be stronger than required.

This rather labored analysis has been included in this discussion to point up the fact that prediction of suitable current levels in irregular cross-sections becomes very complicated, and at best lacks much in accuracy. Results just as accurate and with much less time and effort can be obtained by an experienced operator by cut-and-try methods, starting at high currents and working down, or in other cases, starting with low currents and working up.

One side comment should be made. With strong circular magnetization of the cross-section of Fig. 108, local poles would appear at the inner corners of the two inch edge areas, marked a, b, c and d in the figure. Also a leakage field would develop at the *inside* angles, e, f, g and h, between the edge areas and the web. To avoid confusing indications from these local leakage fields, it would probably be desirable to work with as low a current value as seems feasible, with perhaps a still lower value for the inside fillets. In critical cases two inspections may therefore be in order—one at a low current value for the web section, and another at a higher current value for the square end sections.

6. THE CASE OF PROD CONTACTS ON LARGE OBJECTS. When magnetizing a large part locally by means of prods, using direct current, the problem of determining actual field strengths and distribution is again very difficult since many variables are involved. The current passes into the part or plate at one contact and leaves it at the other. It will spread out non-uniformly between the two contact points, both parallel to and normal to the plate surface. The pattern will depend on prod spacing, dimensions of the part and its electrical conductivity. Obviously the over-all pattern of field density will depend on the pattern of current distribution, and obviously also, the current will be strongest on a direct line between the two points of contact. Naturally, in making the inspection, the prods are so positioned that the area along this line and the area closely adjacent to it is the area in which indications of surface cracks are expected to occur.

The prod contact technique has a special ability to produce indications of discontinuities which lie wholly below the surface, often quite deep within the part. Thus far, attempts to plot the current and field distribution deep within a relatively thick section when prods are applied to one surface have led to somewhat inconclusive results.

7. LABORATORY TESTS FOR FIELD STRENGTH WITH PROD MAGNETIZATION. Laboratory tests have been made to give some information regarding field strength and current distribution, and directions, in the surface layers of the metal by measuring the tangential component of "H" using the Hall Probe.

One series of tests was made to determine the variation in field strength midway between the prods, when prod spacing and current (D.C.) were varied. Prod spacing was varied from 2 inches to 18 inches, and for each spacing, current was varied from 200 to over 2000 amperes. Figure 109 gives the curves for these data, with prod spacing plotted against field strengths, measured by the tangential component of H, each curve representing a different current value. The curves shown are for one inch thick plate. Data for one quarter and one half inch plate are altogether similar. It is evident from these curves that field strength falls off rapidly as prod spacing increases, and that increases in current produce the greatest increase in field strength when the prod spacings are small. The

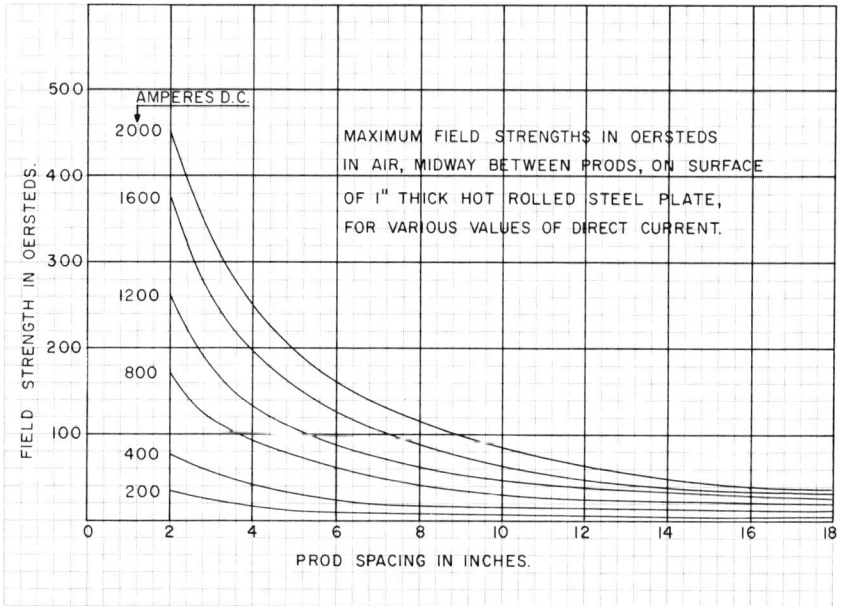

Fig. 109—Field Strength, H, Plotted Against Prod Spacing for Various Current Values, on One Inch Plate.

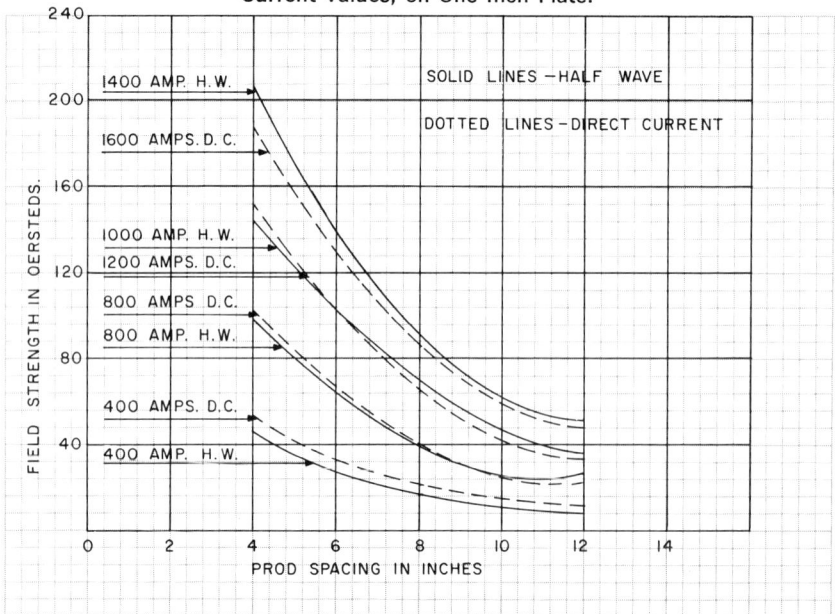

Fig. 110—Comparison of the Effectiveness of Direct Current and Half Wave Current when Used with Prod Magnetization.

experiment of course gives no evidence whatever as to what goes on below the surface of the plate.

8. CORRELATION OF THE TESTS WITH PRACTICE. The values obtained by these tests are not of much use, in themselves, for the purpose of determining how much current and what prod spacing are best to use for any given testing problem. However, many years of experience in steel casting and weld inspection with prods has evolved a pattern of prod spacing and current values for various plate thicknesses. The A.S.T.M. in 1958 published tentative Standard E 109-57 T, giving recommended prod spacings and currents for plate thicknesses under three quarters of an inch, and for thicknesses three quarters of an inch and over. These are quoted in Table I.

TABLE I.

A.S.T.M. Recommended Prod Spacings and Current Values.

Prod Spacing, Inches	Section Thickness, Inches	
	Under $3/4$ inch	$3/4$ inch and over
2 to 4	200 to 300 Amperes	300 to 400 Amperes
Over 4 to less than 6	300 to 400 Amperes	400 to 600 Amperes
6 to 8	400 to 600 Amperes	600 to 800 Amperes

Using the recommended currents in these A.S.T.M. specifications and the field value for the mid-point between the two prods, taken from the plot for one inch plate thickness as shown in Figure 109, we can re-write Table I in terms of magnetizing force, H, expressed in Oersteds at the surface. The values so obtained are shown in Table II.

TABLE II*

Magnetizing Force in Oersteds for Various Prod Spacings, Using A.S.T.M. Recommended Currents.

Prod Spacing, Inches	Section Thickness, Inches	
	Under $3/4$ inch	$3/4$ inch and over
2 to 4	40 to 25 Oersteds	50 to 45 Oersteds
Over 4 to less than 6	35 to 20 Oersteds	45 to 40 Oersteds
6 to 8	30 to 20 Oersteds	40 to 35 Oersteds

*It would seem that the A.S.T.M. values for prod spacing and current (Table I) might well be revised as to current specified, so as to produce more nearly 40 to 50 Oersteds at all spacings.

The magnetizing forces as measured with the Hall probe at the mid-point between prod contacts, can now be read for the A.S.T.M. recommendations. The Table shows the magnetizing forces to vary from 20 oersteds for 6 to 8 inch spacing on plates less than ¾ inch thick, to 50 Oersteds for 2 to 4 inch spacing on ¾ inch plate and thicker. It therefore becomes possible, by taking Hall probe readings, to space the prods and adjust the current so that the minimum magnetizing force of 20 Oersteds (or any other value that may be determined to be desirable) can be obtained.

Also, since field strengths as measured in these tests are almost directly proportional to current, prod current magnitudes can be established in terms of amperes per inch of prod spacing, for various thicknesses of plate. It is also possible by means of the Hall probe measurement of H, to determine at what other points, outside the prod line, the 20 Oersted minimum value of magnetizing force has been developed.

For example, along the prod line or very near it, laboratory tests have shown that 60 amperes per inch of prod spacing will produce the required 20 Oersteds minimum for plate thicknesses under ¾ inch, and 100 amperes per inch will produce the required 35 Oersteds minimum for thicknesses over ¾ inch. As one moves away from the prod line a greater number of amperes per inch are required to produce these minimum values. For example, at a perpendicular distance out from the prod line of one half of the prod spacing, 150 amperes per inch were required to produce the 20 Oersteds minimum for thicknesses under ¾ inch. 250 to 350 amperes per inch were needed to produce the 35 Oersteds minimum for thicknesses over ¾ inch.

It is evident that current values rapidly rise above those permitted at distances of only a few inches away from the prod line. Maximum current permitted is determined by the heating effect at the prod contacts which can damage the steel; by the interference of the powder pattern produced by the field *external* to the plate and around the prods themselves, which becomes pronounced at high currents; by the limitations of the output of the current source; and finally by experience as incorporated in the A.S.T.M. specification.

Experience has also shown that, except in special cases, practical prod spacing should be limited to 8 inches if maximum sensitivity is required. Greater spacing requires higher currents and the fields produced are "wasted" in spreading out into the plate beyond the prod line.

9. PROD INSPECTION USING HALF WAVE CURRENT. Similar experiments to those outlined above, but using half wave instead of direct current were also carried out. Curves comparing the magnetizing force produced on a ¼ inch thick plate, with various current strengths, for direct and for half wave current are shown in Figure 110. It is evident from these curves that field strength per ampere is somewhat higher for half wave than for D.C. at the smaller prod spacings, but is almost the same for either direct current or half wave at prod spacings greater than 8 inches. Since half wave consumes less power and produces lower heating effects in equipment and at the prod contact points, than does direct current, it is the preferred method when prod inspection is indicated. Also, better powder mobility is obtained with half-wave.

10. OVER-ALL MAGNETIZATION OF LARGE OBJECTS. When the over-all method of inspecting large and complex bodies, such as castings, is used to replace inspection with prods, the matter of field direction and the amount of current to use becomes still more complex. Actually there is no method by which the points of contact or the current necessary can be determined except by experience, general analysis, and some cut-and-try tests.

For this type of inspection, large units, delivering from 6,000 to over 20,000 amperes D.C., have been used. In numerous instances, multiple circuits, each capable of delivering the maximum current, have been employed. Two or three such output circuits may be required. By a special switching circuit, current is passed in quick succession through each of the output circuits. Current is carried to the castings or weldments to be inspected, by means of contacting jigs or by heavy flexible cables, which are attached directly to the part. Or, they may be run as central conductors through openings in the casting to form a coil, or wrapped around the casting for longitudinal magnetization.

The exact arrangement of circuits must be determined experimentally for each size and shape of casting. The operation is rewarding, however, because more thorough inspection is achieved than is possible with prods and the man-hours of time for the inspection of large and complex-shaped castings can be reduced by up to 80%. The over-all method will be discussed in greater detail in Chapters 19 and 24.

11. SUMMARY. Before leaving the entire subject of magnetizing means, field strength and distribution, current distribution and strength requirements, and going on to the actual techniques of applying magnetic particle testing, it is perhaps worthwhile to set down a few of the general principles and rules which an operator should remember:

(1) Fields should be at 90° to the direction of defects.

(2) Fields generated by electric currents are at 90° to the direction of current flow.

(3) When magnetizing with electric currents, *pass the current in a direction parallel* to the direction of expected discontinuities. The field will be at right angles to the current and, therefore, also at right angles to the discontinuity.

(4) Circular magnetization has the advantage over longitudinal in that there are few, if any, local poles to cause confusion in particle patterns, and is to be preferred when a choice of methods is permissible.

(5) For circular magnetization, use 1000 amperes per inch of diameter of the part. This often produces *too strong* a field, but *if* used, the field so produced is *sure* to be as strong as necessary.

(6) For coil magnetization use the formula given in Chapter 9 to arrive at the ampere turns necessary for any given set of conditions. This formula is:

$$NI = \frac{45,000}{L/D}$$ where NI is the ampere turns required, and L/D is the length to diameter ratio of the part.

(7) For prod magnetization with direct current, a minimum of 60 amperes per inch of prod spacing will produce the minimum magnetizing force of 20 Oersteds at the midpoint of the prod line for plate $\frac{3}{4}$ inch thick or less. A safer figure to use, however, is 200 amperes per inch, unless this current strength produces interfering surface powder patterns. Prod spacing for practical inspection purposes is limited to about 8 inches maximum, except in special cases.

CHAPTER 11

MAGNETIC PARTICLES—THEIR NATURE AND PROPERTIES

1. GENERAL DESCRIPTION. There are two essential components of the magnetic particle testing process, each of equal importance for reliable results. The first is the proper magnetization of the part to be tested, with fields of the right strength and in the right direction for the detection of the particular type of defect being sought. The second ingredient of a successful test is the use of the proper type of magnetic particles to secure the best possible indications of these defects under the conditions prevailing in any given case.

The particles used are in all cases finely divided ferromagnetic material. The properties of this material vary over a very wide range for different applications—including magnetic properties, size, shape, density, mobility and visibility or contrast. Varying requirements for varying conditions of test and varying properties of suitable materials have led to the development of a large number of different types of materials now available on the market. The choice of which one to use is an important one, since the appearance of the particle patterns at discontinuities will be affected by this choice, even to the point of whether or not a pattern is formed at all.

Since the results of magnetic particle tests usually depend on the interpretation of the particle pattern by the inspector, the appearance of this pattern is of fundamental importance. For this same reason, the reproducibility of results by different operators in different locations depends on the use of the same type of particles by each operator, as well as on the use of the same magnetizing procedure.

A vast amount of investigational work has been done with a view to determining what the desirable properties of magnetic particles are. The values for the properties which were listed above have been determined for a large number of finely divided ferromagnetic materials. But knowing these properties is not the same as knowing what the effect of each of the individual characteristics is on the patterns produced. The very complex inter-relation of these properties, plus the influence of the numerous types and methods of magnetization, makes selecting the optimum properties of magnetic particles for any particular application a matter of experience. Experience can and does lead to the choice of the most suitable of available particles in any given case. The development of the various types of magnetic particles was dictated by this very need, and the formulations were worked out experimentally to produce the best results. Although experience does not necessarily tell us *why* certain combinations perform better than others, demonstrable performance comparisons in actual use in finding defects is the most satisfactory way of determining which material is best for any given set of test conditions.

In the following sections we will discuss the various properties which are considered to affect the over-all results, but no attempt will be made to assign to each property, in any quantitative way, the role that it plays, or its relative importance to the end results.

2. EFFECT OF SIZE. It is self-evident that size must play an important part in the behavior of magnetic particles when in a magnetic field, which *can* be quite weak at a discontinuity. A large heavy particle is not likely to be arrested and held by a weak field when such particles are moving over a part surface. On the other hand, *very* fine powders *will* be held by very weak fields, since their mass is very small. But extremely fine particles may also adhere to the surface where there are no discontinuities, especially if it is rough, and form confusing backgrounds.

A. *Dry Powders.* In general, for the dry powders, sensitivity to very fine defects increases as particle size decreases, but with definite limitations. If the particles are extremely small, of the order of a few microns, they behave like a dust—that is, they accumulate and adhere in depressions on even very smooth-looking surfaces. They will "develop" fingerprints and adhere at any damp or slightly oily areas, whether or not leakage fields exist. These extremely fine powders, though undoubtedly sensitive to very weak fields, are not desirable for general use, because they do leave this heavy, dusty background.

In some special applications, particles of a specific size range are used. For instance, where it is desired to detect only rather large, coarse discontinuities, and not produce indications of very fine ones, only large sized particles are used.

However, most dry ferromagnetic powders used for detecting discontinuities are careful mixtures of particles of all sizes. The smaller ones add sensitivity and mobility, while the larger ones not only aid in locating large defects, but by a sort of sweeping action, counteract to some extent the tendency of the fines to leave a dusty background. Thus, by including the entire size range, a balanced powder with sensitivity over most of the range of sizes of discontinuities is produced.

B. *Wet Method Materials.* When the ferromagnetic particles are applied as a suspension in some liquid medium, much finer particles can be used. The upper limit of particle size in most commercial wet method materials used for magnetic particle testing purposes, is in the range of 60 to 40 microns (about .0025 to .0015 inch). Particles larger than this are difficult to hold in suspension, and even the 40 to 60 micron sizes settle out of suspension rather rapidly. Also, the large particles have another bad feature. When such a suspension is applied over a surface, the liquid drains away, and as the film remaining over the surface becomes thinner, the coarse particles are quickly stranded and immobilized. Such stranded particles often line up in what are called "drainage lines" to form a "high water mark" of particles that can be confused with indications of discontinuities.

In the case of the finer particles, the stranding due to the draining away of the liquid occurs much later, giving the particles mobility long enough to reach the influence of leakage fields and accumulate to form true indications. The *minimum* size limit for particles to be used as liquid suspensions is indeterminate. Commercially available ferromagnetic materials commonly used for this purpose include some exceedingly fine particles. The electron microscope has shown the presence of a large proportion of particles as small as one-eighth of a micron (0.000005, or 5 millionths inch) in diameter.

In actual use, however, particles of this size never act as individuals. Due to the fact that they are magnetized in use, they become actual tiny magnets, because the material used has appreciable retentivity. Under conditions of quiet settling in a suspension, these particles are drawn together as a result of their retained magnetism to form clumps or aggregates of particles. These aggregations of fine particles then tend to act as a unit when they are applied to the surface of parts for magnetic particle testing. The speed and extent to which this process takes place increases with the rententivity of the particle material. Agitating the suspension breaks up the aggregates, but they begin to form again as soon as agitation ceases. This happens as soon as the suspension has been applied over the surface of the part. Therefore, since the particles act as agglomerated units of varying size, and not as individual particles, there is no meaningful lower limit for the size of the individual particles themselves for the wet method.

C. *Advantages of Agglomeration of Fine Particles.* This agglomeration of fine particles into larger clumps is advantageous rather than otherwise, as long as the size of the aggregates does not become larger than the 60 micron "limit" mentioned above. Individual particles of exceedingly small size, and therefore mass, move very slowly through the liquid of the suspension under the influence of leakage fields at discontinuities. Unless special techniques are used, individual particles of the size indicated by the electron microscope are not particularly useful for the location of exceedingly fine cracks until the process of agglomeration into somewhat larger units has taken place. In practical applications this process takes place while drainage of the suspension from the surface of the part is occurring. As the agglomeration proceeds, the clumps formed will vary in size, and, since these clumps act as individual units, the effect is that of a particle size range from very fine to relatively coarse.

D. *Fluorescent Particles.* The above discussion applies primarily to magnetic particles that have not been treated with fluorescent pigments. The treated particles have fewer of the *very* fine particles, though when they *are* present they tend to act in the same manner as those not so treated but with somewhat less tendency to agglomerate. The same comment applies to all other types of wet method materials in which a bonding material is incorporated, such as the newer concentrates. Therefore, a meaningful lower limit for "particle size" cannot be set in these cases either.

3. EFFECT OF DENSITY. Most ferromagnetic materials have fairly high densities—that is, they are heavy in terms of weight per unit volume. The densities of the materials in common use vary from around 5.0 to nearly 8.0 times the density of water. Such heavy materials tend to settle rapidly whether suspended in liquids or in air. This is the principle reason for ruling out large-sized particles, since the smaller sizes, having less mass per particle, as well as more surface area in proportion to mass, settle out much more slowly. The density of many ferromagnetic powders is lowered somewhat by compounding or coating them with pigments whose densities are lower than that of the particles. This is true of both the dry, pigmented powders, and the fluorescent particles in liquid suspension.

4. EFFECT OF SHAPE. The shape of the magnetic particles used for magnetic particle testing has a strong bearing on their behavior in locating defects. When in a magnetic field the particles tend to align themselves along the lines of force, as illustrated in a magnetograph. This tendency is much stronger with elongated or rod-like particles than with more compact or globular shapes for the reason that the long shapes develop stronger polarity. Because of the attraction exhibited by opposite poles, the pronounced north and south poles of these tiny magnets arrange themselves into strings of particles, north pole to south pole, much more readily than do globular shapes. The result is the formation of stronger patterns in weak leakage fields, as these magnetically formed strings of particles bridge the discontinuity.

The superior effectiveness of the elongated shapes over the globular shapes is particlularly noticeable in the detection of wide, shallow discontinuities, or of those discontinuities which lie wholly below the surface. The leakage fields at such defects are more diffuse, and the formation of strings due to the stronger polarity of the elongated shaped magnetic particles makes for stronger, readable patterns in such cases.

In the case of the dry powders, there is another effect due to the shape of the particles which must be taken into account. Dry particles are applied to the surfaces of parts by means of plastic powder bottles, or by rubber squeeze bulbs, or by the use of air-stream-operated powder guns. The ability to flow freely and to form uniformly dispersed clouds of powder that will spread evenly over a surface is a necessary characteristic for rapid and effective dry-powder testing. A powder composed only of elongated shapes tends to "felt" together in a container, and to be ejected in uneven clumps. When a powder behaves in this manner, the inspection becomes extremely slow and difficult. Globular-shaped particles on the other hand, flow freely and smoothly under similar conditions.

Clearly, then, a dry powder must have free-flowing properties for easy application, yet have optimum shape for the greatest sensitivity for the formation of strong indications. These two opposing needs are met by blending particles of different shapes. A fair proportion of rod-like particles must be present for a sensitive blend. To have powder that will flow well for easy and uniform application, a sufficient proportion of more compact shapes must also be present. Since there are considerable variations in the shapes of the different types of powders as received, production of a finished magnetic particle testing powder requires close control to insure uniformity.

In the case of particles for the wet method of inspection, the individual particles are kept dispersed by "mechanical" agitation until they are flowed in suspension over the surface of the magnetized part. There is therefore no need to incorporate unfavorable shapes merely for the purpose of improving the flow of the particles. Long, slender particles, with otherwise desirable characteristics, could be used exclusively. Unfortunately, exclusive use of such particles would be especially costly, and the advantage to be gained from their use would not appear to warrant the added expense to the user. Experience over the years has shown that satisfactory sensitivity is obtainable with particles of less than optimum shape in the wet version of the method.

Because wet method particles are suspended in a liquid medium which is very much denser and more viscous than air, they move in the leakage fields much more slowly than the dry powders, and they therefore accumulate much more slowly at discontinuities. In the vicinity of leakage fields, they can be seen to line up to form minute elongated aggregates. Even the unfavorable aggregate shapes, formed by simple agglomeration in suspension, will line up into magnetically-held, elongated aggregates under the influence of local, low-level leakage fields. This effect contributes to the really remarkable sensitivity of the fine particles comprising wet method materials, a sensitivity which has not been equalled by "magnetic perturbation" detection instrument probes.

5. PERMEABILITY. In theory, magnetic particles used for magnetic particle testing should have as high a permeability as possible. This is for the reason that they must be readily magnetized by the low-level leakage fields that occur in the vicinity of a discontinuity, so that they will be drawn by these fields to the discontinuity itself to form a readable indication, even though these fields are sometimes very weak at very fine discontinuities.

As a practical matter however, there is little connection between permeability and sensitivity for magnetic powders. For instance, the iron-based dry-method powders have permeabilities that are higher, by an order of magnitude or more, than those of the oxides used in the wet method. Yet a typical dry powder has little value for detecting the extremely fine surface cracks that the wet method particles find so well.

One reason for this seeming paradox is that permeability is just *one factor* in the desirable properties of a suitable powder, scarcely more important than size, shape or any other. Unless all other factors are in the proper range for the application at hand, high permeability alone is of little value. Another reason is that published values of permeabilities are not adequate guides for judging the behavior of particles. Most such values are in terms of *maximum material permeability*, whereas in the low level fields at discontinuities it is the *initial* permeability that would seem to be the determining factor. (See the discussion of permeability in Chapter 9, Sections 3 to 5.) Of two powders, A and B, even if powder A has a much higher *material* permeability than powder B, it does not follow that the initial permeability of A will be higher than that of B. It might just as easily be the other way around.

6. COERCIVE FORCE. As a general principle, low coercive force and low retentivity are desirable in magnetic particle testing powders. If these values were high in a dry powder, for example, the particles would become magnetized during manufacture, or in first use, and thus become small, yet strong, permanent magnets. Once magnetized, their tendency to be controlled by the weak fields at discontinuities would be over-shadowed by their tendency to stick magnetically to the test surface wherever they first touch it. This acts to reduce mobility of the powder, and also to form a high level of background which not only reduces contrast and makes indications harder to see, but also wastes powder.

In the case of wet method particles that would become strongly magnetized because of high coercive force, this same objectionable background will form for the same reason. In addition, such particles would stick to any iron or steel in the tank or plumbing of the unit, and cause heavy settling-out losses which would have to be made up by frequent additions of new particles to the bath. Another bad feature shown by highly retentive wet method particles is their tendency to clump together quickly in large aggregates on the test surface. As previously stated, excessively large clumps of material have low mobility and indications are distorted or obscured by the heavy, coarse-grained backgrounds. Therefore, particles having high coercive force and retentivity are not desirable for wet method use.

Experience and extensive tests have shown, however, that coercive force and retentivity of a *low order* are advantageous. In the case of dry powders, a low order of residual magnetism appears to increase sensitivity, especially in the diffuse leakage fields formed by defects lying wholly below the surface. This may be for the reason that the small amount of polarity already established in elongated particles aids in lining them up into strings when the weak leakage fields act upon them. The action is similar to that of the compass needle swinging in the very weak field of the earth.

Wet method particles are actually chosen to have higher than minimum values of retentivity and coercive force. In this case the reason is clearly understood. As described in Section 2 above, these ultra-fine particles begin to collect at discontinuities as soon as they are applied to the test surface, when the agitation which had been present in the bath, ceases. With insufficient retained magnetism the particles remain fine and migrate very slowly through the liquid, due to their small magnetic moments and small mass, and the viscosity of the liquid suspending medium. The indications of discontinuities will build up, of course, but very slowly, taking as long as five to ten seconds. On the other hand, if magnetically "hard" particles are used, the test surface is covered with large immobile clumps as soon as the bath is applied.

Particles having intermediate magnetic properties collect into clumps more slowly while the indications are forming. The leakage field, strongest at the actual discontinuity, draws particles toward it, while the particles themselves are constantly enlarging due to agglomeration. At the same time they sweep up the ultra-fine particles as they move toward the defect. In this way all the magnetic fields present work together.

7. HYSTERESIS CURVES. The overall magnetic properties of the various types of magnetic particles can best be shown by their hysteresis curves. In making these curves, the powders are compressed into tubular holders, either rod-shaped or toroidal. The filled holder is then tested as a solid rod or ring would be, making allowance for voids remaining in the packed specimens. Special operating procedures are necessary to make the test reproducible. The resulting hysteresis curves obtained in this manner are those for the packed, shaped specimens, and not those for the individual magnetic particles or small agglomerations of particles.

Some typical hysteresis curves are shown in Figures 111 and 112. Figure 111 shows the loops for a typical dry powder, (a), when the maximum magnetizing force, H_{max}, was 500 Oersteds, and (b) when H_{max} was only 50 Oersteds. The shapes of these curves show that the material has a high permeability, shown by the height of the curve, B_{max}, a low coercive force, H_c, and a low retentivity, B_r—an excellent combination of properties for this type of powder.

Figure 112 gives curves for typical wet method materials again at 500 Oersteds maximum magnetizing force (a), and at 50 Oersteds maximum (b). In this case the much lower permeability is clearly shown by the much lower height of the curve B_{max}; and the much higher retentivity, B_r, and coercive force, H_c, is shown by the way the curve spreads out. Also, the striking difference in the proportion of retentivity as between the high (500 Oersteds) and the low (50 Oersteds) levels of magnetization is clearly indicated. $\dfrac{B_r}{B_{max}}$ $= 0.38$ at the high level, and $\dfrac{B_r}{B_{max}} = 0.057$ at the low level. No such difference is shown by the two curves of Fig. 111 for the dry method powder, the proportions being $\dfrac{370}{5610} = 0.065$ at 500 Oersteds, and $\dfrac{80}{690} = 0.115$ at the 50 Oersted level.

However, the actual values for B_{max} and H_c for the two types of powders are not by any means a measure of their relative sensitivities for the location of defects, because these powders are used under vastly different conditions, and factors other than the magnetic properties shown by their hysteresis curves are controlling. Nevertheless, the magnetic properties do play an important part under each of the conditions of use.

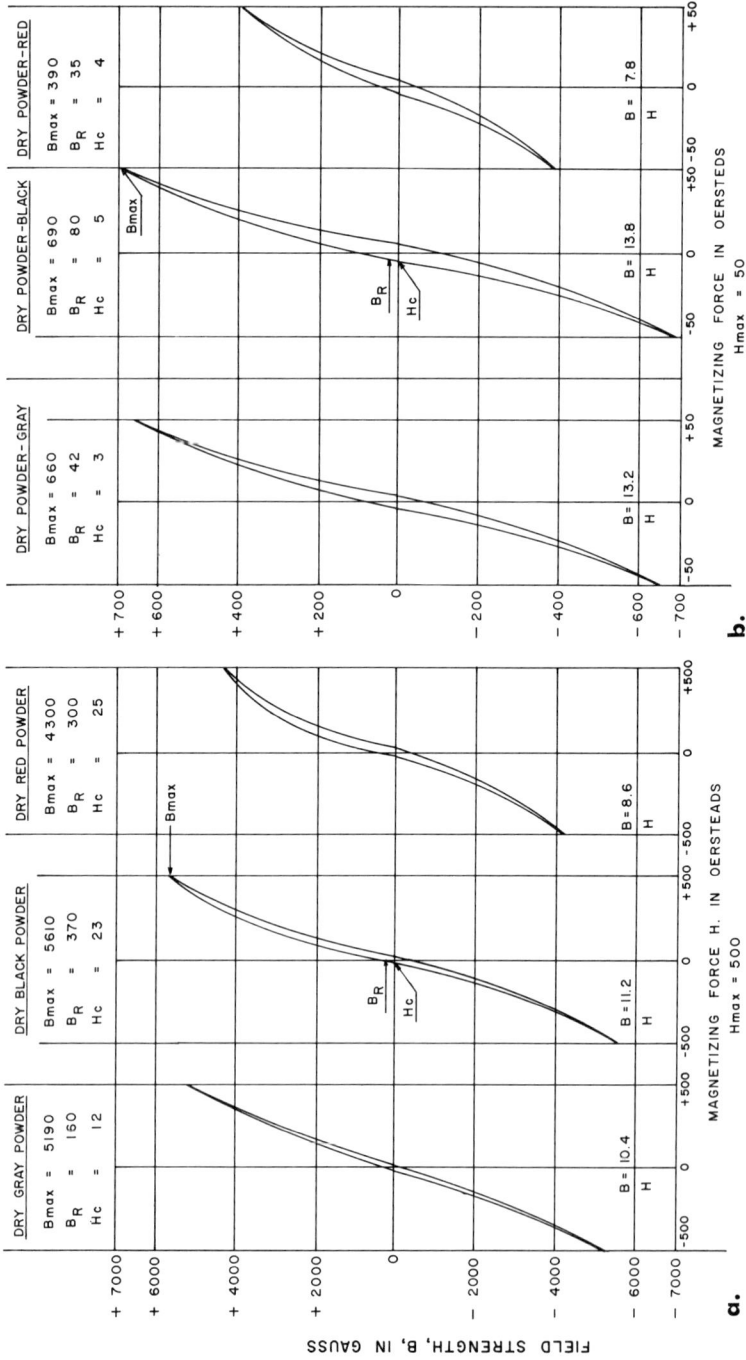

Fig. 111—Hysteresis Curves for Typical Dry Powders.
(a) When H_{max} was 500 Oersteds.   (b) When H_{max} was 50 Oersteds.

218

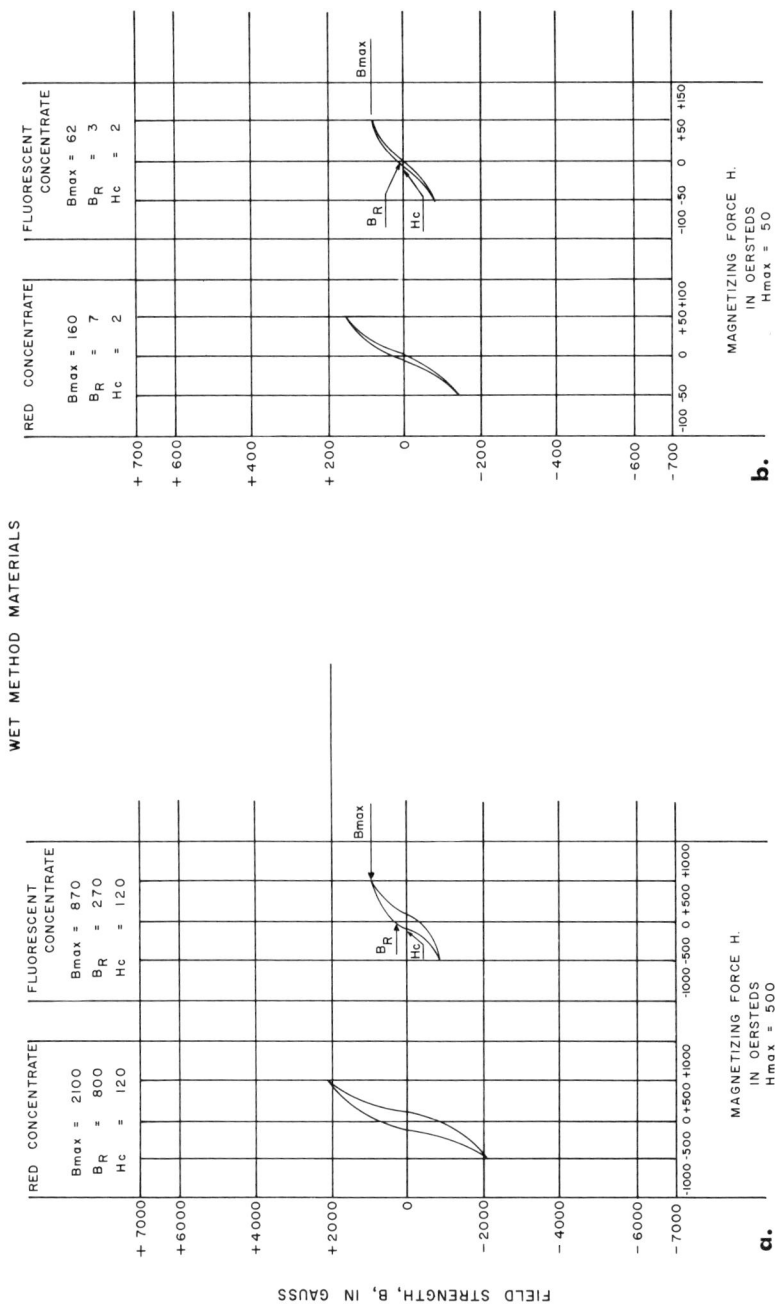

Fig. 112—Hysteresis Loops for Wet Method Materials.
(a) When H_{max} was 500 Oersteds. (b) When H_{max} was 50 Oersteds.

219

8. MOBILITY. It is obvious that when the magnetic particles are applied over the surface of a magnetized part, they must move and gather at a discontinuity under the influence of the leakage field to form a readable indication. Any factor that interferes with this required movement of the particles will have a direct effect on the sensitivity of the powder and the test. Conditions promoting or interfering with mobility are different for dry and wet method materials.

(a) *Dry Powders.* Ideally dry powders should be applied in such a way that they reach the magnetized surface in a uniform cloud with a minimum of motion. When this can be done, the particles come under the influence of the leakage fields while suspended in air, and have three-dimensional mobility. This condition can be approximated when the magnetized surfaces are vertical or overhead.

When the particles are applied on a horizontal or sloping surface they settle directly to the surface and do not have the same degree of mobility. Mobility can be achieved in this case, however, by tapping or vibrating the part, which jars the powder loose from the surface and permits it to move toward the leakage fields.

When A.C., or half-wave rectified A.C. or pulsating D.C. are used for magnetization, the rapid variation in field strength while the current is on, imparts motion to the magnetic particles on the ·surface of the part. The "dancing" of the particles gives them excellent mobility for the formation of indications.

The coatings applied to some of the dry method powders to give color to the indications serve a double purpose in that they also reduce friction between particles and the surface of the part, thus contributing another factor aiding mobility.

(b) *Wet Method Materials.* The earliest device for imparting mobility to magnetic particles was that of suspension in a liquid. Hoke's initial discovery of the magnetic particle principle was made when the fine metallic grindings were washed over the surface of the magnetized steel blocks which were being ground, by the liquid coolant used for grinding.

The suspension of particles in a liquid, which may be water or an organic liquid such as a petroleum distillate, allows mobility for the particles in two dimensions when the suspension is flowed over the surface of the part; and in three dimensions when the magnetized part is immersed in the suspension.

Wet method particles have a tendency to settle out of suspension, either in the tanks of the unit or somewhere on the test surface short of the defect. To be effective, the magnetic particles must move along with the liquid and reach every surface that the liquid covers, without "running aground" or settling out somewhere along the way. According to Stoke's Settling Law, particles settle out of suspension at a rate that is directly proportional to their size and their density—that is, the density in excess of that of the liquid medium—and inversely proportional to the liquid's viscosity. From this one might conclude that to slow down settling and improve the mobility of the particles one need merely use low density particles that are as small as possible, and use a more viscous liquid for the suspending medium.

None of these devices is very useful, either alone or in combination, to improve or control the behavior of particles when used in suspension for wet method defect detection. None of the properties of particles or liquids can be changed as suggested above without sacrificing some other equally important property. Ferromagnetic particles are by their nature of high density. Particle density may be changed to a slight extent by the addition of lower density pigments, but this lowers the magnetic properties of the compound particles in direct ratio. Therefore, this scheme can be used only where the need for color and contrast is great enough so that a small magnetic loss can be tolerated. Particle size has already been discussed (Section 2). Lowering it too far results in slow indication build-up. Raising the liquid's viscosity produces similarly undesirable results—that is, reduces the speed of indication buildup by slowing the movement of particles through the liquid.

As a result of all this the mobility of wet method particles is never ideal, nor can all possible means for improving it be used. But, while it must be balanced off against many other properties, mobility certainly is *one* of the factors which are important to wet method results.

9. VISIBILITY AND CONTRAST. These are additional important properties that have a great deal to do with making a magnetic powder suitable for its intended purpose. Size, shape and magnetic properties of a particle may be in all respects favorable toward making a perfect magnetic particle test medium, but if, having formed an indication, the inspector does not see it, the whole inspection fails.

Visibility and contrast are promoted by choosing colors of particles that are easy to see against the color of the surface of the test part. The natural color of the metallic powders is silver-gray, and we are limited to the colors black and red in the iron oxides commonly used as the base for the wet method materials. Visibility against certain colors of test surfaces can be increased by coloring the powder particles in some way. By use of pigments the silver-grey iron particles are colored white, black, red or yellow, all with comparable magnetic properties. One or another of these colors gives good contrast against the surfaces of most of the parts that are tested.

Among the dry powders, the gray-white powder was the first to be used, and it gives good contrast against the surfaces of many test parts. It fails to give good visibility, however, against the silver-grey of a sand- or grit-blasted surface, or against bright machined or ground surfaces. Choice of colors should be made by the operator to provide the best possible visibility against the surfaces of the test part under the conditions of shop lighting that prevail. In similar fashion, choice of either the black or the red wet-method material is made to suit particular lighting conditions.

In some cases it has been found advantageous to coat the *part being tested* with a color to improve contrast. Chalk or whiting, in alcohol, has been so used in the past for the inspection of large castings and weldments, when lighting conditions were poor in the areas where the inspection was being made. Aluminum paint has been similarly used. During World War II many small parts were given a flash coat of cadmium, principally for the purpose of making the black particle indications of the wet method more visible. The extra cost of coating the test surface was justified only when conditions were such that good contrast could not be obtained by use of the powders and pastes that were then available. The device of color-contrasting the part is rarely used today, because the fluorescent materials now available solve the problem in a much better way.

The ultimate in visibility and contrast is achieved by coating the magnetic particles with a fluorescent pigment, and conducting the search for indications in total or semi-darkness, using black light to activate the fluorescent dyes used. When indications glow in darkness with their own light, it is almost impossible for an inspector not to see them. Magnetically, these fluorescent materials are theoretically less sensitive than uncoated particles, but this reduction in magnetic sensitivity is more than off-set by the fact that patterns of particles can be readily seen even when only a few such particles go to make up the indication. A fluorescent indication easily visible under black light is often quite impossible to see when viewed in white light. The advantage in visibility and contrast of the fluorescent materials is so great that they are being used in a very high percentage of all applications. This is an example of magnetic particles in a particular use, in which one property supersedes in importance not only the magnetic properties, but also some of the other properties which we have been discussing.

Fluorescent magnetic particles are usually available in wet method materials only. It is quite feasible to make fluorescent dry particles, but the conditions under which dry particles are used are usually such that satisfactory visibility under test conditions is achieved without the expense of using the fluorescent version of the method. Today, however, in many applications for which dry powders were formerly used, the fluorescent *wet* method has been adopted. This is true for the inspection of large castings, steel billets and other products, for which the wet fluorescent method gives quicker coverage and better and quicker viewing for indications than the dry powder can do. Development of the over-all method of magnetization for such large objects has had much to do with making the wet method, with its quick, complete coverage of large surfaces, attractive for such tests. This type of inspection was formerly conducted by the dry powder method, almost without exception.

10.   WET METHOD MATERIALS. Wet method magnetic particles are fundamentally similar to each other, once they are dispersed in the suspending liquid. But before this step they appear in a variety of forms. In the past, the most common form, in this country at least, was a paste, made up of the powder and a liquid vehicle, and having a consistency much like that of a paint pigment when ground in oil.

Many of the raw materials as received are in a rather grainy or lumpy condition. When added to an oil or water bath in this condition they do not disperse, thus making a mixture wholly unsuited for magnetic particle inspection purposes. As in the case of paint pigments, a pre-dispersion by grinding the raw material into a paste with a small amount of oil or other liquid, produces a material more readily dispersed by dilution with the liquid of the bath.

However, the thick paste produced by this process is difficult to handle and requires a special procedure before it will blend properly into the bath. More recently the original pastes have been remade as dry powder concentrates. These powders are much easier to use, as they need merely to be measured out and added directly to the agitated bath. The agitation system of the modern magnetic particle units will pick up the powder and quickly disperse it in the bath in the ordinary process of circulation and agitation.

11. PRE-MIXED BATHS. To a small extent in this country, and to a great extent in England, pre-mixed, ready-to-use baths are sold and used. In England they are called "inks". These materials are manufactured in the usual manner and then made up into a bath of proper strength in the supplier's plant. Though more bulk must be shipped and paid for, the user is sure of getting a well proportioned bath of correct strength, provided it is properly mixed before using. This is especially attractive to the small user. Such pre-mixed baths are often used on a completely expendable basis. They are sprayed onto the test surfaces from a pump-agitated drum and not collected for re-use as they are in the ordinary testing unit. In many cases small pressurized aerosol cans containing the bath are furnished. These are particularly convenient for quick "spot" tests, such as for maintenance inspection.

12. THE SUSPENDING LIQUIDS. Wet method particles may be suspended either in water or in a petroleum distillate, or in any other liquid having suitable properties. Water is initially cheaper, but additions must be made to it before it is a suitable medium for suspending the wet magnetic particles. Wetting agents, anti-foaming materials, corrosion inhibitors, suspending and dispersing agents are all necessary. Dry material concentrates to be used for water suspension must contain all the extra ingredients necessary to make the finished suspension. Cost of the concentrates is comparable for water or oil suspension. The cost advantage of water-base baths lies in the difference in cost between water and the oil used. Another advantage of water for operations requiring large volumes of bath is the freedom from fire hazard. This hazard exists in small installations also, but the few gallons of distillate in the ordinary

224

testing unit have, over the years, not shown flammability of the bath as a serious problem.

The need to incorporate into the powder all the special ingredients which are needed for water suspension or for oil suspension, makes necessary two separate and distinct products. A water-suspendable powder cannot be used in oil. This is because the various additives are insoluble in oil and will not disperse the particles in an oil bath. The additions made to the powders intended for oil suspension are, similarly, not soluble in water. However, with suitable water conditioners, some of the oil-suspendable powders *can* be used in water.

13. AVAILABLE MATERIALS. As was stated earlier, the need to meet a variety of conditions for successful magnetic particle testing has resulted in the development of a large number of different materials to accomplish this result. Below is a list of many of these materials, with the special characteristics of each. Their specific applications and methods of use will be discussed in detail in subsequent chapters, especially in Chapters 13, 14, and 15. In order better to identify the several types listed, current designations used by Magnaflux Corporation are given.

Commercially available powders for the dry method are:

#1. *Grey Powder.* This is a general-purpose, high contrast powder, by far the most widely used of the dry powders. It is effective on dark surfaces, whether black, grey or rust colored.

#2. *Yellow Powder.* A pale yellow color, featuring low cost with fair sensitivity, and good contrast on dark colored surfaces.

#3A. *Black Powder.* Especially designed for use on light collored surfaces. This is also the cleanest powder to use, as it is dust-free. It is also the most sensitive of the dry powders. Its higher sensitivity is due to the fact that it contains the highest proportion of magnetic material of all the dry powders.

#8A. *Red Powder.* This dark reddish powder is used on light colored surfaces as is the black powder. However, since the black powder on a silvery surface is sometimes hard to see, the red color often offers a better contrast, particularly under incandescent lighting where the red color stands out.

#11. *Black Powder.* This powder is similar in color to the one listed above, under #3, but features a coarse, insensitive particle size and regular particle shape. It flows easily and is used for application with automatic equipment in circumstances where discrimination for only the larger defects is sought.

Commercially available wet method materials are:

A. *Non-Fluorescent Materials.*

#7C. *Black Powder Concentrate.* This is an oil-suspendable dry powder. It is especially suited for finding fine cracks on polished surfaces, such as bearings or crankshafts. It is the *most sensitive* of the non-fluorescent wet method powders for such applications, though indications may be hard to see. This is similar to the older 7B black paste, which was difficult to put into suspension. A very limited amount of this paste (7B) is still made for use in equipment with the older air-agitation systems. This powder, made for oil suspension, becomes water-suspendable when used with the proper proportion of water-conditioners listed below as WA-2A, or WA-4, in section C.

#9C. *Red Powder Concentrate.* This is a reddish brown oil suspendable powder. It is fully the equivalent of the black powder, 7C above, for all applications except the very fine cracks in polished surfaces. The red color gives improved contrast and visibility in situations where the black-on-silver contrast of the black powder is poor. Also, this color tends to be more visible than the black under incandescent light. This red powder is similar to the old 9B red paste, which shares the limitations of the black paste described above.

As in the case of the black powder, 7C, this red oil-suspendable material becomes suspendable in water when used with the proper amount of water conditioners listed below as WA-2A, and WA-4, in Section C.

#27-A. *Black Powder Concentrate.* This item is similar in all respects to the black powder 7C listed above, except that it is compounded for water suspension, and is for use with water only.

#29-A. *Red Powder Concentrate.* This item is similar in all respects to the red powder 9C listed above, except that it is compounded for water suspension, and is for use with water only.

B. *Fluorescent Materials.*

#10-A. *Fluorescent Powder Concentrate.* A blend of fluorescent magnetic powders and suspending agents, this powder concentrate is for use with oil only. It fluoresces a blue-green color, and features high fluorescent brightness, for tests where lighting conditions are not the best.

#12. *Fluorescent Paste.* This is an obsolescent blue-green-fluorescing paste composed of low fluorescent brightness magnetic particles and dispersing materials. It is for use with water only. In those places where it is still used it is favored because of its low tendency to foam, and its weak fluorescent background.

#14-A. *Fluorescent Powder Concentrate.* This is the most widely used of the fluorescent materials. It fluoresces a bright yellow-green and features high sensitivity, high (but not extreme) fluorescent brightness, and easy handling properties. It is designed for suspension in oil, but can be used for water suspension when combined with the proper amount of water conditioners litsed below as WA-2A, and WA-4 in Section C. It stays in suspension well and does not give objectionable fluorescent background.

#15. *Fluorescent Powder Concentrate.* This powder concentrate is identical in composition with 14-A above, but is made up entirely of larger individual particles. It is usually used as a water suspension but it may be used in oil also. This material finds its greatest application suspended in water for steel billet inspection, where only the larger defects are sought. It is also most effective for the inspection of steel castings and similar "toothy" surfaced parts, which would present a background problem with fine-sized particles. Because of its large particle size it is difficult to keep in suspension. The water conditioner for use with it is extra-low foaming to prevent the bath from foaming excessively during the more violent agitation required to keep the particles suspended.

#20-A. *Fluorescent Powder Concentrate.* This powder is identical to 14-A above, but is compounded to include the proper amount of water conditioner. It is for use with water only.

#24. *Fluorescent Powder Concentrate.* This is a special iron-base powder of extreme fluorescent brightness. It fluoresces a bright yellow-green. Its high density makes it more difficult to keep in suspension. Some users prefer it because they have found it more sensitive in some applications, though its superiority over 14A above is hard to demonstrate. It has the highest magnetic permeability of any of the fluorescent materials.

C. *Water Conditioners.*

WA-2A. This material is a powder designed to help suspend the wet method materials in water—particularly the 14A powder listed above, which is somewhat difficult to wet with water. This wetting agent features excellent wetting and spreading on oily surfaces, moderate to low foaming, good corrosion inhibition and a tolerance for oil contamination in water baths.

WA-3. This material is designed for use with the coarse particle fluorescent material listed as #15 above. In extremely well agitated inspection units, this conditioner features minimum foaming together with ability to suspend the difficult-to-wet coarse powder. It has *no* corrosion inhibiting properties.

WA-4. The conditioner listed as WA-2A above is slow to dissolve in water. This product (WA-4) is a liquid and was developed to eliminate that feature. Besides being rapidly soluble in water, this product shows strong corrosion inhibiting properties, although it foams slightly more than the WA-2A above. Like all water conditioners, this product is not affected by the hardness of the water used to make up the bath.

CHAPTER 12

BASIC VARIATIONS IN TECHNIQUE

1. INTRODUCTION. In the application of magnetic particle testing there are a considerable number of possible variations in procedure which critically affect the results obtained. These variations are made necessary by the many types of defects that are sought, and the many types of ferromagnetic materials in which these defects must be located. The need to assure adequate sensitivity, and maximum reliability *at an optimum cost,* has also caused many special techniques to be developed.

In order to make the proper choices among these variables for any given testing problem, the operator must know what the possible variations are, and how each affects the end result. A given technique may be either favorable or unfavorable, depending upon the specific case he is considering.

2. LIST OF VARIATIONS IN TECHNIQUE. The following list gives the most important variations in the techniques used to apply magnetic particle testing:

(a) *Type of Current.* Possible current types are A.C., D.C., half wave rectified single phase A.C., full wave rectified single phase A.C., and full wave rectified three phase A.C.

(b) *Type of Particles and Method of Application.* Magnetic particles may be applied dry or in suspension in a liquid—that is, the Dry Method or the Wet Method. Particles may be any of several colors, or may be fluorescent.

(c) *Sequence of Steps.* The part may be magnetized first and particles applied later after the magnetizing force has been turned off (the residual method) ; or the part may be covered with particles while the magnetizing force is acting (the continuous method). With parts having high retentivity, a combination of these methods is sometimes used.

(d) *Direction of Field.* Circular, longitudinal or other methods of magnetization are involved in this choice, as well as methods of making contact or applying a coil.

(e) *Sensitivity Level.* This choice has to do principally with the amount of current or other magnetizing force and the test medium, as well as proper application by whatever method has been decided upon.

(f) *Equipment.* Stationary or portable units, use of prods or yokes, or special-purpose or automatic equipment may be selected, depending on the conditions involved.

3. CHARACTERISTICS OF DEFECTS AND PARTS WHICH INFLUENCE THE PROPER CHOICE AMONG THE SEVERAL VARIABLES. In deciding how best to proceed to apply magnetic particle testing and what choices to make among the possible technique variations, the characteristics of the sought-for or expected defect, and of the parts in which they occur, must be surveyed in order to arrive at the optimum in sensitivity, reliability and cost for the test.

Following are a list of considerations with respect to the characteristics of defects and parts which determine the proper choice of procedure:

(a) Are the defects open to the surface or wholly below the surface?

(b) Are they fine and sharp, or are they wide open?

(c) Are they shallow or do they go deep into the metal?

(d) What is their physical size and shape?

(e) If they have direction, are they longitudinal or transverse to the axis of the part?

(f) What is their location and direction with respect to the stresses to which the part will be subjected in service?

(g) What is the service for which the part is intended—is it critical as, for instance, an aircraft engine or landing gear part—or would its failure involve no drastic consequences, as for example, breakage of a tool?

(h) What is the size and shape of the part in which the defects occur?

(i) What are the magnetic characteristics of the part in which they occur?

(j) What are the overall economic considerations? Test costs vs. costs of failure, such as loss of goodwill, damage to customer's equipment, etc.?

4. PRIMARY METHOD CHOICES. There are two basic decisions which must be made before the details of operating procedure can be decided upon. These are:

(a) *Type of Current to be Used.* This choice is dictated by the location of the defects, whether they are open to the surface of the part, or whether they are located wholly below the surface. Choice of current lies between A.C. and some form of D.C. If the defect is open to the surface, either A.C. or D.C. may be suitable and the choice is determined by other considerations. If the defect lies wholly below the surface A.C. cannot in general be used. Instead, D.C., either straight D.C. or some type of rectified A.C. is indicated.

(b) *Type of Magnetic Particles to be Used.* This choice is primarily between the dry and the wet method, and secondarily among the various colors that are available, including flourescent particles. The decision is influenced principally by the following considerations:

> (1) Whether the defect is *on the surface* or *wholly below the surface.* For deep-lying defects below the surface the dry powder is usually more sensitive.

> (2) The *size* of the defect if *on the surface.* The wet method is usually best for very fine and shallow defects.

> (3) Convenience. Dry powder with a portable half wave unit for instance, is easy to use for occasional large parts in the shop or foundry, or for field inspection work.

5. CHOICE OF TYPE OF CURRENT—A.C. VS. D.C. This is the first basic choice to be made, since the skin effect of A.C. at 50 or 60 cycles per second frequency limits its use to the detection of defects which are open to the surface, or only a few thousandths of an inch below it. However, the skin effect of A.C. is less at lower frequencies, resulting in deeper penetration of the lines of force. At 25 cycles the penetration is demonstrably deeper, and at frequencies of 10 cps and less, the skin effect is almost non-existent.

If the defects sought are at the surface, A.C. has several advantages. The rapid reversal of the field imparts mobility to the particles, especially to the dry powders. The "dancing" of the powder helps it to move to the area of leakage fields and to form stronger indications. This effect is less pronounced in the wet method.

Alternating current has another advantage in that the magnetizing effect, which is determined by the value of the *peak* current at the top of the sine wave of the cycle, is 1.41 times that of the current read on the meter. Alternating current meters read more nearly the average current for the cycle rather than the peak value. To get equivalent magnetizing effect from straight D.C., more power and heavier equipment is required. Thus A.C. equipment for a given output of *magnetizing force* can be lighter and less costly, and better adapted for portability.

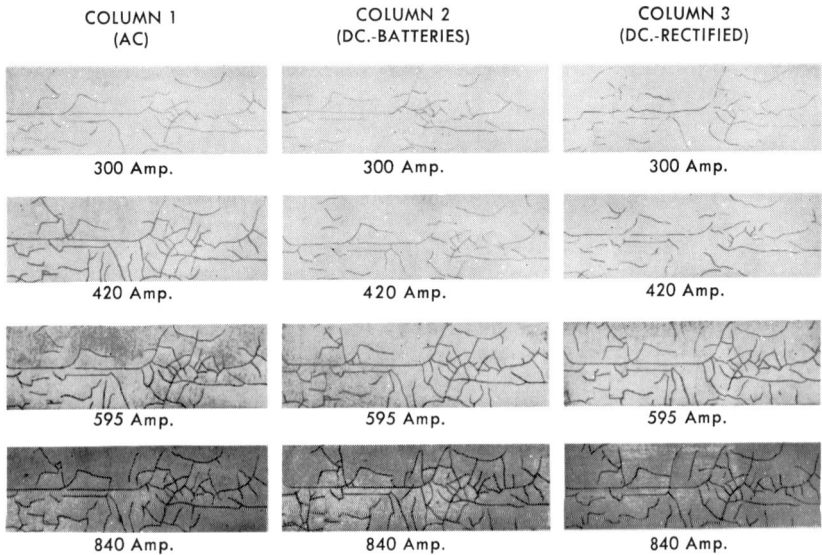

COLUMN 1 (AC)	COLUMN 2 (DC.-BATTERIES)	COLUMN 3 (DC.-RECTIFIED)
300 Amp.	300 Amp.	300 Amp.
420 Amp.	420 Amp.	420 Amp.
595 Amp.	595 Amp.	595 Amp.
840 Amp.	840 Amp.	840 Amp.

Fig. 113—Comparison of Indications of Surface Cracks on a Part Magnetized with A.C., Straight D.C. and Three Phase Rectified A.C., Respectively.

D.C., on the other hand, magnetizes the entire cross-section more or less uniformly in the case of longitudinal magnetization, and with a straight line gradient of strength from a maximum at the surface to zero at the center of the bar in the case of direct contact (circular) magnetization. See Chapter 9, Sections 20 and 25, for field distribution in and around a magnetic bar carrying A.C. and D.C. respectively. Figure 113 is a comparison of indications of the same set of fine surface cracks on a ground and polished piston pin, obtained by the use, respectively, of 60 cycle A.C., D.C. from storage batteries (straight D.C.) and D.C. from rectified three phase 60 cycle A.C. Four values of current were used in each case, using a central conductor to magnetize the hollow pin. The indications produced with A.C. are heavier than the D.C. indications at each current level, although the difference is most pronounced at the lower current values. Straight D.C. and rectified A.C. are comparable in all cases. The A.C. currents are meter (R.M.S.) values, so that peak-of-cycle currents, and therefore magnetizing forces, are 1.41 times the meter reading shown.

Another comparison, which graphically shows the advantage of direct current for the detection of defects lying wholly *below* the surface, and the limitations of A.C. for this purpose, is shown in

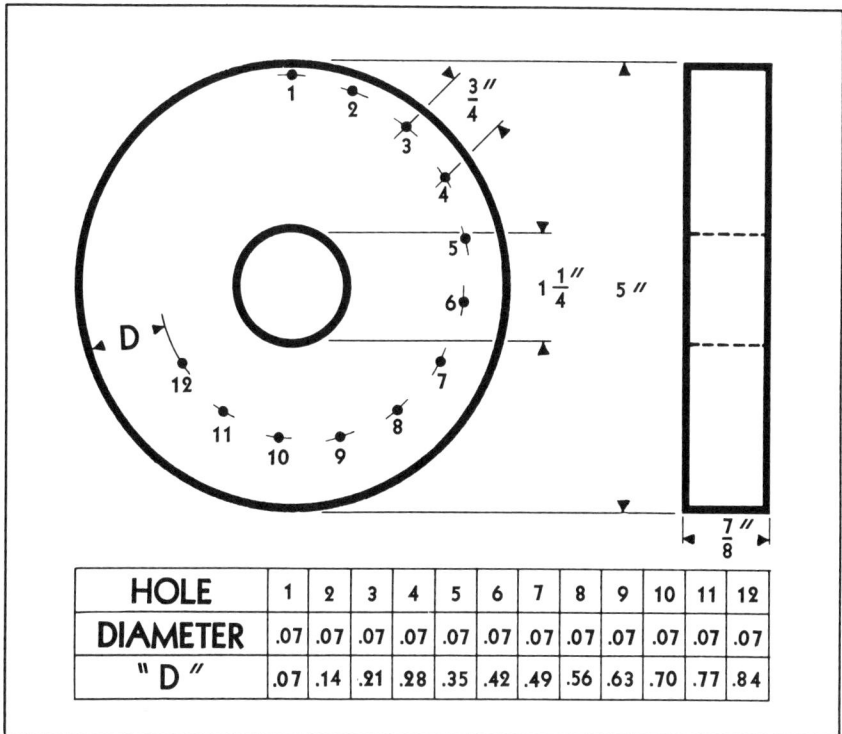

HOLE	1	2	3	4	5	6	7	8	9	10	11	12
DIAMETER	.07	.07	.07	.07	.07	.07	.07	.07	.07	.07	.07	.07
" D "	.07	.14	.21	.28	.35	.42	.49	.56	.63	.70	.77	.84

Fig. 114—Drawing of a Tool Steel Ring Specimen
with Artificial Sub-Surface Defects.

Figs. 114 and 115. Figure 114 is a drawing of a ring specimen of unhardened tool steel (.40 carbon), $\frac{7}{8}$ inch thick. Holes parallel to the cylindrical surface have been drilled, 0.07 inch in diameter, at increasing depths below the surface. The depths vary from 0.07 to 0.84 inch in 0.07 inch increments, from hole #1 to hole #12.

Figure 115 is the plot of the threshold values of current necessary to give a readable indication of holes in this ring, by the dry continuous method with central conductor, using 60 cycle A.C. and three forms of D.C. The three types of D.C. are straight D.C. from batteries, three phase rectified A.C. with surge, and half wave rectified single phase 60 cycle A.C. Currents as read on the usual meters are varied from the minimum necessary to indicate hole #1

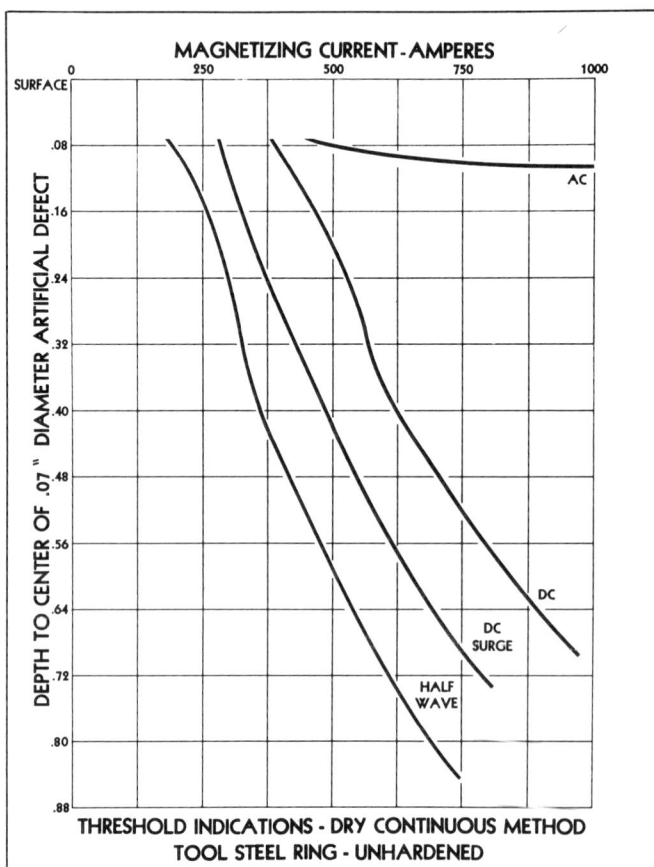

Fig. 115—Comparison of the Sensitivity of A.C., D.C., D.C. with Surge, and Half Wave, for Locating Defects Wholly Below the Surface.

in the case of each type of current, up to a maximum of over 1000 amperes. Alternating current required about 475 amperes to show hole #1 and over 1000 amperes to show hole #2. Hole #3 could not be shown with A.C. at any current value available. Hole #2 was shown with 475 amperes straight D.C., with 275 amperes D.C. preceded by a surge of double this amount, and by 175 amperes of half wave. Seven hundred and fifty amperes of half wave showed hole #12, while 975 amperes were needed to show hole #10 with straight D.C.

For the inspection of finished parts, such as the machined and ground shafts and gears of fine machinery, direct current is frequently used. Although A.C. is excellent for the location of fine cracks that actually break the surface, D.C. is better for locating very fine non-metallic stringers lying just under the surface. It is usually important to locate such stringers in parts of this type, since they can initiate fatigue failures.

These comparisons point up the importance of choosing the right current type to give the best indications possible, and show how the choice will vary, depending upon the nature and location of the defects sought.

6. CHOICE OF TYPE OF MAGNETIC PARTICLES. DRY VS. WET METHODS. This is the second basic method choice to be made. The dry powder method is superior for locating defects lying wholly below the surface because of the high permeability and the favorable elongated shape of the particles. These form strings in a leakage field and bridge the area over a defect. A.C. with dry powder is excellent for surface cracks which are not exceedingly fine, but as shown in the comparisons of Fig. 115, is of little value for defects lying even slightly below the surface.

Figure 116 is a comparison of the effectiveness of the dry method and the wet method for detecting defects lying wholly below the surface, using the same unhardened tool steel ring described in Figure 114. It is clear from this comparison that the dry method is superior to the wet for this purpose at any value of direct current used.

A further comparison of the wet and dry methods is shown in the graph of Fig. 117.

However, when the problem is to find very fine surface cracks, there is no question as to the superiority of the wet method, whatever form of magnetizing current is used. In some cases, direct

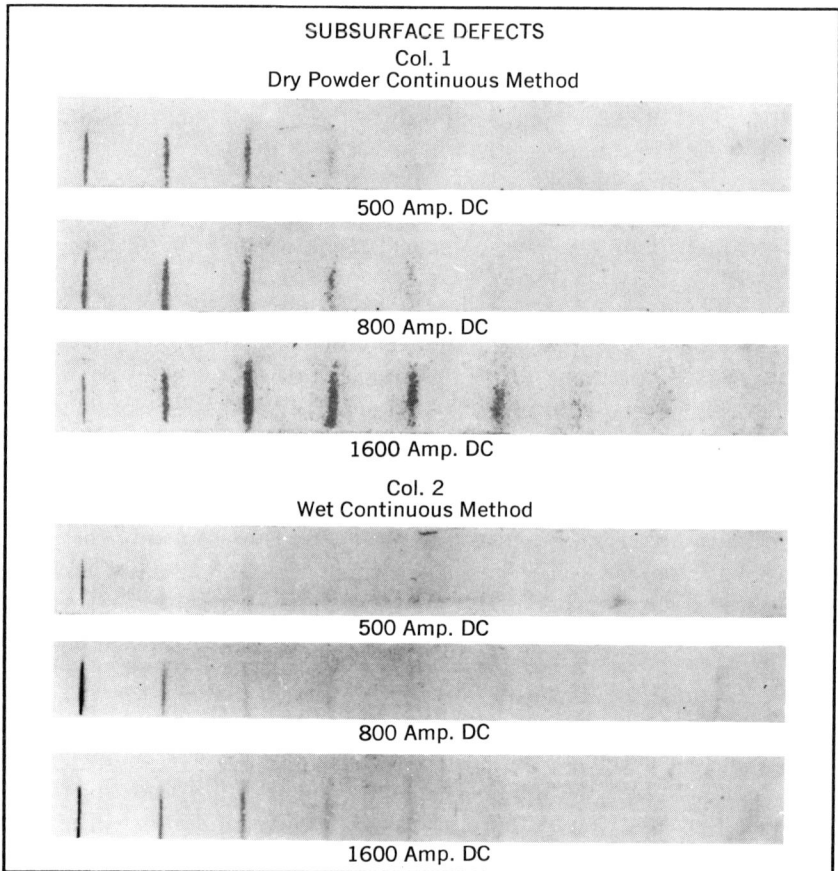

Fig. 116—Comparison of the Dry and the Wet D.C. Method
for Defects Lying Below the Surface.

current is selected for use with the wet method so as to get the advantage of better indications of discontinuities which lie just below the surface, especially on bearing surfaces and aircraft parts. The wet method offers the advantage of easy complete coverage of the surface of parts of all sizes and shapes. Dry powder is often used for very local inspections.

Selection of the color of particles to use is essentially a matter of securing the best possible contrast with the background of the surface of the part being inspected. The differences in visibility among the black, gray, red and yellow particles are considerable on backgrounds which may be dark or bright, and which may be

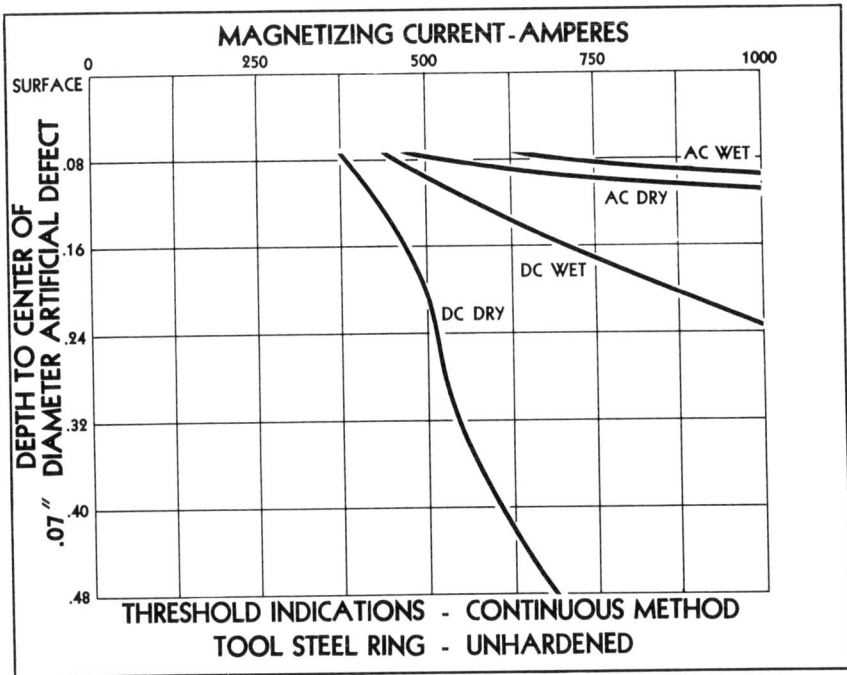

Fig. 117—Comparison of the Dry and the Wet Methods for Defects Lying Below the Surface, when A.C. and D.C. are used for Magnetization.

viewed in various kinds of light. Black stands out against most light-colored surfaces, gray against dark-colored ones, like those of castings. Red is more visible than either against silvery and polished surfaces, especially if the light by which the inspection is made is from incandescent lamps. The yellow-colored powder is not used very extensively, but has some advantages under some certain conditions of lighting. If difficulty is had in seeing indications the operator should try some color of powder other than the one he is using.

In the case of the wet method, as was stated in Chapter 11, the ultimate in visibility and contrast is obtained by the use of fluorescent particles. The fluorescent wet method has been used in constantly increasing numbers of inspection applications for many years, principally because of the ease of seeing even the faintest indications.

7. FIRST OPERATING DECISION: RESIDUAL VS. CONTINUOUS METHOD. This choice between the residual and the continuous method is a relatively easy one. In the residual method, parts are magnetized and subsequently the magnetic particles are applied. The method can be used only on parts having sufficient retentivity. The permanent field they retain must be sufficiently strong to produce leakage fields at discontinuities which in turn will produce readable indications. The method in general is reliable only for the detection of surface discontinuities.

Since hard materials which have high retentivity are usually low in permeability, higher-than-usual magnetizing currents may be necessary to obtain a sufficiently high level of residual magnetism. Some idea of the effect which the lower permeability of highly retentive steels has on necessary magnetizing current is shown in Fig. 118. The tool steel ring of Fig. 114 was hardened to Rockwell C-63, and again tested with the dry continuous method using several forms of direct current. The curves indicate that 4000 amperes D.C. was needed to show hole #5. Even half wave at 1500 amperes showed only hole #4.

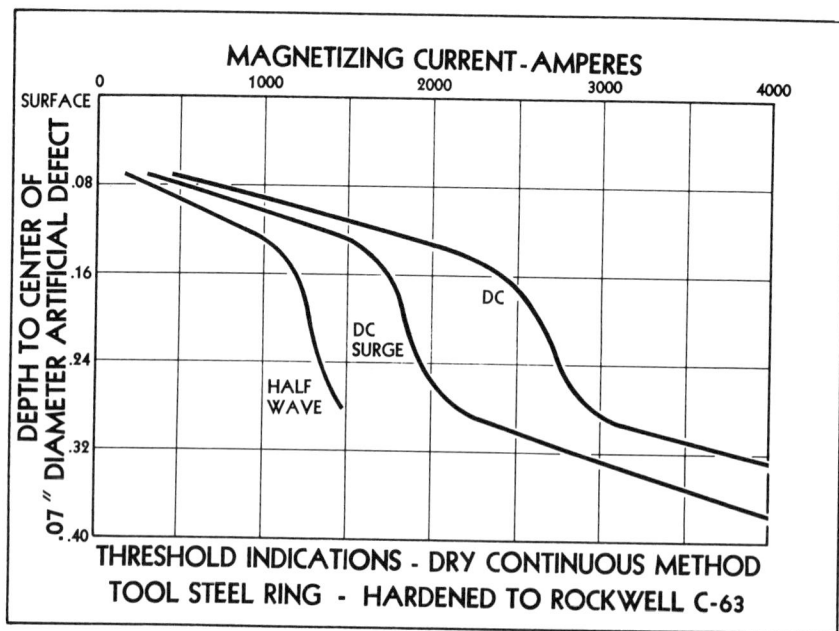

MAGNETIZING CURRENT-AMPERES

THRESHOLD INDICATIONS - DRY CONTINUOUS METHOD
TOOL STEEL RING - HARDENED TO ROCKWELL C-63

Fig. 118—Effect of the Hardness of a Specimen on the Current Required to Locate Defects Lying Below the Surface.

The difference in the behavior between hard steels and soft steels is usually not very serious if only surface discontinuities are sought.

It becomes important when interest lies in defects below the surface. Then the figure of 500 for the minimum value of material permeability for a metal or alloy to be suitable for magnetic particle testing may become a limitation for some hard steels and alloys. (See Chapter 9, Sections 9 to 11.)

Either the dry or the wet method for particle application can be used in the residual method. With the wet method, the magnetized parts may be immersed in an agitated bath of suspended magnetic particles, or they may be flooded with bath by a "curtain spray". In these circumstances a favorable factor, affecting the strength of indications, enters. This is the *time* of immersion of the part in the bath. By leaving the magnetized part in the bath or under the curtain spray for a considerable time, the leakage fields, even at fine discontinuities, can have time to attract and hold the maximum number of particles. This can mean an increase in sensitivity over the mere flowing on of the bath over the surface of the part as it is being magnetized by the continuous method. It should be noted, however, that the location of the discontinuity on the part as it is immersed affects particle build-up. Build-up will be greatest on horizontal *upper* surfaces, and less on vertical surfaces or horizontal lower surfaces. Also, rapid withdrawal from the bath or spray can wash off indications held by extremely weak leakage fields, and care must be exercised in this part of the process.

The residual method, either wet or dry, has many attractive features and finds many applications, even though the continuous method has the *inherent* advantage of greater sensitivity.

This advantage for the continuous method is simple, but basic. When the magnetizing force is applied to a ferromagnetic part, the field rises to a maximum on the hysteresis curve, its value or intensity deriving from the strength of the magnetizing force and the material permeability of the part. But when the magnetizing force is removed, the *residual magnetism* in the part is *always* less than the field present while the magnetizing force was acting. The amount of difference depends on the retentivity of the material. The continuous method, for a given value of magnetizing current, is therefore *always* more sensitive than the residual, at least so far as sensitivity is determined by the strength of field in the part.

Also, techniques have been worked out for the continuous method which make it faster than the residual. The indication is produced at the time of magnetization, whereas the residual method requires two steps—magnetization and application of particles—plus the added time for indications to build up if the immersion method is used. The choice therefore, as between residual and continuous methods, should be for the continuous method unless special circumstances make the residual method more desirable. For example, the residual method is frequently used in special purpose and automatic equipment because the timing of magnetization and application of particles is not critical and the entire process is therefore more easily controlled.

The continuous method is, of course, the only possible one to use on low carbon steels or iron having little or no retentivity. It is frequently used with A.C. on such materials because the alternating current field produces excellent mobility of the particles. With the wet method the usual practice is to flood the surface of the part with the bath, then simultaneously terminate bath application and close the magnetizing circuit switch momentarily. Thus the magnetizing force acts on the particles in the film of bath as they are draining over the surface. A "shot" of $\frac{2}{5}$ to $\frac{1}{2}$ of a second—either A.C. or D.C.—is sufficient time for the current to be on. Strength of the particle bath has been standardized to supply a sufficient number of particles in the film to produce good indications with this technique. It should be noted, however, that the continuous method requires more attention and alertness on the part of the operator than does the residual method. Careless handling of the bath-current sequence can interfere seriously with reliable results.

Probably the highest possible sensitivity obtainable for very fine defects is achieved by immersing the part in the wet bath and passing the magnetizing current through the part for a short time *while immersed;* and then leaving the current on while the part is removed from the bath and while the bath drains from the surface. This technique is used today for the inspection of jet engine compressor blades and vanes, using fluorescent magnetic particles.

8. SECOND OPERATING DECISION. CIRCULAR VERSUS LONGITUDINAL MAGNETIZATION. This decision is determined by the shape and orientation of the defect in relation to the shape and principal axes of the part. The rule of thumb, that the current must be passed in a direction parallel to the defect, is controlling. This rule may require circular magnetization in some sections of the part, and longitudinal magnetization in others. Of course, if the principal direction of discontinuities is unknown both circular and longitudinal magnetization must be used, in order that all possibly-present discontinuities be located.

When feasible, circular magnetization is preferred, because it produces a minimum of external polarity that will attract and hold particles where no discontinuity exists. Particles so held may confuse or prevent the operator from seeing the actual indications of defects.

9. THIRD OPERATING DECISION. AMOUNT OF CURRENT REQUIRED. The amount of magnetizing current, or the number of ampere turns to use for optimum results, is the final factor in determining the sensitivity level of the process. The types and minimum dimensions of the defects that must be located—or the size and kind of defects that may be tolerated—are the major considerations in fixing current levels. The type of service for which the part is intended, the importance of the part in its assembly, and the effect on the continuing performance of the assembly if this part fails, are further important factors in determining the sensitivity level required. For highly critical parts maximum sensitivity would be called for in order that all minute discontinuities be located. Since the most critical defects are practically always on the surface, this situation would probably call for the wet continuous method, with circular magnetization if possible, and the use of fluorescent particles.

In some primary metal and industrial applications, sensitivity is deliberately limited and controlled so as to detect only gross defects and to ignore the small ones. This is true in the inspection of steel billets for seams. Here only defects over a certain depth are of interest, since shallow ones will scale off in subsequent re-heating and rolling of the billets. Sensitivity here is controlled by the amount of current used, by the character of the magnetic particles, and by sequencing of magnetization and bath application.

The amount of current is initially determined by the rules of thumb given in Chapter 9. The formula for longitudinal magnetization with a coil is: Ampere Turns $= \dfrac{45,000}{L/D}$, where L is the length and D the diameter of the part. (See Chapter 9, Section 8.) For circular magnetization the rule is to use 1000 amperes per inch of part diameter. This rule cannot, however, be followed in all cases. For small diameters a somewhat smaller current is usually sufficient.

Current must not be so high as to heat the part enough to affect its heat-treated structure, or to cause burning due to the resistance at the points of contact. On the other hand, for large diameter parts the level of current called for by the rule is extremely high and in many cases not available. Experience has shown that for most steels, for the location of surface cracks only, considerably lower current values than the rule calls for are sufficient. On large parts, if the purpose of the inspection is to locate defects lying wholly below the surface, either local magnetization with prods must be used, or else high current for overall magnetization must be provided.

If the current to be used is A.C. or half wave, the above rules of thumb for current strengths will call for higher currents than needed, since the fields produced by A.C. and half wave tend to be stronger at the surface than in the case of similar values of D.C. However, these rule-of-thumb current values are only guides to start from. Experience with the results produced will show quickly whether the chosen value is higher than required.

10. THE FOURTH OPERATING DECISION. EQUIPMENT. This decision depends on the size, shape, number and variety of parts to be tested. For production testing of numerous parts which are relatively small, but not necessarily identical in shape, a bench type unit with clamping head contacts for circular magnetization, and a built-in coil for longitudinal magnetization is commonly used. In such applications the wet continuous method is generally selected, although some dry powder units of this type have been in service.

If the part or parts are large, portable units using prods or C-clamp contacts and hand-wrapped coils may be most convenient. Half-wave and dry powder are often used with such portable equipment, as in the inspection of welds and large castings. The wet method is also used with such portable equipment, in which case the bath usually is not recovered and re-used, but is allowed to drain away—the "expendable" technique. This system, using fluorescent particles is being used increasingly today, especially with the overall method of magnetization.

When the number of identical or closely similar parts is large, as in mass production operations, single-purpose magnetization and inspection units or jigs on multi-purpose units, usually automatic or semi-automatic, are often designed and built. In such cases the extra cost of the unit is justified by greater output per inspector, and by more uniform and reliable inspection. Reproducibility of results is assured since each part receives exactly the same preselected treatment during magnetization and bath application. Automatic units of this type invariably use the wet method, because it is easier to cover all surfaces of parts with a liquid suspension of particles than with a spray or cloud of dry powder. (See Chapter 19.)

11. OTHER OPERATING DECISIONS. There are a number of modifications and variations of the standard techniques for magnetization and particle application that apply to certain sizes and shapes of parts, and specific inspection conditions. When any of these variations are applicable, special tooling is usually necessary. The choice of one of these special methods is dictated by their ability to do the testing job more reliably and at lower cost than can be done by more standard procedures. Some of these variations are:

(a) Multi-directional magnetization.

(b) The overall method.

(c) The induced current method.

(d) The use of automatic units.

In some cases no power is available to operate magnetizing equipment in the area where the inspection must be carried out, or the nature of the inspection does not justify the purchase of more expensive equipment. In such cases permanent magnet yokes may be used; or electromagnet yokes operated either by A.C. or rectified A.C. if some current source can be had. Yokes using automobile or other storage batteries as a source of energizing current are another solution of the no-power problem. In explosive areas, permanent magnet yokes furnish a safe way to secure local magnetization for maintenance inspection. Inspection with these devices is usually limited to spot checking or occasional testing of miscellaneous parts.

CHAPTER 13

THE DRY METHOD—MATERIALS AND TECHNIQUES

1. HISTORY. When A. V. de Forest first applied magnetic particle testing to the drill pipe problem, he used the dry powder method. At that time there was no record of performance for either the dry or the wet methods. In Hoke's original observation of magnetic particle patterns on steel blocks, the fine grindings from the blocks themselves provided the magnetic particles and they were carried by the grinding coolant that he was using. This was, therefore, the wet method. But for de Forest's work some form of "iron filings" was doubtless readily at hand—the material may even have been mill scale, plentiful around a pipe mill. As it served his purpose, he continued to use the dry form of magnetic particles.

Having visualized the value of the new method in broader applications, he undertook to improve the powder as a means of securing reliable and reproduceable results. He recognized that control of particle size and shape, as well as permeability, were important for a satisfactory powder. Furthermore he saw the need for a colored powder to make a better contrast against the surfaces of the part being tested, and early devised a means for coating the particles to give them a gray-white color.

The material for the dry method was thus fully developed at the outset of commercial use of the magnetic particle testing method, and there has been little basic change since that time. Much more precise control of particle size and shape, better coating methods and additional colors have improved the performance of the dry powders, but essentially they are the same as originally worked out by Doane and de Forest in 1929. The use of particles in liquid suspension, now more prevalent than the dry method, was not begun until the middle 1930's.

The dry method was first used extensively by the railroads, the automotive industry and foundries, and during World War II for the inspection of welds and steel castings on all manner of war materials from gun mounts to submarines. It is still used somewhat in similar fields today, although in many instances, even for these purposes, the wet method has taken over. This appears to be due largely to the development of fluorescent particles for the wet method, with their many advantages. Nevertheless the dry method retains a field of importance of its own, as in the inspection of welds and castings where the detection of defects lying wholly below the surface is considered important.

2. ADVANTAGES AND DISADVANTAGES OF THE DRY METHOD. In the previous chapter (12) the good points and the drawbacks of the dry method were quite thoroughly discussed, and comparisons drawn with the wet method. This discussion may be briefly summed up in the following list:

Advantages:

(a) Excellent for locating defects wholly below the surface and deeper than a few thousandths of an inch.

(b) Easy to use for large objects with portable equipment.

(c) Easy to use for field inspection with portable equipment.

(d) Good mobility when used with A.C. or Half Wave.

(e) Not as "messy" as the wet method.

(f) Equipment may be less expensive.

Disadvantages:

(a) Not as sensitive as the wet method for very fine and shallow cracks.

(b) Not easy to cover all surfaces properly, especially of irregular shaped or large parts.

(c) Slower than the wet method for large numbers of small parts.

(d) Not readily useable for the short, timed "shot" technique of the continuous method.

(e) Difficult to adapt to a mechanized test system.

3. MATERIALS. Chapter 11 contains a description of the dry magnetic particles and discusses their various characteristics, contrasting them with the wet method materials. Outstanding properties of the dry powders are their more favorable shape and higher permeability in comparison with the wet particles. It is these two characteristics more than any other that are responsible for the

245

good performance of these dry powders within their field of application.

The availability of four colors—grey, red, black and yellow—increases the chance of getting good visibility under almost any conditions of background or lighting.

4. STEPS IN APPLYING THE DRY METHOD. There are five essential steps in applying the dry method of magnetic particle testing. These are:

(a) Preparation of the surface.

(b) Magnetization of the part.

(c) Application of the powder.

(d) Blow-off or removal of excess powder.

(e) Inspection for indications.

In a few instances some cleaning of the parts after testing is required to remove adhering powder.

5. SURFACE PREPARATION. In general, the smoother the surface of the part to be tested and the more uniform its color, the more favorable are the conditions for the formation and the observation of the powder pattern. This statement applies particularly to inspections being made on horizontal surfaces. For sloping and vertical surfaces, the dry powder may not be held on a very smooth surface by a weak leakage field. The surface should be clean, as far as possible, and dry and free of grease. The dry particles will stick to wet or oily surfaces, and not be free to move around over the surface to form indications. This may completely prevent the detection of significant discontinuities. On surfaces that have been cleaned of grease by wiping with a rag soaked in naphtha, a thin film of oil often remains that is sufficient to interfere with the free movement of the powder. This film can be removed by dusting the surface with chalk or talc from a shaker can, and then wiping the surface with a clean dry cloth. An initial application of the dry magnetic powder itself, followed by wiping, often will give a surface over which a second application of powder will move readily. If it is feasible to use it, vapor degreasing will give a dry, oil-free surface.

Any loose dirt, paint, rust or scale should be removed with a wire brush, or by shot or grit blasting or other means. If cleaning is done with shot or grit blasting, there is a peening effect, especially on softer steels, which may close up fine surface discontinuities. The effect is more pronounced with shot than with grit, but if these cleaning methods are used the operator should be aware of the danger of missing very fine cracks. A thin, hard, uniform coating of rust or scale will not usually interfere with the detection of any but the smallest defects. The inspector should know the size of the smallest defect he is to consider significant, in order to judge whether or not such a coating of rust or scale should be removed.

Paint or plating on the surface of a part has the effect of making a surface defect behave like a sub-surface one. The relative thickness of the plating or paint film and the size of the defects sought determine whether or not the coatings should be stripped. The dry method is more effective in producing indications through such non-magnetic coatings than the wet method, but if fine cracks are expected the surface should be stripped of the coating if its thickness exceeds 0.005 inch. Most coatings of cadmium, nickel or chromium are usually thinner than this, and the plating makes an excellent background for viewing indications. Hot galvanized coatings, however, are thicker, and in general should be removed before testing, unless only gross discontinuities are important. Broken or patchy layers of heavy scale also tend to interfere, by their tendency to hold powder around the edges of the breaks or patches, and should be removed if they are extensive enough to interfere seriously with the detection of genuine discontinuities.

6. MAGNETIZATION. All the usual methods of magnetization are applicable to the dry powder method of testing. Although the residual method is sometimes used with dry powder, the continuous method is generally employed. This is especially true when A.C. or half wave is being used because these types of current give mobility to the powder. The dry method is often used when large objects are being inspected, and portable equipment is the power source.

7. CIRCULAR MAGNETIZATION. Stationary magnetizing units, having a grill top and adjustable clamping contact heads, are used for magnetizing small parts. These units are similar to those used for the wet method, but do not have the tank and circulating pumps. A pan to catch the dry powder for re-use is under the grill table top. This provision for catching the dry powder for re-use does not imply that the same batch of powder should be used indefinitely. In time, with repeated dusting over magnetized surfaces, the powder loses most or all of the fine particle sizes, and therefore loses sensitivity to fine defects. Color coatings of some of the powders also tend to be lost with prolonged use. The powder should therefore be watched and discarded and replaced with fresh material before this loss becomes serious.

For the testing of large objects such as large castings and weldments, when portable units are used, contact for circular magnetization is usually made with hand-held prods, and powder applied while the current is on. The current switch is built into the handle of the prod. For applications in which the holding of the prod contacts by hand is difficult or tiring for the operator, magnetic "leeches" are available, to hold the contacts to the work magnetically. Figure 119 illustrates this device. The electrodes carrying the magnetizing current are held firmly to the work by strong permanent magnets. Both electrodes may be attached, or one leech and one hand-held prod can be used. The latter arrangement is particularly advantageous

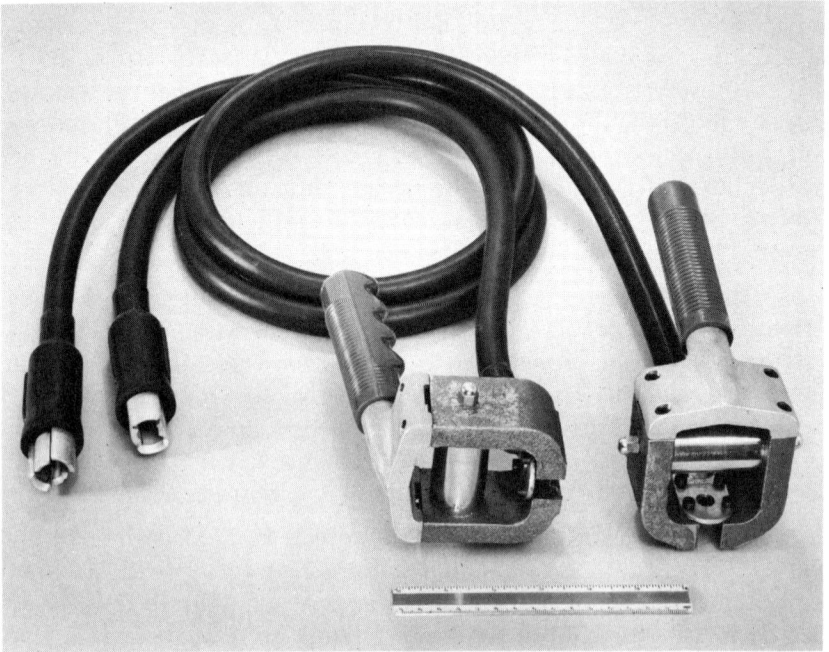

Fig. 119—Magnetic Leech Contacts.

in weld inspection. With hand held prods two inspectors are required, one to hold down the two prod contacts, and one to manipulate the powder gun. With leech contacts one inspector can handle one prod contact and the powder blower, thus making it a one man operation.

Another system uses a double prod contact. The two prods are mounted on a common handle, spaced at the proper distance. The operator can handle the prods with one hand and apply the powder with the other. The current control switch is in the handle on which the prods are mounted. This system is flexible and easy to operate, and requires less equipment than the leeches. However, holding the double prod properly in place requires more effort on the part of the operator and is more tiring than the leech system.

Current values for the dry powder method of testing are the same as indicated by the rules of thumb for other magnetic particle testing methods. These are, for circular magnetization, 1000 amperes per inch of diameter, and for prod contacts, 60 amperes *minimum* per inch of prod spacing. Somewhat larger current values are usually advisable, rather than the minimum. Up to some 200 amperes per inch of prod separation are frequently used.

8. LONGITUDINAL MAGNETIZATION. When the bench types of stationary units are used for the dry powder method, a fixed coil for longitudinal magnetizing is permanently mounted on the unit. When portable units are used, heavy flexible cable leads are usually wound into coils of three or four turns to suit the size and shape of

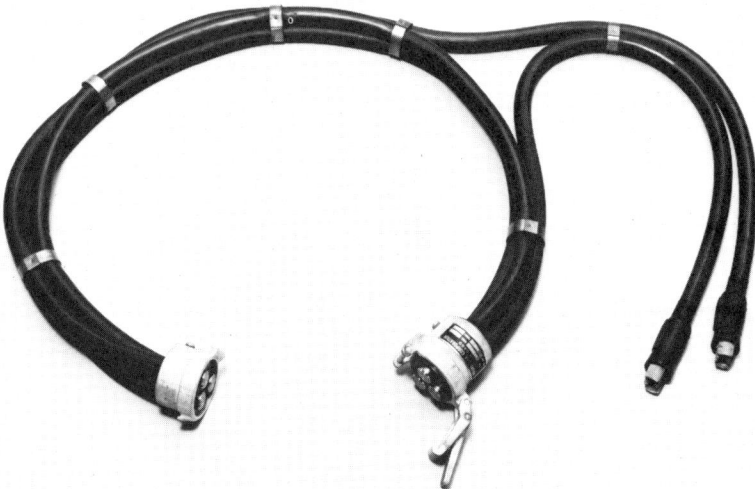

Fig. 120—Pre-formed Split Coils of Flexible Cable.

the part. Fixed coils are seldom used with portable units although pre-formed split coils are sometimes convenient, especially for threading through openings in parts to form central conductors. Figure 120 illustrates a pre-formed split coil for this purpose. When alternating current or half wave is used for magnetizing, the current is usually turned on in the coil, the powder applied and the coil moved about to cause the particles to "dance" under the effect of the alternating or pulsating field, and to move over the surface to form indications at defects.

9. APPLICATION OF THE POWDER. A few rules for the application of the dry powders will, if followed, make the process of testing easier and more effective. It should be remembered that the dry particles are heavier and individually have a much greater mass than the very fine particles of the wet method. If, therefore, they are applied to the surface of a part with any appreciable velocity, the fields at the discontinuities may not be able to stop them and hold them. This is especially true when vertical or overhead surfaces are being examined. The powder should reach the surface of parts as a thin cloud with practically zero velocity, drifting to the surface so that leakage fields have only to hold it in place. This technique has been referred to as the same as that "for gently salting a very good steak". For vertical and overhead surfaces, the fields must overcome the pull of gravity which tends to cause the particles to fall away. But, since the dry particles have a wide range of sizes, the finer particles will be held under these conditions, unless the leakage fields are extremely weak.

On horizontal surfaces this problem is minimized. The usual mistake, however, is to apply too much powder. Since, once on the horizontal surface of a part, the powder has no mobility (unless A.C. or half wave is being used) too heavy an application tends to obscure indications. If the part can be lifted and tapped, the excess powder will fall away and indications will be more readily visible; or, the excess powder can be gently blown away with an air stream not strong enough to blow off magnetically held particles forming an indication.

Various devices have been used to make proper powder application easy. Two of the most widely used are shown in Fig. 121. The "squeeze bottle" is light and easy to use. With some practice, by a combination of shaking as with a salt shaker, and a squeeze on the bottle, powder can be ejected with minimum velocity. Practicing with the bottle on a sheet of white paper will rather quickly enable the operator to produce an even, gentle overall coverage. With the

Fig. 121—a) Squeeze-Bottle Applicator for Dry Powder.
b) Air Operated Powder Gun.

powder gun or blower a better job of application can be done, especially on vertical and overhead surfaces. The powder gun throws a cloud of powder at low velocity, much like a very thin paint spray. Held about a foot distant from the surface being inspected, a very light dusting of powder permits easy observation of the formation of indications. On horizontal surfaces the excess of powder is blown away with a gentle air stream from the blower. Two push-button valves on the blower gun control the flow of powder or clean air. Less powder is used with the gun, which is a saving in cost, and better inspection is accomplished. Figure 122 shows the inspection

Fig. 122—Inspection of a Weld, Using One Leech Contact and One Prod, with a Powder Gun.

of a structural weld in the shop. The operator is using one magnetic leech contact and one hand held prod, which he can move from point to point, operating the current on-and-off switch by means of a push-button in the handle of the prod. At the same time, with the other hand, he operates the gun to apply powder.

As was said earlier, clean, smooth surfaces are usually best for successful dry powder testing, provided the surface is horizontal when tested. If the surface is rough, powder tends to gather and be held mechanically by gravity in depressions of the rough surface. A stronger stream of air than should normally be used may be required to blow off this loose powder. Care must be taken in the inspection of such rough areas (as, for example, a rough weld bead) that weakly-held indications are not also blown away. By watching the area very carefully while applying powder and while blowing off the excess, weak indications can often be seen as the powder shifts.

For very critical inspections, the weld bead is sometimes machined away. Indications of discontinuities which are wholly below the surface are more readily formed and seen by the inspector on the smooth machined surface of the weld. If the surface being tested is vertical or even at an angle to the horizontal, an extremely smooth surface becomes a disadvantage, since the dry powder tends to slide off easily, and weak leakage fields may not be able to hold it in place. Under these circumstances the rough surfaces give better results.

10. INSPECTION. A good light and good eyesight are the principal requirements for observing the presence of indications on the surface of parts. Choice of the best color powder for contrast against the surface is an aid to visibility. On the large discontinuities, powder build-up is often very heavy, making indications stand out clearly from the surface. For finer cracks the build-up is less, since only the smaller particles are held by the leakage field in this case. For exceedingly fine cracks it may sometimes be better to go to some form of the wet method, which is more sensitive to very fine discontinuities.

In the case of discontinuities lying wholly and deep below the surface, experience and skill, or carefully established, controlled practices for repetitive tests are required to secure the best results of which the method is capable. The depth below the surface and the size and shape of the discontinuity (see discussion in Chapter 20) determines the strength and spread of the leakage field. A really experienced inspector will observe the surface as the powder is allowed to drift onto it, and can see faint but significant tendencies of the powder to gather. Often indications are seen under these conditions, but are no longer visible when more powder has been applied, the excess blown off, and the surface *then* examined for indications. Standardized techniques for careful and proper application of the powder can provide excellent sensitivity where similar assemblies are repetitively tested.

In the case of steels having appreciable retentivity, indications are held at the defect by the remanent field, making the inspection somewhat easier. In low carbon steels, such as low to medium strength plate and structural steels, the retentivity is very low. On these it is important to make the inspection while the magnetizing current is on and the powder is being applied, since indications may not remain in place after the current is turned off. This is particularly true on vertical and overhead surfaces, where gravity plays a part in causing particles to fall away if loosely held. Harder new high strength steels are being used more and more for pipe, pressure vessels, and structural shapes. Smaller discontinuities become more important in this type of steel, since it is used at higher stress levels, often approaching the yield point. Although retentivity is high, great care must be taken in the inspection of these steels, especially for fine cracks open to the surface.

CHAPTER 14

THE WET METHOD—MATERIALS AND TECHNIQUES

1. HISTORY. The original de Forest development of the Magnetic Particle Testing Method (1929) employed the dry powder technique. Hoke's first observations of the magnetic particle principle back in 1918, however, were made with the magnetic particles (grindings) suspended in a liquid (grinding coolant). The advantages and disadvantages of the two methods of bringing the particles to the leakage field at defects were, at that stage, neither realized or even considered.

It was not until the middle 1930's that liquid suspensions were first used. At the Wright Aeronautical Company, at Paterson, N.J., the black magnetic oxide was first suspended in a light petroleum oil similar to kerosene. And about the same time, the General Electric Co. at Schenectady, N.Y., started using finely ground mill scale suspended in a similar light oil. The idea was also quickly taken up at Wright Field, Dayton, Ohio, by O. E. Stutsman, who was in charge of inspection of aircraft engines and parts in manufacturer's plants and in modification centers.

Initially, trouble was experienced in properly dispersing the fine black magnetic oxide in the oil bath. This problem was solved in the Magnaflux Corporation laboratory by pre-dispersing the oxide in a light oil by grinding it as paint pigment or ink is ground, with a small amount of oil vehicle. A suitable dispersing agent was incorporated in the mixture to prevent excessive floculation in the bath. The pre-dispersed oxide prepared in this fashion resulted in a concentrate of paste consistency, and was referred to as "Paste" for the wet method.

The paste was made into a bath of suitable concentration by diluting it with petroleum distillate. This was rather a tedious and messy procedure, since it involved first diluting the paste with bath liquid to a thin slurry before adding it to the bath. Nevertheless, the paste form of wet method particles, with oil for a bath, was used exclusively until 1956.

However, a few incidents involving fires in equipment focused attention on this hazard. These were due to electric sparks at contact heads, which ignited the bath vapors. This led to the development of particles that would be compatible with water as a liquid for the particle bath. It was some time before a really satisfactory formulation was achieved, since there were problems in attempting to use water for a bath liquid. These had to do with wetting of oily surfaces, rusting, electrolysis in equipment, etc. Today, water is used extensively, both in tank-type units and in testing of large objects by the expendable bath technique.

With the use of water as a suspensoid, the paste form of particles became even more difficult to disperse, and this led to a reformulation which produced the magnetic particles in the form of a dry powder concentrate. The powder incorporated the needed materials for dispersion, wetting, rust inhibiting, etc. Today, the dry concentrate has almost completely replaced the paste form for both oil and water suspensions. Use of water instead of oil for the bath is also increasing, especially for very large installations where many gallons of bath are required, as in the large billet testing systems.

2. GOOD POINTS OF THE WET METHOD. As is true of every process, the wet method has both good points as well as less favorable characteristics. The more important of the good points of the wet method, which constitute the reason for its extensive use, are the following:

(a) It is the most sensitive method for *very fine* surface cracks.

(b) It is the most sensitive method for *very shallow and fine* surface cracks.

(c) It quickly and thoroughly covers all surfaces of irregular shaped parts, large or small, with magnetic particles.

(d) It is the fastest and most thorough method for testing large numbers of small parts.

(e) The magnetic particles have excellent mobility in liquid suspension.

(f) It is easy to measure and control the concentration of particles in the bath, which makes for uniformity and accurate reproducibility of results.

(g) It is easy to recover and re-use the bath.

(h) It is well adapted to the short, timed "shot" technique of magnetization for the continuous method.

(i) It is readily adaptable to automatic unit operation.

3. LESS FAVORABLE CHARACTERISTICS. Some of the less attractive characteristics of the wet method are the following:

(a) It is not usually capable of finding defects lying wholly below the surface if more than a few thousandths deep.

(b) It is "messy" to work with, especially when used for the expendable technique, and in field testing.

(c) When oil is used for a bath and direct contact for circular magnetization, it can present a potential fire hazard.

(d) A fairly critically designed recirculating system is required to keep the particles in suspension.

(e) Sometimes it presents a post-inspection cleaning problem to remove magnetic particles clinging to the surface.

4. BATH CONSTITUENTS. The magnetic particles used in the wet method were discussed and described in detail in Chapter 11. Briefly, their outstanding characteristic is their extremely small size—down to one-eighth of a micron. These very fine particles do not act as individuals, however, but agglomerate into groups which have diameters of some 5 microns or more. Two colors of particles are available, red and black. In addition (and most widely used) fluorescent particles give maximum visibility (Chapter 15). The more modern dry concentrates are formulated to include other bath constituents, with the fine magnetic particles already bonded together in optimum sizes.

The bath liquid may be either a light petroleum distillate of specific properties, or water. Both require conditioners to maintain proper dispersion of the particles and to permit the particles freedom of action in forming indications on the surfaces of parts. These conditioners are usually incorporated in the powders.

5. OIL AS A SUSPENDING LIQUID. Oil was a natural first choice as a bath liquid, since most machine parts that are inspected tend to have an oily film on their surface. Gross amounts of oil or grease should be removed of course, but any film remaining is readily wetted and dissolved by the light oil of the bath.

The oil should have very definite properties to be suitable for bath purposes. It should be a well refined light petroleum distillate of low viscosity, odorless, with a low sulphur content, with a high flash point and a close-cut, fairly high boiling range. The usual specifications for a suitable oil are given in Table III.

TABLE III

PROPERTIES OF OILS RECOMMENDED FOR MAGNETIC PARTICLE
WET METHOD BATH.

Viscosity, Kinematic at 100°F (38° C.)	3 Centistokes (max.)
Flash Point—Closed Cup	135° F. (57° C.) Min.
Initial Boiling Point	390° F. (199° C.) Min.
End Point	500° F. (260° C.) Max.
Color (Saybolt)	Plus 25.
Sulphur—Low Available	Pass the Copper Test ASTM-D129-52.

Of these properties, viscosity is probably the most important from a functional standpoint. It should not exceed three Centistokes as tested at 100° F., and must not exceed five Centistokes at the point and temperature of use. High viscosity slows the movement of particles under the influence of leakage fields. Laboratory tests have shown that at viscosities above five Centistokes the movement of magnetic particles in the bath is sufficiently retarded to have a definite effect in reducing the build-up, and therefore the visibility, of an indication of a small discontinuity. Heavy oil from the surface of parts tends to build up in the bath and increase its viscosity. This is the main reason for pre-cleaning parts to remove oil and grease.

Addition of light non-volatile oils to the bath as an "automatic" rust preventive has been suggested. This must be done with caution, if at all, because tests have shown that even small additions can raise the bath viscosity above a safe unit. The addition of ten percent of an S.A.E. #10 lubricating oil will raise the viscosity of the bath from three Centistokes to above five. If the bath already has a viscosity of four Centistokes, as little as two percent of #10 oil will raise it above five.

Much lighter distillates would have a much lower viscosity than those usually used, but they would have other properties undesirable in a magnetic particle bath. Lighter distillates would have an initial boiling point lower than that specified, and therefore a lower flash point, making them a greater fire hazard. Also, evaporation losses from the tank would be greater with a lighter oil. Breathing of unpleasant fumes from a light distillate leads to operator discomfort. Odors from distillate are objectionable from the operator's standpoint. Odor of distillate is associated with color and sulphur content, which is the reason for specifying that these properties be within limits. The approved, highly refined oils are nearly odorless, and meet the color and sulphur specifications.

6. WATER AS A SUSPENSOID. The great attractions of the use of water instead of oil for magnetic particle wet method baths are lower costs and the complete elimination of bath flammability. It should be pointed out here that, although there have been some instances of oil fires in magnetic particle testing units, they have been relatively few over more than thirty years of using oil for this purpose. In some of the earlier automatic or semi-automatic units, the oil bath was sprayed onto the part, creating a mist of oil which was more easily ignited than the liquid oil. Better means of applying the bath—such as flowing or "dumping"—have almost completely eliminated this source of trouble. But, as units became larger and the volume of oil needed for the bath became much greater, practical considerations indicated a change from an oil to a water bath as an overall economy and safety measure.

The use of water as a suspensoid is quite extensive today. But the use of water is not a panacea for all the problems which the oil bath presents, because, although they are different from those of the oil bath, water has problems of its own. The cost advantage of water base baths lies entirely in the cost of the oil, about 40 cents/ gallon, as against water, the cost of which is nominal. This is because of the complex formulation necessary in the way of wetting agents, dispersing agents, rust inhibitors, anti-foam agents, etc., which run up the cost of water-suspendable concentrates. Consequently, water-suspendable concentrates are comparable, or even slightly higher, in cost than those used for oil suspension.

Water baths must be used in shop areas where the temperature is above freezing. Use of anti-freeze liquids is not feasible because the quantities needed to be effective would raise the viscosity of the bath above the maximum allowable. Because the use of detergents to assure the wetting of oily surfaces causes foaming of the bath, circulation systems must be designed to avoid air entrapment or other conditions that produce foam. Anti-foaming agents are used to minimize this tendency, but are not 100% effective.

Since water is a conductor of electricity, units in which it is to be used must be designed to isolate all high voltage circuits in such a way as to avoid all possibility of an operator receiving a shock, and the equipment must be thoroughly and positively grounded. Also, electrolysis of parts of the units can occur if proper provision is not made to avoid this. When water was first being tried, some users simply changed the bath in their units from oil to water. Rapid corrosion and occasional electric shock to the operator resulted. Consequently such conversions are not to be recommended. Units *designed* to be used with water as a suspensoid are, however, safe for the operator and minimize the corrosion problem.

There is no critical requirement as to the water that is used for the bath, as there is in the case of oil. Ordinary tap water is suitable, and hardness is not a problem since the mineral content of the water does not interfere with the conditioning chemicals necessary to prepare the bath.

7. THE MAGNETIC PARTICLES. The second ingredient of the bath for applying the wet method is of course the magnetic particles. These have already been described. Three basic types of materials are available:

(a) The oil base pastes, either black, red or fluorescent. These are primarily for use with oil suspensoids, though with suitable water conditioners they can be used with water. Their use is rapidly being abandoned in favor of the newer dry concentrates.

(b) Dry concentrate powders, either black, red or fluorescent. One type is used with water as the suspensoid, and a second type for suspension in oil.

(c) Water conditioners. These are used to suspend oil pastes in water, and to make up the quantity of conditioner in the water bath when needed.

As has been said, the dry powder concentrates are now preferred for both oil and water baths. The new self-dispersing particles are readily picked up by the pump circulating system, and are easily maintained in suspension. Some of the oldest wet method units used air for agitation of the bath. This form of agitation is not satisfactory for the dispersion of the dry powder concentrates, or maintaining bath suspension, and has not been used in new equipment for a good many years.

8. STRENGTH OF THE BATH. It should be obvious that the strength of the bath is a major factor in determining the quality of the indications obtained. Too heavy a concentration of particles gives a confusing background and excessive adherence of particles at external poles, thus interfering with clean-cut indications of very fine discontinuities, so that there is danger of their being missed.

The guide, therefore, to proper bath concentration, is the formation of satisfactory indications with the particular technique being used. For the wet continuous 1/2-second-contact process used on most small stationary units, a different concentration is required from that which may be used in the tank for dipping parts for the residual method, where the exposure time of the magnetized part to the bath is much longer.

Experience has produced some bath concentration rules for the 1/2-second-contact system which are satisfactory for most applications. Table IV lists these concentrations for the usual wet method particles. (For fluorescent types, see Chapter 15.) But the best method of assuring optimum bath concentration for any given combination of equipment, bath application, type of part and defects sought, is to make use of parts with known defects. Bath strength can be adjusted by cut-and-try methods until satisfactory indications are obtained. This concentration of bath can then be adopted as standard for those conditions.

It is important that the proportion of magnetic particles in the bath, once the satisfactory concentration has been arrived at, be maintained uniform. If the concentration varies, the strength of the indication will also vary, and interpretation of the meaning of an indication may be erroneous. Fine indications may be missed entirely with a weak bath.

It must be remembered that the important consideration in bath strength is the proportion of active ingredient — that is, the actual amount of magnetic oxide per unit of volume. Since the proportion of oxide varies in the different pastes and powders, the amount of these materials required to produce the necessary concentration of magnetic particles in the bath will vary also. To determine the actual amount of oxides in the bath, a measured sample of the bath is filtered, the oxide purified from the various compounding materials, and the separated oxide then dried and weighed. This is, however, a laboratory test and not readily used by the operator at the unit. Since the bath strength should be checked no less often than at the beginning of every shift, when the unit is in continual use, some quick and easy test method is required.

The settling test has served this purpose very well, and is easily and quickly performed at the unit. It is not as accurate as is desirable, but is reasonably quantitative and is reproducible. It can be easily standardized with the material in use, and is quite satisfactory as a daily guide for the operator. It is less satisfactory on some of the newer, more complex formulations than on the older types of bath materials. In these cases another form of test (which will be described) can be used if preferred. (See Chapter 26, Section 9.)

In the settling test, 100 ml. of well agitated bath are run from the bath applicator or taken from a well agitated tank, into a pear-shaped centrifuge tube, as illustrated in Fig. 123. The volume of solid material that settles out after a pre-determined interval (usually 30 minutes) is read on the graduated cylindrical part of the

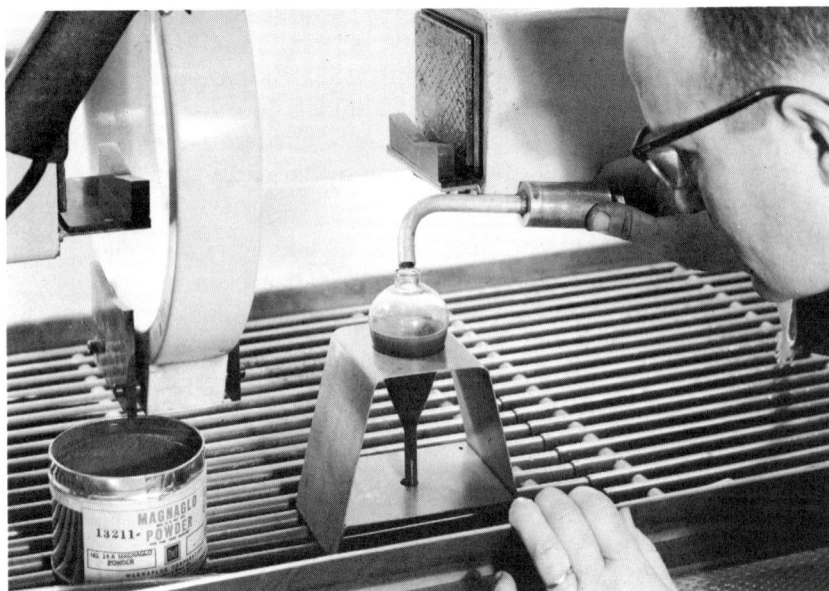

Fig. 123—Making the Settling Test for Wet Bath Concentration.

tube. Dirt in the bath will also settle out and usually shows as a separate layer on top of the oxide. The layer of dirt and lint is usually easily distinguishable, as it is of a different color from the oxide. See Chapter 26, Section 8 for details of this test.

Table IV shows the suggested amounts of the several concentrates to be added per gallon of bath, and the volume of solid material which settles out when the bath is made up with these amounts of concentrates. The black and red pastes, 7-A and 9-B are included in the Table, since they are still used to a slight extent, although they are rapidly becoming obsolete.

TABLE IV

BATH STRENGTH CHART.

Magnaflux Corporation Designation of Material	7-C Black Powder	9-C Red Powder	27-A Black Powder	29-A Red Powder	7-A Black Paste	9-B Red Paste
Oz. of Concentrate per gallon of bath	$1\frac{1}{4}$	$1\frac{1}{4}$	$2\frac{1}{3}$	$2\frac{1}{3}$	$1\frac{1}{2}$	$1\frac{1}{2}$
Settling Test. * Volume Range, in ml.	1.3 to 1.5	1.7 to 1.9	1.1 to 1.2	1.4 to 1.5	1.4 to 1.6	1.3 to 1.6
Liquid to be used for the bath	Oil Water **	Oil Water **	Water	Water	Oil	Oil

*Settling time is 30 minutes, using a 100 ml centrifuge tube.
**Requires use of water conditioner WA-2A to disperse in water.

9. MAKING UP THE BATH. When a new bath is to be made up, for a new unit or after dumping a dirty bath from a unit in use, it is important first to make sure that the agitation system is clean, and not clogged by dried particles or accumulated dirt such as lint or shop dust. Next fill the tank with oil or water as required, and operate the agitation system to make sure it is functioning properly.

(a) *For Dry Powder Concentrate.* Measure out the required amount of powdered concentrate in the graduated cup and pour it directly into the bath liquid in the tank. The agitation system should be running and the concentrate poured in at the pump intake, so that it will be quickly drawn into the pump and dispersed. See Figure 124. After 10 minutes of

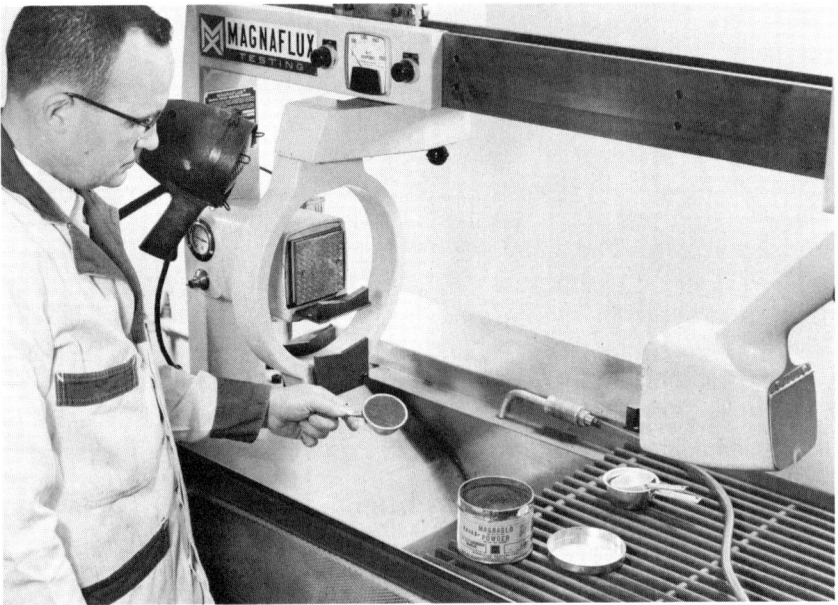

Fig. 124—Mixing the Bath Using Dry Concentrate.

operation the bath strength should be checked with a settling test. The amount of settled material should check approximately with the figures given in Table IV., if the recommended amounts of concentrate given in the table have been used. If the volume is less than called for add concentrate— if more, add liquid. A better plan for using the settling test as a bath check is to observe the actual amount of settling for a carefully made up bath of the desired strength, record it, and thenceforth using this figure for the standard of bath strength for subsequent tests.

The new pre-wet concentrates will disperse very quickly even through the large volume of bath in large units.

(b) *For Oil Paste and Oil Bath.* The procedure is similar to that followed in the case of the dry powder concentrates, except that the paste must be weighed out instead of measured. It is transferred to a mixing cup or bowl. Add bath liquid, a little at a time, and mix, until a smooth thin slurry has been produced. This slurry is then poured into the liquid in the tank at the point where the agitation system will pick it up and disperse it. After agitating ten minutes the strength should be checked by a settling test as in the case of the dry powder concentrate.

10. MAINTENANCE OF THE BATH. As the bath is used for testing it will undergo changes due to use. Some of these changes are:

(a) Drag-out of magnetic particles, both by mechanical and magnetic adherence to parts, thus tending to reduce particle concentration in the bath.

(b) Loss of liquid due to the film which adheres to the surface of parts.

(c) Loss of liquid by evaporation, tending to increase particle concentration.

(d) A gradual accumulation of shop dust, dirt from parts not properly cleaned, lint from wiping rags, and oil from parts that carry a residual film of oil.

(e) Miscellaneous scraps of foreign material which shop workers sometimes toss into the tanks!

It is important, therefore, that the bath be frequently checked and corrections applied as needed. Frequent settling or otherwise suitable tests are desirable, if the unit is in constant use. Concentrate should be added when the oxide concentration is low. Evaporation or liquid drag-out should be watched, and volume maintained when the level drops appreciably. Since loss of liquid may be either by drag-out or by evaporation, and the corrective measures are different for these two types of loss, it may be difficult to keep the bath in balance over long periods. This is especially true of water baths, since water has a higher evaporation rate than the oil. To make up for evaporation loss, addition of oil or water is required. Conditioning materials are not volatile. They will therefore build up in the bath as the liquid evaporates. Bath volume lost by drag-out, however, includes conditioner in solution, and proper correction should then include the right amount of conditioner.

It is difficult to know what the cause of volume loss actually is in any given case. However, for a unit in constant use it can be assumed that more than 50% of the loss is due to drag-out. For a unit used only occasionally, loss by evaporation is likely to be the major one. Actually the problem is not a serious one, because with constant use the accumulation of dirt, scraps, lint, etc. requires the dumping of the tank and a new bath before loss of liquid becomes serious. But *magnetic* particle content is of most critical importance and should be carefully watched at all times.

Dirt accumulation in the bath can usually be easily determined, as it shows up in the settling test for magnetic particles. When the settling test is run, the oxide, being heavy, settles out first. Dirt, lint, etc. is lighter and settles more slowly. It is seen as a second layer on top of the oxide. For oxide determination, this layer of dirt must be carefully excluded from the volume read. When the layer of dirt approaches the volume of oxide, formation of proper indications will be impeded, and the bath should be dumped and a new one made up. This may occur as often as once a week when a unit is in constant use.

11. STEPS IN THE APPLICATION OF THE WET METHOD. The steps in applying magnetic particle testing by the wet method consists of 1) surface preparation; 2) magnetization; 3) application of the magnetic particles; and 4) inspection for indications.

12. PREPARATION OF THE SURFACE. In general the same requirements apply for the wet method as for the dry technique. Dirt, rust, loose scale and oil or grease should be removed. The oil bath will dissolve oil or grease, but this builds up the viscosity and fluorescence of the bath and shortens its useful life. With a water bath, oil on the surface of the part makes wetting more difficult, although the conditioners in the bath are usually sufficient to take care of a moderately oily surface. But excessive oil on part surfaces contaminates even the water bath in time. Paint and plated coatings, if over 0.005 inch thick, should be stripped. Tests have shown that non-magnetic coatings of any kind, in excess of 0.005 inch in thickness, seriously interfere with the formation of magnetic particle indications of small discontinuities.

13. THE CONTINUOUS WET METHOD. MAGNETIZATION AND BATH APPLICATION. All the usual methods of magnetization are used in the wet method—longitudinal, circular, multi-directional, induced current and over-all magnetization are employed, with either A.C. or D.C., as are the continuous and residual methods of bath application.

The term "continuous method" implies that the magnetizing force is acting while the magnetic particles are applied. While the current in on, maximum flux density will be created in the part for the magnetizing force being employed. In some cases—usually when

A.C. or half wave D.C. is the magnetizing current being used—the current is actually left on, sometimes for minutes at a time, while the magnetic particles are applied. This is more often needed in dry method applications than in the wet.

To leave the current on for long durations of time is not practical in most instances, nor is it necessary. The heavy current required for proper magnetization can cause overheating of parts and contact burning if allowed to flow for any appreciable length of time. Over-heating of the power equipment could also result. If such truly continuous currents are required, power supplies must be built much heavier and they become much more costly. In practice the magnetizing current is normally on for only $\frac{1}{2}$ of a second at a time. All that is required is that a sufficient number of magnetic particles are "in the zone" and free to move while the magnetizing current flows. The bath ingredients are so selected and formulated that the particles can and do move through the film of liquid on the surface of the part and form strong, readable indications in the time of one-half of a second. This is, of course, one reason why the viscosity of the bath and bath concentration are so important, since anything that tends to reduce the number of available particles or to slow their movement tends to reduce the build-up of indications.

The procedure for this method of magnetization and bath application is simple and easy to perform, once the operator understands the importance of the sequence and timing of the two operations. Bath from the nozzle of the unit or other applicator should be flowed liberally over all surfaces on the part. The magnetizing current button should be pressed at the instant the source of bath flow is withdrawn. The timer in the circuit of D.C. units gives a $\frac{1}{2}$ second "shot" of current of pre-selected strength. With other units, (A.C.) the operator controls the time the current is on. There should be no interval of delay between the withdrawal of the flow of bath and the current "shot". It is most important that the maximum film of bath be on the surface of the part when the current flows. On the other hand, it is of equal importance that the flow of bath from the nozzle or applicator be stopped *before* the current button is pushed, else indications may be washed away. Many operators will give two or even three shots of current in quick succession to strengthen the indication somewhat—but *not* with any re-application of bath.

For small parts the usual equipment is a bench-type unit with

Fig. 125—Simple Wet Method Unit Setup for Circular Magnetizing.

contact heads between which parts are clamped for circular magnetization. Under the table of the unit is the bath tank, the control
equipment, power supply and agitator pump. The bath in the tank
is kept constantly agitated by pump circulation. Bath is available
from the pump system through a hose and nozzle for flooding the
part. Current values are determined basically by the 1000 ampere
per inch of diameter rule, modified by experience for very small or
very large parts.

For longitudinal magnetization, the fixed coil, which can be seen
in Fig. 126 is used. The sequence of the magnetizing and bath application operations is the same as for circular magnetization. The
amount of current to pass through the coil to produce the proper
number of ampere-turns, is determined by the formula (See Chapter 9) $NI = \dfrac{45,000}{L/D,}$ where N = the number of turns in the coil, I
the current, L the length of the part and D its diameter.

The magnetic particles in the bath could be either black or red,
depending on which color gives the best contrast with the background of the surface of the particular parts being inspected. Or,
more often, fluorescent particles are used for maximum visibity.
(See Chapter 15.)

Fig. 126—Wet Method Unit Setup for Longitudinal Magnetizing.

14. THE RESIDUAL METHOD. In the residual method the parts are first magnetized, and the bath is *subsequently* applied, either by flowing-on or by immersion of the part in a tank containing the bath. Obviously, satisfactory inspection by the residual method can only be attained on parts having a relatively high retentivity. High carbon or alloy steels are usually highly retentive of magnetic fields, even in the unhardened condition. Hardened steels of this type have even higher retentivity. On the other hand the material permeability of such steels may be quite low.

With usual amounts of magnetizing current, residual fields in such steels can be established which very satisfactorily give indications of surface cracks—even very shallow ones. However the location of defects wholly below the surface in such steels is not usually possible by *either* the continuous or the residual methods, if the discontinuities are more than a few thousandths below the surface. Although the residual method is not as widely used today as the continuous method, it does have some advantages which make it attractive in some circumstances.

269

The residual method is applied in two different ways:

(a) Magnetizing and applying the bath as a "curtain" (not a spray) which may be left on to flow gently over the magnetized part long enough to develop good indications.

(b) Magnetizing and immersing the part in an agitated bath of magnetic particles for the length of time needed to produce good indications.

The "curtain flow" method is used largely on automatic or special-purpose units. Parts are usually mounted in jigs or fixtures, given a shot of current and the curtain flow turned on. The system avoids the need for critical timing of the magnetizing-bath application sequence.

In the immersion method as usually practiced, mostly on relatively small parts, the parts are magnetized rapidly, one at a time, then placed in a tray and immersed in a tank containing an agitated bath of magnetic particles. The parts must be so disposed in the tray that they do not touch one another, else non-relevant indications may be produced at such points of contact. Haphazard loading into a basket for immersion should not be permitted. Strength of bath and immersion time both have an effect on the size of the indications produced. If the leakage field at a shallow crack is weak, for example, prolonged immersion permits more particles to come into the influence of the field and makes the indication more prominent and visible.

The residual method is capable of close control, and of giving very uniform results—possibly to a greater degree than the continuous method as usually practiced. The fact that it is applicable only to parts having relatively high retentivity is probably the reason, as much as any other, that the method is not used more extensively.

15. CLEANING AFTER TESTING. After testing, a film of bath remains on the surface of parts, which may be objectionable. The bath liquid dries rather rapidly, whether it is oil or water. In the case of oil, a film of oxide particles remains after drying, which may be difficult to remove. If removal is important the parts should be cleaned at once after inspecting. They should first be demagnetized if they contain remanent field, then washed in solvent spray or by some other means. Where water is the bath liquid, a water spray after demagnetization does a good cleaning job, especially if the bath has not been allowed to dry on the parts.

If parts have been allowed to dry before cleaning, removal of the bath residue may be more difficult. Mechanical scrubbing or detergent washing may be necessary. Sometimes vapor de-greasing may remove the residue, and in some cases the use of ultrasonic cleaning has been successful.

Cleaning is in many cases not necessary, especially on unfinished parts which will be subsequently machined or otherwise processed. If fluorescent particles have been used, the residue left on the surface is very slight since particle concentration is low in fluorescent particle baths. On the finished parts, however, when inspection is the final process, cleaning is very often, if not always, required. On highly polished surfaces, residual powder from the bath can be unsightly, and may in fact, contribute to early rusting.

16. RUST PREVENTION. After testing by the wet method using oil as the bath liquid, the surfaces of parts are left vulnerable to rusting. The bath oil is by specification free of any residual non-volatile material, and when it dries it leaves no protective film. Some sort of rust preventive should be applied soon after testing. A light coating of mineral seal oil is often used as a temporary preventive—or some other form of more permanent anti-rust compound can be applied. Addition of a small amount of mineral seal oil to the bath itself has been used by some operators. For the reasons discussed in Section 5 of this chapter, this procedure is *not* recommended.

When water is the vehicle of the bath, the dried film on the surface of parts consists of the various conditioners that have been used in the bath formulation, in addition to residual magnetic particles. One of the conditioners is a rust inhibitor, so that some rust protection is afforded by this inhibitor after testing. However, this is by no means permanent, and a rust preventive should be applied unless the parts are soon to be processed further.

17. SKIN PROTECTION. Continuous exposure of the hands and arms of the operator to the bath of the wet method may cause the skin to dry due to the removal of the natural oils of the skin by the bath liquid. This can result in drying and cracking of the skin, which in turn may open the way to secondary infection. In most cases the liberal use of some good protective hand cream—preferably one that contains lanolin or silicones—will minimize such effects. Operators should avoid oil-soaked sleeves which are constantly in contact with the skin, as this intensifies the effect.

Cleanliness is, however, a *must* if secondary infection is to be avoided. Thorough washing of the hands and arms followed by a thoroughly rubbed-in application of hand cream will usually provide protection for some hours. After work a thorough washing with soap and warm water will remove the hand compound and dirt, and a fresh application of cream will prevent further drying and chapping.

The practice of adding liquid skin protector to the bath is not a good solution to the problem. In the first place such additions are ineffective in avoiding skin drying, and in the second place, additions of any kind to the bath may raise its viscosity and affect the proper inspection of parts. Synthetic rubber gloves are sometimes used, but these are clumsy and slow up the inspection. Synthetic rubber aprons are, however, recommended to keep the bath liquid from saturating clothing. Ordinary rubber gloves or aprons are not suitable for use when the bath is oil, because oil softens and destroys rubber. They are satisfactory, however, when water is the bath liquid.

In some cases individuals have a true allergy toward petroleum oils or other ingredients of the bath, especially very light-complexioned persons. If the usual preventive measures just described are ineffective to prevent a rash, there is usually no remedy but to transfer the person involved to other work.

18. PREPARED BATH. In this country wet method magnetic particle baths are in almost all cases made up as needed by adding the concentrate to the bath liquid. In some operations abroad, it is the usual practice for the supplier to prepare the bath ready to use, and ship it in cans or drums to the user. For large users this increases shipping costs, although the custom does have some attraction for small users.

19. PRESSURIZED CANS. Prepared bath is widely sold in this country in pressurized cans for spraying. Such cans, usually containing oil-based baths, are very convenient to use for spot-checking or small area tests in the field. They are often furnished in kits, including a permanent magnet·or electro-magnet yoke, which makes an extremely portable package for small field testing jobs, or for maintenance testing around the shop.

20. LACQUER METHOD. An interesting variant of the usual wet method is a bath in which the magnetic particles are suspended in a thin quick-drying lacquer, either clear or with a small amount of white pigment. Such a lacquer bath is sprayed or painted onto the surface of a magnetized part. At leakage fields, magnetic particles are drawn to form indications while the lacquer is still liquid. When the lacquer dries, the indication is fixed in place. If a strippable lacquer is used, the film containing the indication can be peeled off, and transferred to a report or other record. Or it can, if transparent, be projected or reproduced by photographic printing. If the method is used, care must be taken that the viscosity of the lacquer bath does not exceed the maximum allowed for oil baths. If it does, loss of sensitivity for the inspection results. The method is not very practical for general use, but it is one technique used to advantage if a removable transfer of a specific indication is desired.

CHAPTER 15.

FLUORESCENT MAGNETIC PARTICLES—THEIR NATURE AND USE

1. HISTORY. Fluorescent magnetic particles were first offered to industry in the summer of 1942. The method of coating magnetic particles with fluorescent dye was devised by R. C. and J. L. Switzer, who were issued a patent covering the idea late in 1941. Magnaflux Corporation secured an exclusive license under this patent, and after an extensive program in cooperation with the Switzer Brothers to refine the formulation of the new particles and prove their practical application as a real advance in the art of magnetic particle testing, marketed the fluorescent particles under the trade name "Magnaglo"*.

When exposed to near-ultraviolet light—or "black light"—the magnetic particles so treated glow with their own light, having a highly visible yellow-green color. Indications produced are easily seen, and the fluorescent particles give much stronger indications of very small discontinuities than do the ordinary non-fluorescent magnetic particles. This fact caused the new material to be put to use at once, especially in critical tests where very small defects were sought. Tests were faster and more reliable than with the older non-fluorescent particles.

Since the close of World War II, and especially since the early 1950's, the use of fluorescent particles has grown at a steadily increasing rate, and they are most widely used today.

2. PRINCIPLE OF FLUORESCENCE. Many substances have the property of fluorescence. Such materials give off light when energized or exposed to light of a wave length shorter than that of the violet of the visible spectrum. This "black" or invisible light energy is absorbed by the fluorescent material and re-emitted as light of a longer wave length in the visible range. Different substances fluoresce in a variety of different colors, from blue through green, yellow and red. Many minerals fluoresce brilliantly under black or

---

*Magnaglo. Trademark registered in U.S. Patent Office. Property of Magnaflux Corporation. Also registered in Canada, Great Britain, Italy, and Mexico.

ultraviolet light, and prospecting for such ores—as, for example, ores of tungsten and vanadium—is often done at night with a portable black or ultraviolet light.

Specifically, the dyes used to treat the magnetic particles to make them fluoresce, glow with a greenish-yellow color when exposed to black light. The black light used for fluorescent magnetic particle testing is radiant energy having a wave length of 3650 angstrom units. True ultraviolet, of 3000 angstroms and below, also activates many substances to make them fluoresce. However, ultraviolet of this shorter wave length is harmful, causing burns and damaging the eyes. The black light of 3650 angstroms, on the other hand, is entirely harmless, and is used in thousands of locations without any ill effect whatever on the inspector.

Except for a few necessary modifications, fluorescent magnetic particles are used in wet suspensions as in the ordinary wet method.

3. ADVANTAGES OF THE FLUORESCENT PARTICLE METHOD. Fluorescent particles have one tremendous advantage over the untreated or "visible" particles. This is their ability to give off a brilliant glow under black light. This brilliant glow serves three principal purposes:

(1) In semi- or complete darkness even very minute amounts of the fluorescent oxide are easily seen, having the effect of increasing the apparent sensitivity of the process tremendously, even though, magnetically, the fluorescent particles are not superior to the untreated oxides.

(2) Even on discontinuities large enough to give good visible indications, fluorescent indications are so much more easily seen that the chance of the inspector missing an indication is greatly reduced even when the speed of scanning parts is increased. To state it simply, the fluorescence of indications takes "the looking out of seeing".

(3) Inside drilled holes or cavities, or in sharp corners such as threads or key ways, the fluorescent indications are clearly and readily seen, while visible color indications can be easily obscured.

The fluorescent particle method then, is faster, more reliable and more sensitive to very fine defects than the visible colored particle method in most applications. Indications are harder to overlook, especially in high volume testing. In addition, the fluorescent method has all the other advantages possessed by the liquid suspension technique.

4. DISADVANTAGES OF FLUORESCENT PARTICLES. Fluorescent magnetic particles used in suspension in liquids have the same unfavorable characteristics which go with the usual wet method techniques; and there is the additional requirement for a source of black light, and an inspection area from which at least most of the white light can be excluded. Experience has shown, however, that these added special requirements are more than justified by the gains in reliability and sensitivity.

5. MATERIALS. There is no difference between the fluorescent and non-fluorescent materials so far as the liquids and bath requirements are concerned. Oils must meet the same specifications as listed in Table III, with one additional requirement. The oil itself must not fluoresce strongly, nor with any color other than the usual bluish-white of most petroleum products. Most of the commercial distillates approved for the regular wet method are also satisfactory for use with the fluorescent particles. The reason for limiting the fluorescence of the bath itself is obvious, since fluorescence of the film of oil on a part would produce a confusing all-over background.

The particles for this method are magnetically the same as the visible type, but they must carry the fluorescent dye and the binding material that holds the dye and particle together as a unit. This loading of the particles would tend to make them less effective in producing indications were it not for the fact that to be easily visible, a fluorescent particle indication requires only a small faction of the particles needed for the non-fluorescent type. Thus the over-all effect is a large *practical* and effective increase in sensitivity.

The fluorescent particles are now supplied primarily as a dry concentrate incorporating all the ingredients necessary for dispersion, and—in the case of concentrates for suspension in water—for water conditioning. Probably a larger proportion of all fluorescent particles are used in water suspension than is the case with the non-fluorescent type. This is partly because the fluorescent particles are favored for large installations, such as for the inspection of steel billets for high volume testing, and for the inspection of large objects by the expendable method, in the foundry and shop, and in the field. These operations inherently consume large quantities of fluorescent particles.

It is, of course, of great importance that the bond between the fluorescent dye and the magnetic particle be able to resist the vigorous agitation it receives in the pump circulation. If dye and magnetic particle can become separated, the dye tends to cling to the surfaces of the part separately, independent of any magnetic attraction, causing a meaningless and confusing background. At the same time the magnetic particles that *are* held magnetically at indications have lost some or all of their fluorescing ability, which results in a net loss in sensitivity for any size of discontinuity.

6. STRENGTH OF BATH. The quantity of fluorescent particles used to make up a suitable bath is very much smaller than for the non-fluorescent type. It is actually less than 1/6th as much. This is because fewer particles are required to form readable indications. Concentrations of the order of those required for the visible type particles would result in an excessive background fluorescence from the particles in the surface film of bath. This would make fine indications difficult to see.

Table V shows the recommended amount of particle concentrate to be added for the several types of powders and pastes for oil and water suspensions. See the discussion of bath strength in Chapter 14, Section 8. The same requirements dictate the proper strength of fluorescent particle baths.

TABLE V.

BATH STRENGTH CHART—FLUORESCENT PARTICLES

Magnaflux Corporation Designation of Concentrate.	10-A	14-A	15**	20-A	24***	Paste 10
Oz. of Concentrate per Gallon of Bath.	0.2	1/6	1/6	1⅓	2.0	0.2
Settling Test* Volume Range in ml.	0.30 to 0.40	0.20 to 0.30	0.10** to 0.15	0.18 to 0.22	0.75*** to 1.00	0.25
Liquid to be used for bath.	Oil	Oil Water	Water	Water	Water	Oil

*Settling time is 30 minutes, using 100 ml centrifuge tube except as noted below.
**Settling time is 5 minutes.
***Settling time is 45 minutes, using 1 liter Imhof sedimentation cone.

The volume of particles that settles out in the test for bath strength is larger for the fluorescent particles in proportion to the quantity of concentrate used, as compared to the non-fluorescent particles (Table IV). This is because the fluorescent dye and binder reduce the density of the particles so that they do not settle so rapidly, nor so compactly, in the standard settling time of 30 minutes. The amount of settled-out material is not uniformly proportionate to the amount of concentrate added, for the double reason that composition of the several concentrates varies to include necessary conditioning materials, and that the specific gravity of the several suspensoids also varies, partly as the result of these additives. These factors affect the rate of settling and the degree of compactness of the settled material in the time allowed for the settling test. The #15 powder is a special large particle powder, which settles more rapidly (5 min.) ; the #24 is a special powder which settles very rapidly and compactly. To get sufficiently accurate readings a 1000 ml sample is used, in an Imhof sedimentation cone, with a 45 minute settling time.

7. MIXING THE BATH. Except for the difference in the quantity of concentrate used, the details of bath mixing are the same as for the non-fluorescent particles. (Section 9, Chapter 14.) Old type oil pastes must be first reduced to a thin slurry by dilution with bath liquid before adding to the bath; and the dry concentrates are measured out by volume, in a calibrated measure, and added directly to the tank at the point where the pump suction line will pick them up at once.

8. MAINTENANCE OF THE BATH. Here, again, the rules are identical with those described in Section 10, of the previous chapter, for non-fluorescent particles. There are three additional sources of deterioration in a bath of fluorescent particles that must be watched for, and that require discarding of the bath when the condition becomes severe. The first is the separation of the fluorescent pigment from the magnetic particles, which occurs in some of the older types of particles. Such separation causes a falling off of fluorescent brightness of indications, and an increase in the overall fluorescence of the background. When this has occurred to a noticeable degree, the bath should be dumped and a new one prepared. This condition cannot readily be detected in the settling test, but must be observed by the operator in the way the bath performs.

A second source of deterioration of the bath of fluorescent particles which does not act in the case of the non-fluorescent type, is the accumulation of *magnetic* dust or dirt in the bath. In foundries and metal working shops there is a considerable amount of finely divided magnetic material in the dust carried by the air, and this material will accumulate in the bath along with other dust and dirt. In a bath of non-fluorescent particles this does no special harm, until the accumulation of total dirt is excessive. In the case of fluorescent particles, however, it tends to decrease the brightness of the indication. This is because the fine magnetic material is attracted to indications along with the fluorescent particles, and it takes very little of such non-fluorescent material to reduce significantly the fluorescent light emitted by the indication.

A third source of deterioration of the fluorescent particle bath is the accumulation of fluorescent oils and greases from the surfaces of tested parts. This accumulation, in time, builds up the fluorescence of the liquid of the bath to a point at which it interferes with the viewing of fluorescent particle indications.

9. STEPS IN THE TESTING PROCESS. Application of the test using fluorescent magnetic particles is identical in every way with the procedure described in detail in Sections 12 through 16 of the preceding chapter, except, of course, for the examination for indications. One precaution in the preparation of the surface of parts before testing must be given special attention. This is the removal of surface oil and grease. Most petroleum distillates, lubricating oils, and greases fluoresce with various degrees of brightness. Such materials must be kept out of the testing bath because of the increase in background fluorescence which they produce.

10. EXAMINATION FOR INDICATIONS. After parts have been processed with fluorescent magnetic particles and are ready for examination for indications, the procedure is entirely different. The inspection for indications must be conducted under black light in an area from which at least most of the white light has been excluded. Under proper circumstances, with adequate black light intensity, the location of discontinuities is easy and rapid, since the indications glow brightly with their own light. But there are certain requirements that must be met for proper inspection, and the test engineer and operator must know about these and understand them.

11. THE INSPECTION AREA. Complete darkness is of course the optimum for maximum visibility and contrast, since even minute points of light emission are readily seen in complete darkness. To secure complete darkness means a light-proof room; but in most practical applications it is not necessary to go this far in the exclusion of white light. As a matter of fact, absolute darkness is never achieved in any case, since the black light lamps give off some visible violet light. This small amount of violet light is actually not a disadvantage, as it makes it possible to see the part being handled. Indications, which glow with a bright yellow-green, are in good contrast with the violet, and are readily seen.

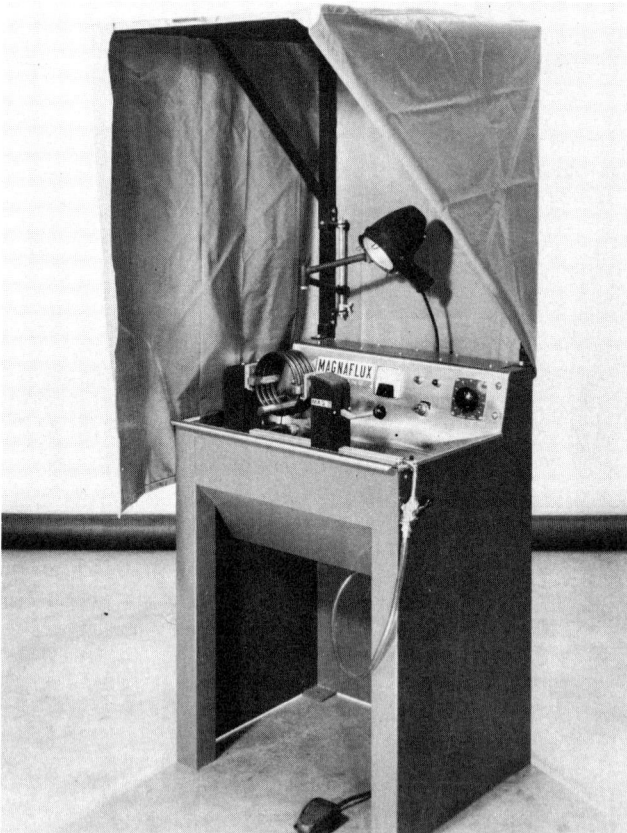

Fig. 127—Simple Inspection Unit for Fluorescent Magnetic Particle Testing.

12. CURTAINED INSPECTION BOOTHS. Nearly all fluorescent magnetic particle inspection is carried out in a curtained enclosure around the magnetizing unit, with suitable black lights mounted inside, usually including at least one hand-held lamp. The curtain black-out is complete at the ends and the back of the unit. Toward the front the curtains extend out at the sides, overhead and across the front, allowing space for the operator to stand in the darkened area before the unit. Usually these curtains do not extend to the floor or even much below the height of the working top of the unit. They do admit some white light, but access to the booth is easier and the ventilation inside is improved by this arrangment. It normally does not admit enough light to interfere seriously with good inspection under the black light on the working top of the unit.

Painting the floor under the booth a dark color helps reduce the light admitted. Bright shop lights in the area of the unit are not desirable.

Since the black light generates considerable heat, small ventilating fans in the booth are often necessary. Usually a white light is also provided to facilitate cleaning of the booth and making up baths, etc., but this light should not be turned on during inspection of parts. Even brief exposure to bright white light destroys the dark adaption of the inspector's eyes. See Section 21, this chapter.

13. SMALL DARK CABINETS. For occasional inspection of small parts with fluorescent magnetic particles where complete facilities are perhaps not warranted, small table top viewing cabinets are sometimes used. These consist of a box-like enclosure in which a black light is mounted. Parts may be held under the black light through an opening at the table-top level, and are viewed by the inspector through a slot in the upper part of the cabinet. See Figure 128.

The device is not very satisfactory for high sensitivity inspection, because full dark-adaption of the inspector's eyes cannot be attained, and very fine indications may not become visible. However, parts can be held quite close to the black light bulb, and a high intensity of black light can be secured—much higher than needed under standard conditions. This compensates to a considerable degree for the lack of full dark adaptation of the inspector's eyes.

When used, the cabinet should be located in a dimly lit area, and the inspector should put his eyes to the viewing slot and wait a short time before attempting to make decisions as to the presence or absence of indications. Gross indications, of course, are easily seen without any dark adaption of the eyes.

14. INSPECTING IN THE OPEN. In some instances it is necessary to inspect objects or parts of an assembly in the field. A super-bright

Fig. 128—Table-Top Inspection Cabinet, with Built-in Black Light.

fluorescent particle is available which can be used more or less in the open, with only a make-shift arrangement to exclude white light. With such an arrangement, and a high intensity of black light radiation, satisfactory tests for all but very fine discontinuities can be carried out.

Field kits are available for this type of work, using portable magnetizing units such as yokes, and pressure spray cans containing prepared fluorescent particle bath. A high intensity black light is also included.

15. INSPECTING LARGE PARTS. Frequently inspection for cracks must be carried out on very large objects or structures. In such cases the inspection is usually made in place, and only portable testing equipment can be applied. Examples are large forgings and castings; the magnetic blades of steam turbines assembled on the spindle; welds in pressure vessels, bridges, buildings, etc.

Since small cracks are usually very important in such inspections, high intensity black lights should be used, and some degree of white light exclusion obtained. Often a make-shift frame, covered with canvas or black plastic sheeting serves excellently for the purpose.

16. THE BLACK LIGHT. Black lights are a critical part of the equipment used in fluorescent magnetic particle testing. In the following chapter (16) the construction and operation of the black light will be discussed in detail. A few practical points should be mentioned here, however, since an understanding of the requirements *at the point of inspection* is important if the operator is to secure good results.

The *intensity* of black light *at the point of inspection* is of the utmost importance. The most common black lights currently in use are 100 and 400 watt quartz mercury arc lights. They are sealed into a housing which includes reflectors, and which is covered with a special filter glass. The filter transmits black light, but excludes nearly all visible light and is opaque to the short wave length ultraviolet light. The 100 watt lights are furnished in two types of re-

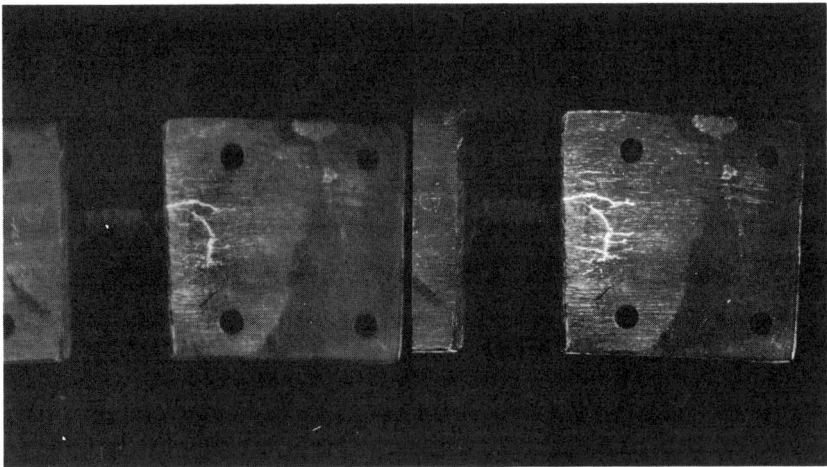

Fig. 129—Effect of Intensity of Black Light on the Brightness of Indications.
Same Indication under Weak (30 Foot-Candles) and Intense
(120 Foot-Candles) Black Light.

flectors, either "spot" or "flood", but only the "spot" type is recommended for inspection work. This light concentrates most of its output in a six inch diameter circle, and gives adequate intensity over this area. It therefore insures a high level of black light intensity at the point of inspection. The 100 watt flood light spreads its output so much that satisfactory intensity at any point is not obtainable. The 400 watt lamps, on the other hand, give satisfactory light intensity over a 16 inch diameter circle. Batteries of these lights are used in installations where a large area must be illuminated, as in the inspection of large steel billets. (See Frontispiece.) The areas of satisfactory black light intensity cited above obtain when the filter glass of the lamp is 15 inches from the work. Much higher intensities result when the lamp is held closer than 15 inches.

17. INTENSITY REQUIRED. The intensity of the black light radiation used to energize a fluorescent material determines the amount of visible light which the material emits. Within limits, doubling the intensity of the black light at an indication will double the brightness of the indication. Therefore black light available *at the indication* must not fall below a safe minimum. Usually 90 to 100 foot-candles is adequate, but for critical work, it should be greater than this (see Chapter 16, Section 15).

This intensity (90 to 100 foot-candles) is obtainable with the 100 watt spot light over a six inch diameter circle when the lamp is held 15 inches from the surface of the part. When a hand-held portable light is being used, much greater intensities are obtainable by holding the light closer to the work—or, a small part can be held closer to the lamp. The operator should remember that very little black light—20 to 30 foot-candles—is sufficient to light up *gross* indications, but without sufficient intensity, a *fine indication may go entirely unseen.*

18. OPERATING CHARACTERISTICS OF BLACK LIGHTS. Mercury arc black lights should be turned on five to ten minutes before starting to use them for inspection. This is because the full output of ultraviolet light is not developed until the mercury arc gets hot, which requires at least five minutes. A certain minimum voltage is required to maintain the arc—about 90 volts—so that a sudden drop in line voltage below this point will cause the light to go out. If this happens, the arc will re-establish itself, but not until the whole lamp has cooled down considerably. This requires usually from five to ten minutes. Lights should therefore not be operated on shop lines subject to severe voltage variations unless constant-voltage transformers are used to avoid such occurrences.

284

The life of the mercury arc capsule is shortened by frequent starting and stopping. It is, therefore, usually more economical to start the light at the beginning of inspection and not turn it off again until its use is ended for the day, even though there may be intervals when it is not actually in use. The output of a black light falls off slowly over the life period of the mercury arc capsule. This life period may be as much as 1000 hours for the 100 watt lamps, and up to 4000 hours for the 400 watt type, but varies greatly from lamp to lamp. As the lamp gets older, therefore, it is most important to check its output at rather frequent intervals, to make sure that the output has not fallen below the minimum permissible intensity. See Chapter 16, Section 15 for the method of measuring black light intensity.

19. BLACK LIGHT FILTERS. The black light filters should be kept clean. Dust and dirt collecting on the filter glass will reduce the black light output significantly. Care should be taken against breakage of the filter. The glass gets very hot from the heat of the mercury arc and even though the glass has been heat-treated to minimize the effect of thermal shock, a splash of moisture or contact with a cold object *can* crack it. A cracked filter should not be used. This is because, even though the visible light that gets through a crack may not be enough to interfere seriously with the fluorescence of indications, the unfiltered white light which may get through such a crack may affect the dark-adaption of the eyes of the inspector.

20. THE INSPECTOR. As in any job of inspection, successful and reliable defect location depends on the inspector remaining alert even though there is a great tendency to get bored with the job when very few indications show up. If an inspector appreciates the importance of the work he is doing—in some applications lives will depend on his effectiveness—he is not likely to become perfunctory in his examination of parts. He must have good eyesight, of course, and if he wears glasses, they must focus on the surface of the part he is examining. This may not always be the case with the two focal distances of bi-focal lenses. A special pair of glasses with an in-between focal distance may be necessary.

21. DARK ADAPTATION. The human eye is a peculiar and complex mechanism which has the power to change its ability to perceive objects and differences in light and colors, depending on the general level of illumination that exists when the observation is being made. In strong white light the sensitivity of the eye to small differences in light intensity is not particularly great, though its ability to differentiate colors and degrees of contrast is at a maximum. When the illumination falls to a low level, an entirely different perceptive mechanism comes into use. In the dim light, the ability to distinguish differences in color and contrasts is poor, but the ability to see dimly-lighted objects and small light sources is tremendously increased. Figure 130 is an attempt to show these two ranges of perceptive ability of the eye in graphic form. No scale is conveniently applicable in such a representation since the two areas of vision are entirely different and operate by entirely different mechanisms. Everyone has experienced the sensation of "not being able to see a thing" when passing from a brightly lighted room to a dark one, but finding after a short time that objects in the dark room become quite readily visible. This increased ability

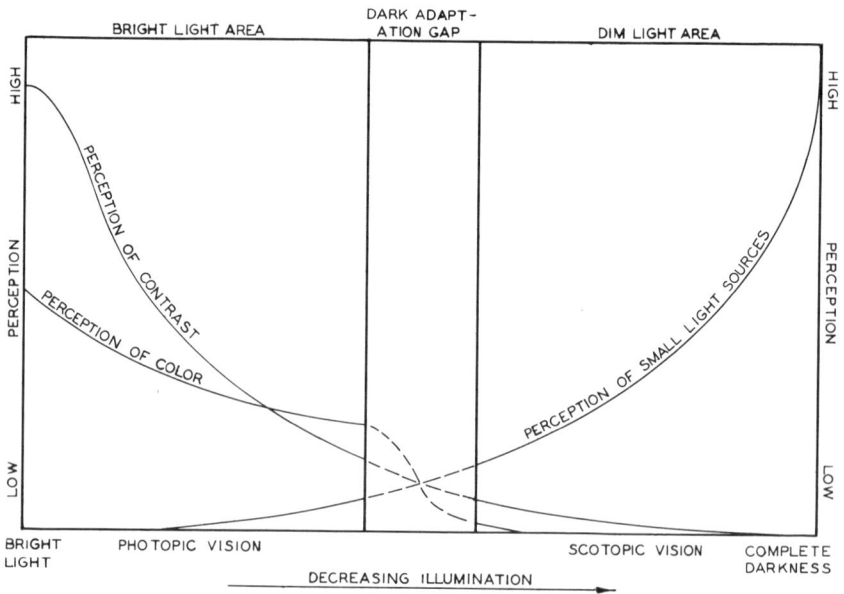

Fig. 130—Chart of Perception of the Human Eye.

to see in dim light is called "dark adaptation". When returning to the bright room, one is at first dazzled until the eyes readjust to this high light level.

This is a most important phenomenon in connection with fluorescent magnetic particle testing. Fluorescent indications are viewed in darkness or dim light where the more sensitive eye mechanism is in action. The dark adaptation of the eyes necessary to the seeing of fluorescent indications requires from five to ten minutes in the dim light before it is well attained. Once the eyes are dark adapted, minute light sources, too small to be seen in a bright light environment, appear relatively brilliant and easily seen. The ability to perceive small light sources such as fluorescent indications is increased by the fact that the eye is drawn to any source of light in a dark background.

The color of the fluorescent light from the fluorescent magnetic particles is a yellow-green, and this color is a peculiarly appropriate one because of another characteristic of vision with the human eye. The eye is not equally sensitive to all colors of the spectrum, but varies quite widely, reaching a maximum in the yellow-green areas, especially for low-level intensities.

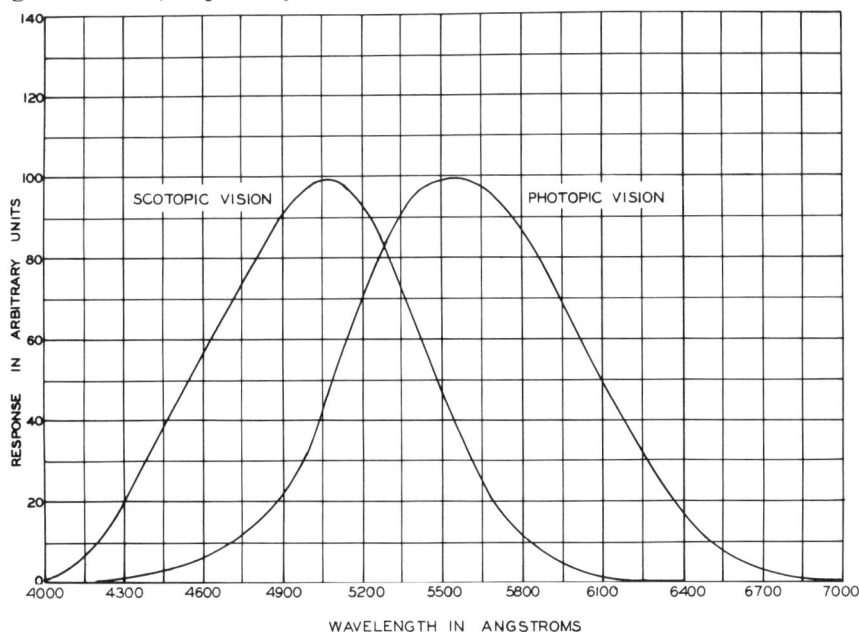

Fig. 131—Chart of Color Response of the Human Eye.

Figure 131 shows the curve of the eye response to the various colors. Use of the yellow-green color for the fluorescence of magnetic particles adds one more plus factor to the sensitivity of this method. Yellow-green light has the added advantage of being "out of context" as to color, compared to the white or blue-white fluorescence of petroleum oils and distillates as encountered in magnetic particle testing. Thus the perceptibility of fluorescent indications is, by their nature and by the environment in which they are viewed, of a distinctly higher order than those of the non-fluorescent type.

22. HEALTH HAZARDS OF MERCURY VAPOR ARCS. Black light is in no way injurious to the operator. The wave length range, peaking at 3650 Å, is below the shortest visible violet light in the spectrum, but is well above the wave length of ultraviolet light which causes sunburn and other injurious effects—3000 Å or less. The mercury arc inside the lamp produces large amounts of this short-wave ultraviolet, but it is completely removed by the envelope and the black light filter combined. The clear glass envelope surrounding the mercury arc capsule is not necessarily entirely impervious to the short wave ultraviolet, and some of it may get through. The lamp should therefore *never* be turned on without the filter in place, and cracked filter glasses or bulbs should be replaced immediately.

23. AVOIDANCE OF OPERATOR DISCOMFORT. When working with black light the operator will quickly notice that many objects and materials have the property of fluorescence. The teeth and fingernails fluoresce with a bluish-white light. Dyes in neckties or other clothing may become startlingly bright, usually not with the same color they have in white light. Men's white shirts usually fluoresce with a quite bright blue-white color, due to daylight fluorescent dyes incorporated in today's detergents to provide the "whiter than white" finishes.

The human eyeball also fluoresces, and when black light is allowed to reach the eyes directly, an unpleasant effect is experienced when this fluorescence is seen, as it were, from the inside. This effect, though it *is* unpleasant, is quite harmless. But black light should not be allowed to reach the inspector's eyes in sufficient amount to create this effect, because he cannot see normally while it is on.

24. EYE FATIGUE. Although working with black light creates no direct health hazard, any operation requiring close attention of the eyes creates fatigue, and if this becomes acute, it can interfere with effective inspection. "Breaks" should be permitted at reasonable intervals. If eye fatigue is complained of chronically, yellow-green tinted glasses are excellent as a means of reducing this effect. These glasses must be of the right composition to pass all the yellow-green light from the fluorescent indications, and to cut out the black light and most of the visible violet that passes the black light filter.

25. POST-INSPECTION MEASURES. After inspection the need to clean parts of residual particles and dried bath ingredients is less than for the non-fluorescent types of particles since bath concentration is so much less. If required, however, the same means are used for the purpose—that is, scrubbing with a detergent or with solvent, washing in a vapor degreaser, or using ultrasonic vibration while parts are immersed in a bath of water or oil. Rust prevention measures must also be taken just as in the non-fluorescent technique.

26. SKIN PROTECTION. The use of fluorescent magnetic particles in liquid suspension does not increase skin irritation or similar hazards above the level existing in the non-fluorescent technique. Extensive medical laboratory skin-patch tests made on the fluorescent dyes have not shown that they have any tendency to affect the skin. But since the materials of the baths used in this testing method are solvents, these solvents do tend to dry the skin, and cause it to chap or crack. Strict cleanliness should be observed around the testing unit to avoid the danger of secondary infections. The use of hand creams to help replace the oils which the solvents have extracted goes a long way toward preventing cracking of the skin. The operator should avoid allowing his clothing to become saturated with the bath liquid, since this accentuates any tendency to skin irritation.

27. PREPARED BATH. Pre-mixed bath, containing the right proportions of fluorescent magnetic particles and conditioners, are available, but are not usually purchased in bulk for filling tanks of testing units. It is more economical to purchase the easily mixed concentrate and buy the oil in bulk. For water based baths the liquid is always at hand for mixing with the concentrate.

Prepared bath is convenient, however, in connection with field inspection or other occasions when portable magnetizing equipment is being used. The bath in such applications is furnished in pressurized spray cans for easy application to surfaces. The kits for these inspections also include spray cans with solvent for surface cleaning before and after inspection.

CHAPTER 16.

BLACK LIGHT—ITS NATURE, SOURCES AND REQUIREMENTS

1. DEFINITIONS. *Black Light,* as the term is used for the purposes of fluorescent magnetic particle testing, is radiant energy having a wave length in the band from 3200 to 4000 Angstrom units, peaking at 3650 Å. This is shorter than the shortest visible violet wave length, and longer than the true "hard" ultraviolet. It is often called "near-ultraviolet" light. Black light has the property of causing many substances, such as certain minerals and dyes, to fluoresce. Though the radiation is not visible to the eye and is therefore characterized as being "black" light, it is produced by a high pressure mercury arc as a by-product of visible white light along with other short-wave radiation.

Fluorescence is defined as the property of certain substances of "emitting radiation as the result of, and *only during* the absorption of radiation from some other source." The terms "phosphorescence" and "luminescence" both refer to other types of light emission. The former is caused by the absorption of other radiation, and continues to glow for some time after the energizing radiation has been removed. The latter is a self-generated light resulting from organic, chemical or electronic reactions within the substance itself. To produce *fluorescent* light, the energizing applied radiation must be acting during the time light is emitted.

In the fluorescent magnetic particle testing process, dyes are used which absorb the invisible short-wave black light, and emit this energy in longer wave lengths in the visible range, giving a bright yellow-green light. Other substances may fluoresce in various colors, from blue through green, yellow, orange and red.

2. ULTRAVIOLET LIGHT. Ultraviolet light is the term applied to radiation of wave lengths shorter than the shortest visible violet wave length—from around 4000 Angstrom units down to 2000 Å. In general, the shorter the wave length the more penetrating and active is the radiation.

However, the band between 4000 Å and 3200 Å is relatively inactive when compared to the band from 3200 Å down to 2000 Å.

Ultraviolet around 2500 Å is very harmful to many forms of life. It will kill bacteria, cause sunburn, generate ozone and can be very injurious to the human eye. The electric arc as used in welding is rich in such short wave ultraviolet, and welders (as well as watchers) must take special care to filter out these short waves before they reach the eye, or else not look directly at the arc.

Welding helmets are fitted with very heavy dark blue or dark purple glass filters to safeguard the eyes. Filters are also used in the case of fluorescent magnetic particle testing, only in this case the filter is a special one, designed to *pass* the maximum amount of wave lengths which activate the fluorescent dyes, and to exclude all other wave lengths to as great a degree as possible, both visible light and short wave ultraviolet. The desired wave lengths are between 3500 Å and 3800 Å.

3. SUNLIGHT AS A SOURCE OF ULTRAVIOLET LIGHT. Radiation from the sun is composed of energy of a very wide range of wave lengths in addition to the visible light which the eye can see. Just below the visible light there is a strong invisible radiation in the ultraviolet. This energy will cause fluorescent dyes to glow, although the strong white light also present does not normally permit this to be apparent.

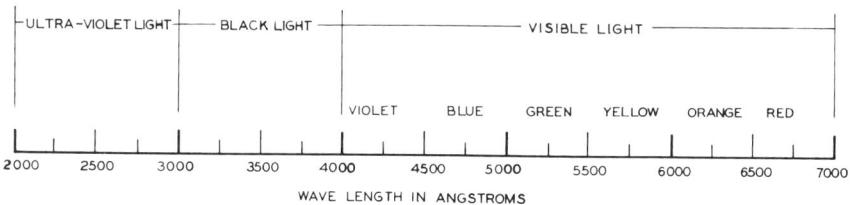

Fig. 132—Spectrum of Light Through the Visible, Black Light and Ultraviolet Wave Lengths.

However the now familiar "daylight fluorescent" colors, paints and printing inks are energized by the natural ultraviolet in sunlight, to make them super-bright even in full daylight. Some of these colors react to short wave ultraviolet as well as to black light. At dusk and dawn there is a difference in light intensity and light distribution in sunlight, due to the longer distance the sun's rays must travel through the atmosphere at those times of day. This results in different amounts of refraction for different wave lengths. There is greater refraction of the shorter wave lengths in sunlight, so that these are bent toward the earth to a greater degree than are the longer visible rays. Thus the actinic energy striking daylight fluorescent materials at dusk and at dawn is proportionately higher than at high noon. This effect causes daylight fluorescent objects to appear especially bright at twilight and just after sun rise.

For the purposes of fluorescent magnetic particle testing, however, it would not be very practical to filter sunlight to secure proper black light intensity. Therefore artificial sources must be utilized.

4. FLUORESCENT DYES. As has been stated, numerous fluorescent dyes are known, and a good many have been used in fluorescent magnetic particle and in penetrant testing. The best dyes for this purpose react strongly when energized with black light of 3650 Å wave length, and the preferred ones emit light in the green-yellow range, the most favorable range for easy "seeing" with the eye. Dyes have been especially developed to give improved performance for the purposes of fluorescent inspection, and those in use today are far superior in fluorescent brilliance and color to those that were available ten years ago.

These dyes are also remarkably stable in their fluorescent light output during prolonged exposure to black light—a property of importance for fluorescent particle testing, and one not possessed by very many of the fluorescent dyes available.

5. SOURCES OF BLACK LIGHT. Although sunlight contains a large amount of ultraviolet, it is not very adaptable for the production of black light for the purposes of fluorescent particle testing. However, the electric arc drawn between two metal or carbon electrodes, is especially rich in energy in the ultraviolet range. The enclosed high pressure mercury vapor arc lamp also offers a convenient source which is high in output of the desired black light wave length. With few exceptions, black light lamps used for fluorescent particle testing employ the mercury arc in one form or another. Figure 133 gives the spectrum of light given out by a high intensity mercury arc lamp, and shows the energy distribution over the range of ultraviolet and visible wave lengths. There are a number of peaks, one of which is at the desired wave-length of 3650 Å.

6. FILTERS FOR BLACK LIGHT. The glass filter almost universally used to separate out the 3650 Å wave-length is a dark red-purple color. It is selected to remove effectively practically all visible light from the energy given off by the mercury arc. At the same time,

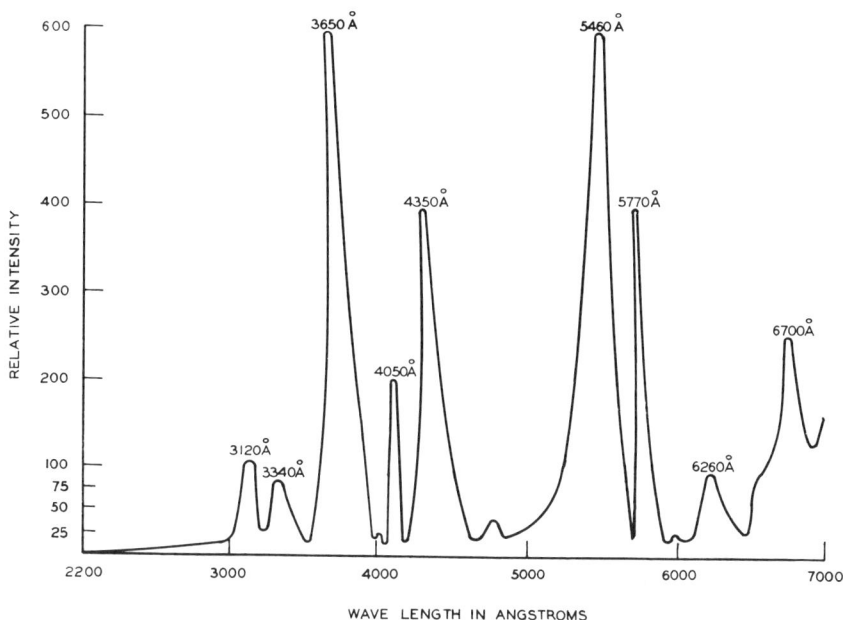

WAVE LENGTH IN ANGSTROMS

Fig. 133—Spectrum of the Output of the High Pressure Mercury Arc.

it also removes all radiation of wave length below 3000 Å—that is, it eliminates all the harmful short-wave ultraviolet. It passes near-ultraviolet radiation in the range from 4000 Å (the lower edge of the range of visible violet) down to 3200 Å. The radiation passed by the filter peaks at 3650 Å, the optimum for energizing most of the fluorescent dyes used for magnetic particle inspection. Figure 134 shows the transmission curve of such a filter (Kopp 41 glass).

It will be noted from the curve that a small amount of visible light is transmitted. This is not altogether undesirable, since it permits the inspector to discern the objects in the immediate vicinity of the black light source, and therefore facilitates handling of parts during inspection. The filter also passes infra-red radiation of wave length longer than the visible red, but this is of no consequence from the point of view of fluorescent particle testing.

7. FLUORESCENT EMISSION FROM DYES. It has been stated that fluorescent dyes are available that emit light in various wave lengths from red to blue. Tests have determined that the eye finds a yellow-green light more easily seen than reds or blues, so that the dyes

293

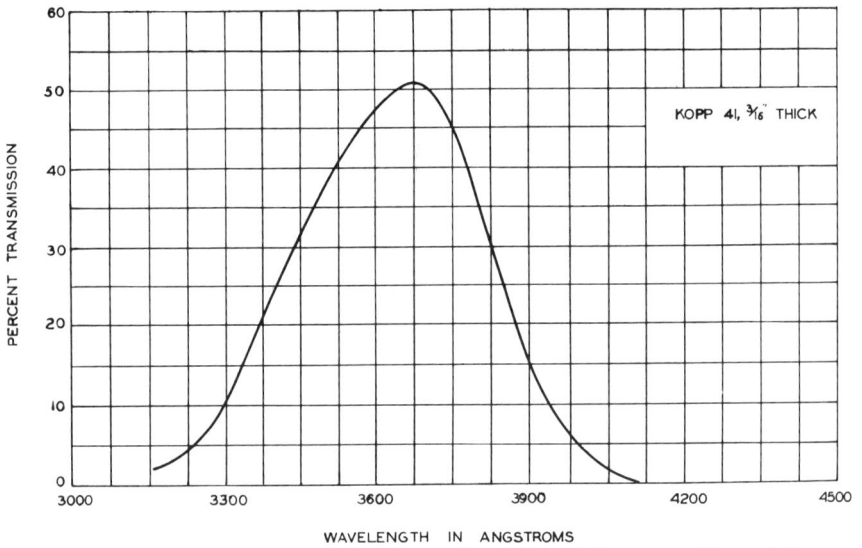

WAVELENGTH IN ANGSTROMS

Fig. 134—Light Transmission Curve of Black Light Filter Glass (Kopp 41).

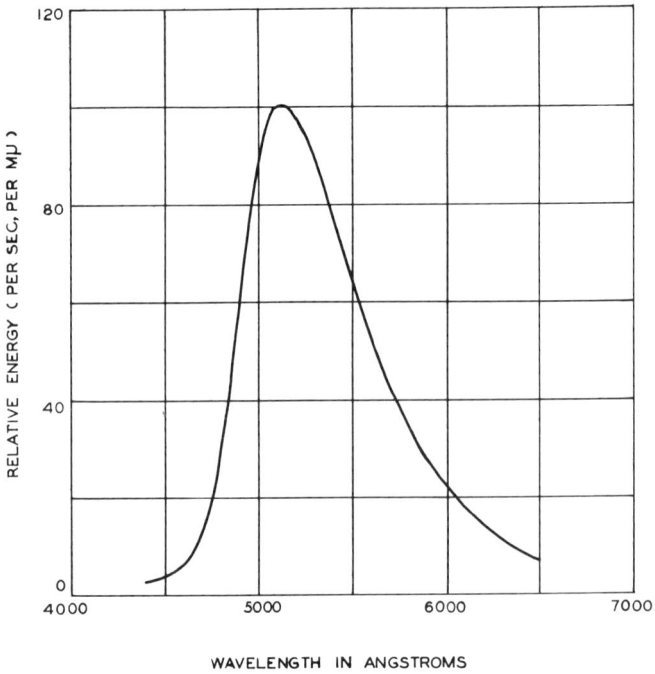

WAVELENGTH IN ANGSTROMS

Fig. 135—Light Emission Spectrum of Yellow-Green Fluorescent Dye.

most widely used for fluorescent particle inspection are those which give off this wave length of light. Figure 135 shows the emission curve of a typical dye when energized with black light of 3650 Å wave length.

8. BLACK LIGHT LAMPS. There are a number of types of black light lamps available commercially, though not all are satisfactory as sources of radiation for fluorescent inspection. Some of these lamps do not give sufficient energy in the 3650 Å range to meet the *minimum* intensity requirements for inspection purposes. All use colored filters of glass intended to remove visible and short wave radiation, but not all are equally successful. Some commercial filters pass excessive amounts of white light.

9. TUBULAR BLACK LIGHT LAMPS. Tubular black lights are similar in construction and operation to the familiar tubular fluorescent lamps for general illumination. These employ a low pressure mercury vapor arc, and the inside of the tube is coated with a phosphor which fluoresces under the energy of mercury vapor discharge. Surrounding these tubes with larger tubes of red-purple filter glass (or using tubes of red-purple glass to manufacture the lamps) results in a black light source.

Small, battery operated tubular black lights and 6 watt tubes operated on 110 volts are used for detecting and identifying fluorescent minerals. They are sometimes used in fluorescent inspection, but only when it is known that the sought-for defects, if present, will produce strong indications. In such cases the lamp should be held very close to the work and the eye also should be brought close to the area being examined.

The larger size 40 and 60 watt tubular lamps have the same limitations as the small ones—insufficient output. These larger tubes will give proportionately more light, but it is produced over a larger area. The tubes are 18 to 36 inches long, whereas the 6 watt tubes and the battery lamps are only 5 to 6 inches long. Since the larger tubes are not portable, as the 6 watt tubes are, they have even less general utility for fluorescent magnetic particle inspection. If mounted in banks of 4 or 6 they may give fairly good results if close enough to the inspection table (15 to 18 inches), especially if only gross defects are being sought. The black light intensity level in even a bank of tubular lights is likely to be too low to ensure the location of very fine cracks. Tubular black lights also put out relatively larger amounts of visible light than do the mercury arc types. One further drawback to the use of these tubular lights is that their output of 3650 Å light drops off rather rapidly with age.

10. INCANDESCENT BLACK LIGHTS. A black light lamp has been on the market which is similar to an ordinary photoflood lamp, except that the envelope is blown out of the red-purple filter glass. This is another low output black light source of little or no value for fluorescent inspection work. The incandescent filament does not put out a great deal of ultraviolet, so that the amount of black light passing through the filter is not adequate. Lamp life is short and heat output great. Like the 6 watt tubular lamp, its only utility is in the intermittent energizing of gross fluorescent indications. It cannot be relied upon for any important flaw-detection work.

11. MERCURY VAPOR LAMPS. The enclosed high pressure mercury vapor arc lamp is by far the most important black light source for fluorescent inspection and is almost universally used. The construction of this lamp is shown in the drawing (Fig. 136). The

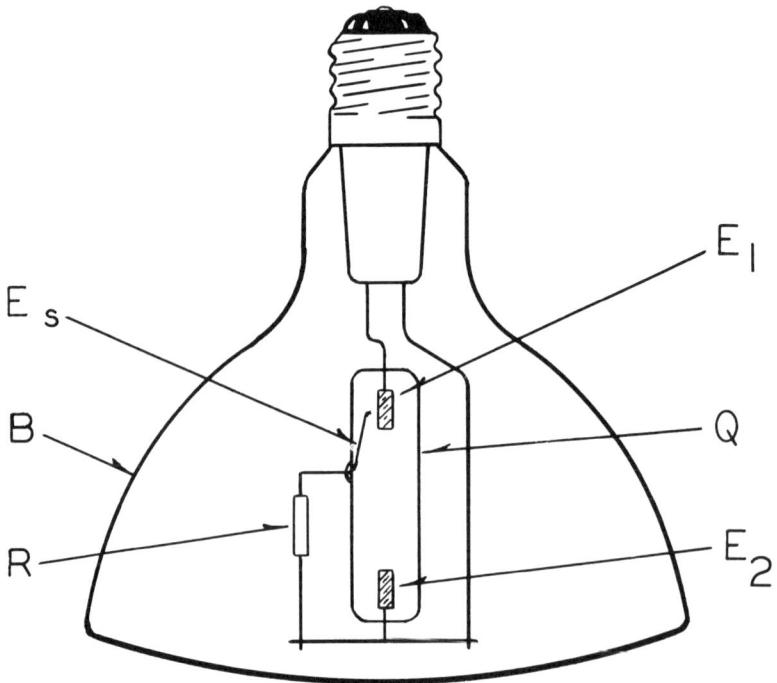

MERCURY VAPOR LAMP CONSTRUCTION
Fig. 136—Construction Drawing of the High Pressure Mercury Arc Lamp.

mercury arc is drawn between electrodes enclosed in a small quartz tube, Q. The current-carrying electrodes are E_1 and E_2, and E_s is an auxiliary starting electrode or heater. The resistor R limits the current in the starting electrode. This arc cartridge is mounted and enclosed in an outer glass envelope, B, which serves to protect the quartz cartridge and also to focus the light emitted. The lamp is supplied current from a special current-regulating ballast transformer, which limits the current the arc can draw.

When current is first turned on, the mercury arc is not set up at once. A small, low-current arc through the gas in the cartridge is first started by the auxiliary electrode, bringing about sufficient vaporization of the mercury in the tube to start the arc between the main electrodes. This process takes about 5 minutes to build to full intensity.

12. LAMP OUTPUT. The quartz inner capsule passes energy of the higher ultraviolet wave lengths, as well as visible light. See Fig. 133. The character of this spectrum is controlled by the design and manufacture of the lamp, principally by selecting the vapor pressure inside the quartz capsule. At very high pressures (100 atmospheres) the spectrum is practically continuous over a wide range, but at a somewhat lower pressure level (10 atmospheres), the output is largely in the visible and ultraviolet range. It is about equally distributed over that range, and includes the desirable black light wave lengths. Lamps of this type yield the maximum energy in the form of black light after filtering, and this is the type used for inspection purposes.

13. COMMERCIALLY AVAILABLE BLACK LIGHTS. The self-contained mercury arc lamps illustrated in Fig. 136 are available in several forms. The 100 watt reflectorized lamp is most commonly used in ordinary work. The 100 watt bulbs are made as either "spot" or "flood" lights. The latter gives a more even illumination over a larger area than the "spot", but is not nearly so desirable for critical inspection use. The illumination level is less than the minimum usually required unless the lamp is held extremely close to the work. The "spot" lamp, on the other hand, concentrates much of its energy in a relatively small area, thus giving maximum illumination on the locations at which the eye of the inspector is focusing in looking for indications. To be more specific, this type of lamp gives adequate intensity of black light for nearly all inspection purposes over a 6 inch diameter circle in a plane 15 inches from the black light filter. In the center of this circle the light intensity is about double the intensity at the edge of the circle.

Fig. 137—The 100 watt Black Light Mounted on a Fixture
on a Magnetizing Unit.

Figure 137 illustrates a lamp of the spot type, mounted in a convenient fixture on a magnetizing unit. The inspector can lift it easily and hold it in his hand to direct the light to the area being examined. Figure 138 shows the 400 watt mercury arc black light, similar in all respects to the 100 watt, except, of course, that its size is unsuitable for portable mountings. This light at 15 inches gives a much larger circle of maximum intensity than the 100 watt. It may be mounted much farther from the work and still give adequate illumination over an area ten times as large as that covered by the 100 watt light. Batteries of the 400 watt lamps are used

Fig. 138—The 400 Watt Black Light.

when, in larger installations, large areas must be covered adequately with black light. Currently even higher wattage lamps are becoming available, which will undoubtedly be used for large test-area illumination.

14. INTENSITY REQUIREMENTS OF BLACK LIGHT. Since the amount of visible light emitted from a fluorescent pigment is directly dependent on the amount of black light energy supplied to it, the intensity of black light at the point of inspection is of the highest importance.

Three factors determine how perceptible an indication is: (a) the *amount of dye* at the indication; (b) the *amount of response* of the fluorescent dye, in the form of emitted visible light in relation to the energy supplied by the black light; and (c) the *actual amount of energy* supplied to the dye by the black light radiation. Given an indication with a fixed amount of fluorescent dye, the brilliance of the indication varies with the intensity of incident black light. This consideration is critical when seeking extremely fine indications.

Experience has shown that 90 foot candles *at the point of inspection* is the minimum black light intensity adequate for the location of fine cracks, such as grinding cracks. Larger or more open cracks, which hold a larger amount of fluorescent particles, may require only 70 foot-candles, and 50 foot-candles may be sufficient for many gross defects. However, unless one knows in advance how fine a crack it may be important to detect, no permanent installation should be set up to provide less than 90 foot-candles at the working point. These levels assume substantial exclusion of white light from the inspection area.

As previously stated, this intensity is provided over a 6 inch diameter circle by a 100 watt spot lamp held 15 inches from the work. At the *center of the spot* the intensity is at least double this amount. Figure 139 shows the distribution of black light intensity,

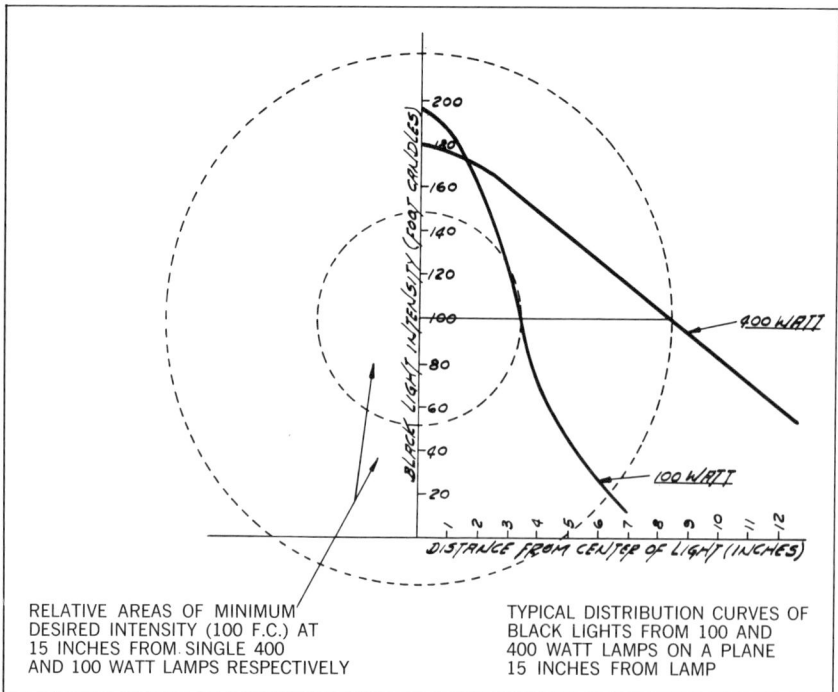

RELATIVE AREAS OF MINIMUM
DESIRED INTENSITY (100 F.C.) AT
15 INCHES FROM SINGLE 400
AND 100 WATT LAMPS RESPECTIVELY

TYPICAL DISTRIBUTION CURVES OF
BLACK LIGHTS FROM 100 AND
400 WATT LAMPS ON A PLANE
15 INCHES FROM LAMP

Fig. 139—Distribution of Black Light from 100 Watt and from 400 Watt Black Light Lamps, Held 15 Inches from the Work Table.

furnished by the 100 watt spot lamp and by the 400 watt lamp, when held 15 inches from the work, from the center of the spot radially to the edges of the illuminated area.

15. MEASUREMENT OF BLACK LIGHT INTENSITY. For the purposes of control of the inspection process, black light intensity should be checked at regular intervals for reasons discussed in the following section. Exact photometric measurements are not required, although reproducible determinations of intensity with reasonable accuracy are essential. Fortunately such a check can be easily and quickly made with very simple equipment.

Standard light meters, consisting of photo-voltaic cells and indicating micro-ammeters, give reasonably accurate readings of light intensity directly in foot-candles. Standard meters of this type are calibrated for white light intensity readings. However, the meters have a reasonable response in the black light wave length range, as can be seen from the curves in Fig. 140. In this figure the response of the photo-voltaic cell is superimposed upon the curve of light transmission by the filter glass. Readings taken under black light will not be numerically accurate, but are reproducible, and

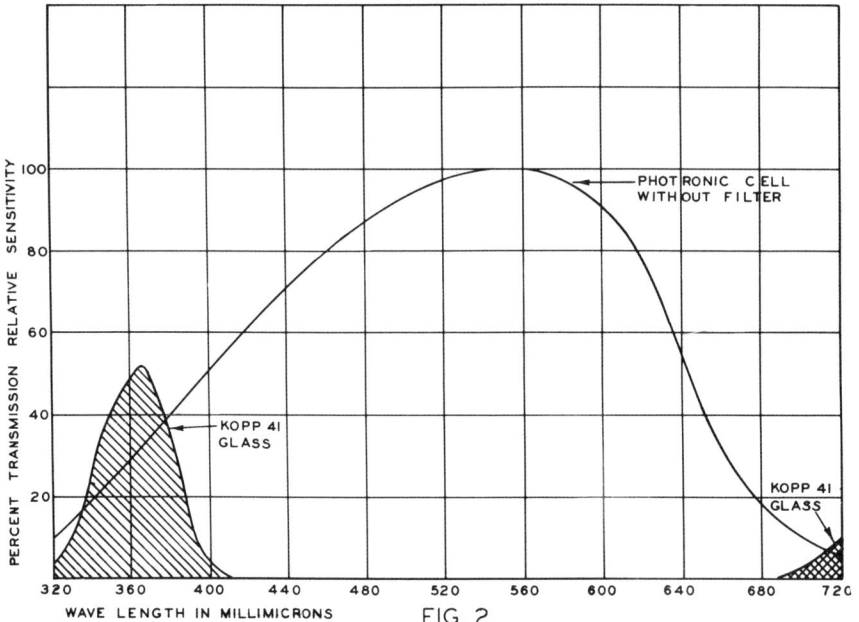

Fig. 140—Transmission Curve for Black Light Filter Glass (Kopp 41)
and Response Curve of the Photo-Voltaic Cell.

therefore serve adequately as a measure of the *relative* black light intensity.

The meters generally used are the Weston Model 703 Type 3, Sight Light Meter, and the General Electric Meter, Model No. 8DW40Y16. The meters are calibrated from zero to 75, or zero to 100 foot candles with multiplier masks to multiply readings by 10 for higher intensities. In use, the meter is exposed to the black light, face up, at the point the measurement is to be made, and foot-candle readings taken directly from the meter scale, using the multiplier mask when needed. "Viscor" or other filters which pass only visible light and filter out black light should of course not be used in making such measurements.

16. CAUSES OF VARIATIONS IN BLACK LIGHT INTENSITY. It has been stated that the intensity of the black light at the point of inspection should be checked at intervals, because a lower-than-optimum intensity may seriously affect inspection results. Some of the causes of intensity variations are the following:

(a) Variations in bulbs. Black light output of mercury arc lamps of the same nominal wattage ratings may vary significantly in lamps from different manufacturers. Even bulbs from the same manufacturer cannot be relied upon to produce uniform amounts of black light. Wattage ratings of bulbs do not, therefore, guarantee the amount of black light that individual bulbs will produce, even when new.

(b) Black light output of a given bulb varies almost directly with applied voltage, and a bulb that gives good intensity at, say, 120 volts will produce much less black light at 105 volts. Light meter checks are the best means for verifying the output on any given electric supply circuit.

(c) Black light output of any bulb falls off with age, and as a bulb nears the end of its life output may drop to as low as 25% of what it was when the bulb was new. The lives of these bulbs may vary widely. Nominal life expectancy is given by the manufacturer (1000 hours for the 100 watt spot), but for various reasons the actual life of a bulb in inspection work is less than this figure. The manufacturer's life tests are based on continuous, steady operation in a fixed and well ventilated position. Black lights used in inspection are subject to numerous starts and shut-offs, and to rough

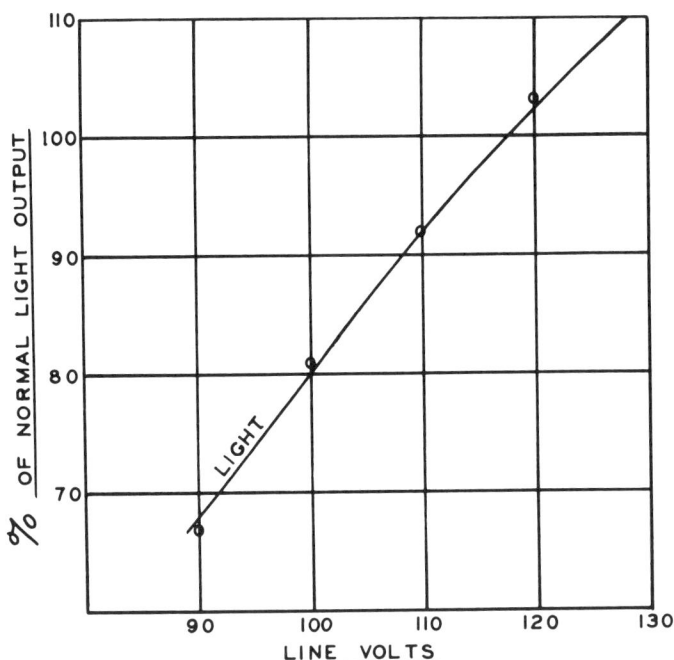

Fig. 141—Output Variations with Varying Voltage. 100 Watt "spot" Black Light.

handling. In addition, due to the needs of filtering and portable housing, such lights may operate at higher temperatures than are desirable. (See also Section 17 below.)

(d) Accumulations of oils films or dust and dirt on the bulb and filter seriously reduce the black light output, sometimes by as much as 50%. This cause of variation can be avoided by keeping the filter clean, but an output check with the meter is the only safe way to make sure that the cleaning has been effective.

17. BLACK LIGHT OPERATING CHARACTERISTICS. As has been stated, when the current to a black light bulb of the mercury arc type is first turned on, it takes five minutes or more for the bulb to warm up to its full output, and no inspection should be started until this time has elapsed. If, for any reason, the arc is extinguished, as by interruption of the current, the bulbs will not respond immediately when the current is again turned on. Time must be allowed for the lamp to cool somewhat and then for the arc to re-establish itself. This takes up to ten minutes. Customarily, a

303

black light, once in operation, is best left on even when not actually in use continuously.

Another reason to leave the bulb turned on is that *each start* affects the life of the bulb, possibly reducing it by as much as three hours per start. It is therefore better, from the standpoint of bulb life, to leave the light on all day, even if it is in actual use for only a few hours.

Low voltage will extinguish the mercury arc, and where line voltage is subject to wide fluctuations, with low points at 90 volts or less, the lamp may go out. Black lights should not be operated on such circuits if it can be avoided, but if it cannot, special constant-voltage transformers for the black lights should be used.

Large line-voltage fluctuations or low voltages may cause annoyance and delay due to bulbs going out, and loss of bulb life due to unnecessary starts, but high voltage surges also decrease bulb life seriously. Line voltages above 130 volts may cause very early burnouts, and the special constant voltage transformers should be used if line voltages are consistently high.

18. ACHIEVING ADEQUATE BLACK LIGHT INTENSITY. The 100 watt spot black light bulb in the hand fixture is a very convenient, reliable and flexible source of adequate black light. It may be held closer to the work than 15 inches, though the amount of visible violet light in the spot may cancel out some of the advantage of doing this. Another advantage is that the inspector's eye can be closer to the point of inspection when the black light is hand-maneuvered to a favorable location.

In some installations, especially where large numbers of parts are being examined, this approach is not so convenient, and it is desirable to have adequate black light intensity over a much larger inspection area. To accomplish this, clusters of black light spot lamps can be directed into the inspection area. In a very few cases banks of tubular black lights have been adequate. Where still larger areas are to be illuminated, a number of the larger 400 watt lamps are used.

19. EYEBALL FLUORESCENCE. Reference has been made several times to the fact that the eyeball fluoresces when black light is directed at it. Although the effect is harmless, it is not only annoying, but interferes with vision while it exists. In the location and manipulation of black light fixtures, this occurrence must be avoided. Black light is reflected from shiny surfaces just as white light is, and if reflected into the eyes will cause such fluorescence to occur. Placement of black lights therefore must avoid reflections as well as direct radiation into the eyes.

When the hand held light is used, the surface of the part being examined should be at an angle so that reflection into the eyes does not occur. Although it is sometimes necessary to bring the eye close to the work when looking for fine indications, this must be done without getting into any direct or reflected black light.

Yellow glasses (such as "Wilsonite" sun glasses) may be worn by the inspector to cut out all black light from the eyes and in cases where the operator is bothered by the fluorescence of his eyes, this is an effective remedy. The lenses of these glasses must pass all of the green-yellow fluorescent light from the indication, but exclude black light. Yellow glasses also exclude most of the visible violet, and therefore many operators do not like to wear them, since they give the effect of working in almost total darkness. Before adopting such protective glasses, the light transmission curves of the yellow glass should be checked against the light emission curve of the fluorescent dye.

CHAPTER 17.

DEMAGNETIZATION

1. INTRODUCTION. After having gone to a great deal of trouble and care to introduce a strong magnetic field of special kind and direction into a part, the operator is frequently faced with the problem of getting rid of this field again, after the test is completed. Demagnetizing after inspection is sometimes as much of a problem as was proper magnetization in the first place.

All ferromagnetic materials, after having been magnetized, will retain a residual field to some degree. This field may be negligible in very soft materials; but in harder materials it may be comparable to the intense fields associated with the special alloys used for permanent magnets. Almost any ferromagnetic material may, for one reason or another, be subjected to magnetic particle inspection, and may have to be demagnetized afterwards. The problem of demagnetization may be easy or difficult depending on the type of material. Materials having a high coercive force are the most difficult to demagnetize. High retentivity is not necessarily related directly to high coercive force, so that the strength of the retained field is not always a good guide as to the probable ease or difficulty of demagnetizing.

It is not always necessary to demagnetize parts after inspection, and since the process involves time and expense, there is no need to apply it unless there is some good reason to do so. In the earlier days of magnetic particle testing demagnetization was almost always carried out as a matter of course and without really considering whether it was actually necessary or not. However, in many cases it *is* essential to demagnetize and the operator should understand the reasons for this step, as well as the problems involved and the available means for solving them.

2. REASONS FOR DEMAGNETIZING. There are many reasons for demagnetizing a part after it has become magnetized from any cause. Demagnetizing *before* magnetic particle testing is sometimes necessary, in cases where highly retentive parts come up to the test with strong residual fields from some previous operation, such as magnetic chucks or cranes. See Fig. 186, Chapter 21.

Demagnetization is necessary when the residual field in a part

(a) may interfere with subsequent machining operations by causing chips to adhere to the surface of the part or of the tool, the tip of which may become magnetized from contact with the magnetized part. Such chips can interfere with smooth cutting by the tool, adversely affecting both finish and tool life.

(b) may interfere with electric arc welding operations to be performed subsequently since strong residual fields may deflect the arc away from the point at which it should be applied.

(c) may interfere with the operation of instruments which are sensitive to magnetic fields, such as the magnetic compass on an aeroplane, or may affect the accuracy of any form of instrumentation incorporated in an assembly which includes the magnetized part.

(d) may interfere with the functioning of the part itself, after it is placed into service. Magnetized tools, such as milling cutters, hobs, etc. will hold chips and cause rough surfaces, and may even be broken by adherent chips at the cutting edge.

(e) might cause trouble on moving parts by holding particles of metal or magnetic testing particles—as, for instance, on balls or races of ball bearings, or on gear teeth.

(f) may prevent proper cleaning of the part after inspection by holding particles magnetically to the surfaces of a part.

(g) is likely to interfere with the magnetization of a part at a lower level of field intensity, not sufficient to overcome the remanent field in the part.

(h) may hold particles which interfere with later applied coatings such as plating or paint.

Although demagnetization is not necessary in practically all other cases, it is often still practiced as a routine step with no particular objective.

3. WHEN DEMAGNETIZATION IS NOT NECESSARY. Demagnetization is not necessary and is not usually carried out when

(a) parts are of soft steel and have low retentivity. In this case the remanent field is low or disappears after the magnetizing force is no longer acting. An example is low-carbon plate such as that used for low strength weldments, tanks, etc.

(b) the material in question consists of structural parts such as weldments, large castings, boilers, etc., even when made of high-strength alloys which retain a considerable field. In such cases the presence of a residual field would have no effect on the proper service performance of the part.

(c) if the part is to be subsequently processed or heat-treated and in the process will become heated above the Curie point, or about 770°C (about 1390°F). Above this temperature steels become nonmagnetic, and on cooling are completely demagnetized when they pass through the reverse transformation.

(d) the part will become magnetized anyway during a following process, as, for example, being held on a magnetic chuck.

(e) a part is to be subsequently re-magnetized in another direction to the same or higher level at which it was originally magnetized as, for example, between the steps of circular and longitudinal magnetizing, for magnetic particle testing purposes.

(f) the magnetic field contained in a finished part is such that there are no external leakage fields measureable by ordinary means as, for example, the testing by circular magnetization of welded and seamless pipe.

The case cited under (e) above is sometimes a cause of confusion. The establishment of a longitudinal field after circular magnetization wipes out the circular field, since two fields in different directions cannot exist in the same piece at the same time. If the magnetizing force is not of sufficient strength to establish the longitudinal field it should be increased, or other steps taken to insure that the longitudinal field actually has been established. For example, a large part having a large L/D ratio may require several longitudinal "shots" along its length to wipe out the circular field. But, once a *truly* longitudinal field is established in the part, the circular field no longer exists. The same is true in going from longitudinal to circular magnetization. If the two fields, longitudinal and circular, are applied simultaneously a field will be established which is a vector combination of the two in both strength and direction.

But, if the fields are impressed successively, the last field applied, if *strong enough to establish itself in the part,* will wipe out the remanent field from the previous magnetization. If the magnetizing force last applied is not of the same or higher order of magnitude as the preceding one, the latter may remain as the dominant field.

4. LIMITS OF DEMAGNETIZATION. When a piece of *unmagnetized* steel is first magnetized, the field within the part increases from zero to the saturation point along the virgin magnetization curve shown as the dotted line on the hysteresis curve in Fig. 56, Chapter 5. Once having been magnetized, the field in the part cannot again be made to traverse this line unless the part is *completely* demagnetized. All steels have a certain amount of coercive force, and it is extremely difficult to bring the steel back to the zero point by any magnetic manipulation. In fact, it is so difficult that for all practical purposes it may be said that the only way to demagnetize a piece of steel completely is to heat it red hot, and cool it with its length directed east and west, to avoid re-magnetization by the earth's field.

When steel is heated it passes through a transformation point, approximately at 770°C., or about 1390°F. for soft steels) at which it becomes nonmagnetic and its permeability drops to 1, the same as air. Above this point the steel becomes austenitic. Fig. 142 illustrates the change that takes place in magnetic properties when iron

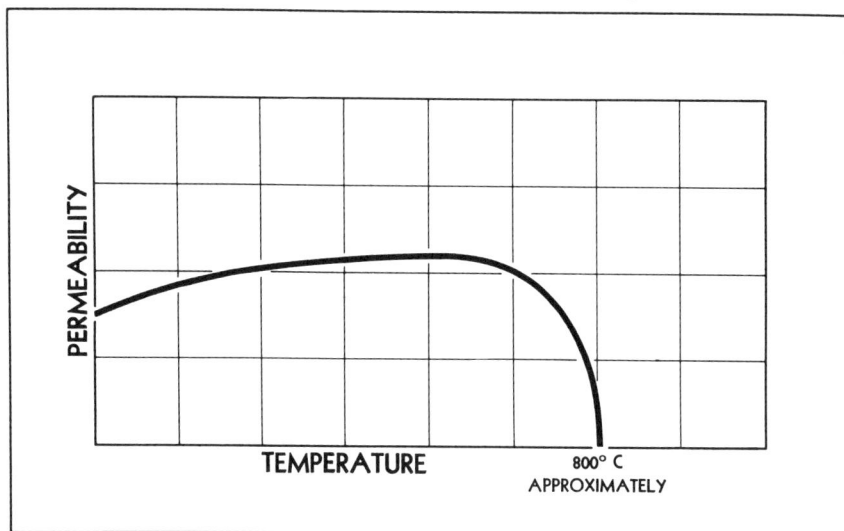

Fig. 142—Effect of Temperature on the Magnetic Properties of Iron.

is heated above the Curie Point. Table VI gives the Curie Point for several materials. When the steel cools down it goes through the reverse transformation, and unless cooled under the influence of a magnetic field, will contain no residual magnetism.

TABLE VI

CURIE POINTS FOR SOME FERROMAGNETIC MATERIALS

Material	Curie Point, °C.
Iron	770
Cobalt	1130
Nickel	358
Iron, 5% Silicon	720
Iron, 10% Chromium	740
Iron, 4% Manganese	715
Iron, 6% Vanadium	815

Other means of demagnetizing almost always leave, for one reason or another, some residual field in the part. As a practical matter, therefore, the process of demagnetization results in either

(a) the best possible job that existing means will produce, or

(b) the level of residual magnetism considered permissible in the particular part involved. "Complete" demagnetization is usually not necessary, even though it is often specified. The principal reason for attempting to secure complete demagnetization is in cases where the remanent field may affect the operation of instruments sensitive to weak magnetic fields.

It must be remembered that the earth's field will always affect the magnetism in a ferromagnetic part and will often determine the lower limit of practical demagnetization. Long parts, or assemblies of long parts, such as welded tubular structures, are especially likely to remain magnetized, at a level determined by the earth's field, in spite of the most careful demagnetizing technique.

Many articles and parts become quite strongly magnetized from the earth's field alone. Handling of parts, such as transporting from one location to another, may produce this effect. Long bars, demagnetized at the point of testing, have been found magnetized when delivered at the point of use. It is not unusual to find that parts of aircraft, or automotive engines, railroad locomotive parts or, in fact, any parts made from steel of fair retentivity, are quite strongly magnetized after having been in service for some time, even though they may never have been near any artificially produced magnetic field. Parts also become magnetized accidentally by being near electric power lines carrying heavy currents, or near some form of magnetic equipment.

Such parts have apparently operated perfectly satisfactorily in this low level magnetized condition, which suggests that the need for extremely thorough demagnetization may often be over-emphasized.

It seems therefore, that the limits of demagnetization may be considered to be either the maximum extent to which the part *can* be demagnetized by available procedures, or the level to which the terrestrial field will permit it to become demagnetized. These limits may be further modified by the practical degree or limit of demagnetization which is actually desired or necessary.

Specifications for demagnetization should be examined to take these facts into account. It is unquestionably true that specifications for demagnetization have called for levels of remanent fields lower than are practicable or even possible of attainment. If demagnetization of parts is called for, the specification should state a limit of permissible residual field that it is reasonably possible to achieve. Unrealistic requirements should be modified in the light of what needs to be or what can be done. See Section 14, this chapter.

5. APPARENT DEMAGNETIZATION. In considering the problem of demagnetization it is important to remember that a part may retain a strong residual field after having been circularly magnetized and yet exhibit little or no *external* evidence of such a condition. On the other hand, parts which have been longitudinally magnetized possess external poles which are easily detected.

Since it may be sometimes only the external evidence of a remanent field that is objectionable, parts may not require demagnetization after being circularly magnetized. The circular step should be conducted *after* the longitudinal magnetization if this is to be done. But circular fields, even if they show no external evidence of their presence at the time of testing, may still be objectionable. As soon as such a part is cut into, or even in some cases if it merely comes in contact with an unmagnetized part, an external pole will be developed. Thus, if a circularly magnetized part is machined, the tool tip may become magnetized and hold chips; or the cut being made may draw a field to the surface and set up an external pole that has the same effect. In such cases, if truly *complete* demagnetization is desired, it is often helpful if the longitudinal magnetization be conducted last, since it is easier to remove this field and to check the extent of its removal.

6. HOW DEMAGNETIZATION IS ACCOMPLISHED. There are a number of ways and means of accomplishing demagnetization, of various degrees of effectiveness. All can be explained by means of the hysteresis loop, and nearly all are based on the principle of subjecting the part to a field continually reversing its direction and at the same time gradually decreasing in strength down to zero.

What happens is illustrated in Fig. 143. The sine wave or curve of a reversing current at the bottom of the graph is used to generate the hysteresis loops. As the current diminishes in value with each reversal, the loop shrinks and traces a smaller and smaller path.

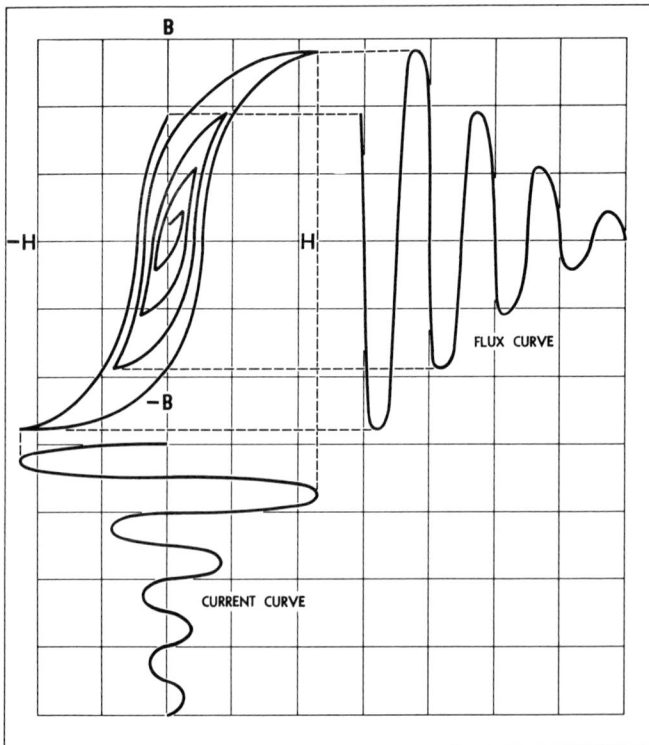

Fig. 143—Flux Curve During Demagnetization, Projected from the Hysteresis Loop.

CHAPTER 17
DEMAGNETIZATION

The curve at the upper right of the drawing represents the flux in the part as indicated on the diminishing hysteresis loops. Both current and flux curves are plotted against time, and when the current reaches zero the remanent field in the part will also have approached zero.

Precautions to be observed in the use of this principle are: first, to be certain that the magnetizing force is high enough at the start to overcome the coercive force, and to reverse the residual field initially in the part; and second, that decrements between successive reductions of current are small enough so that the reverse magnetizing force will be able, on each cycle, to reverse the field remaining in the part from the last previous reversal.

Frequency of reversals is an important factor affecting the success of this method. With high frequency of current reversals the field generated in the part does not penetrate deeply into the section, since penetration decreases as frequency increases. At a frequency of perhaps one reversal per second penetration of even a large section is probably near 100% although for moderate sized parts, the 60 or 50 cycle commercial frequencies of alternating current give quite satisfactory results.

7. REMOVING LONGITUDINAL AND CIRCULAR FIELDS. As has already been said, in most cases it is the external effect of a residual field that is objectionable and that constitutes the reason for demagnetization. The external poles attract ferromagnetic chips and particles and influence the operation of magnetic instruments. A part which is longitudinally magnetized has external poles, and demagnetization aims to reduce these to the lowest possible value. The degree of success attained can be checked by means of a compass needle or field meter. See Section 14, this Chapter and Chapter 8, Section 5.

A part which is circularly magnetized may have no external poles at all and there may be no apparent indication of the presence of an internal field. Such a field is more difficult to remove by any of the usual methods, and there is no easy way to check the success of the demagnetization. There may be local poles on a circularly magnetized piece at projecting irregularities or changes of section, and these can be checked with a field meter. However, in undertaking to demagnetize a circular magnetized part, it is often better first to convert the circular field to a longitudinal field in a D.C.

313

solenoid. The longitudinal field is more easily removed and the extent of removal can be easily checked.

As has been said, although a circular field may give no evidence of its presence, it can still cause trouble, and may require removal. If, therefore, demagnetization is called for, the operator must know the best steps to be taken to secure as thorough a job as possible, whether the existing field be longitudinal or circular.

8. DEMAGNETIZATION WITH A.C. The most common method of demagnetizing small to moderate sized parts is by passing them through a coil through which alternating current at line frequency is passing (usually 50 to 60 c.p.s.). Alternatively, the 60 cycle A.C. is passed through a coil with the part inside the coil, and the current gradually reduced to zero. In the first case the reduction of the strength of the reversing field is obtained by withdrawal of the part, axially, from the coil (or the coil from the part) and for some distance beyond the end of the coil (or part) along that axial line. In the second case the gradual decay of the current in the coil accomplishes the same result.

This is a simple, quick, easy way to demagnetize (when it works), to produce an acceptably low level of residual field. For best results parts should be passed through the coil with their longest dimension parallel to the axis of the coil. The part should be held close to one wall of the coil, or one corner if the coil is rectangular in shape, to take advantage of the stronger field at such locations. Ideally the coil should be of a size to be nearly filled by the part, and for this reason, demagnetizers of this type are furnished with various coil diameters.

Small parts should *not* be piled into baskets and the baskets passed through the coil as a unit, since A.C. will not penetrate into such a mass of parts, and only the few parts on the outer edges will ever be properly demagnetized. Even these will often be only partially demagnetized because of their contact with other parts deeper in the pile. Small parts can be demagnetized in multiple lots only if they are placed in a single layer on a tray which holds them apart and in a fixed position with their long axes parallel to the axis of the coil.

Sixty cycle A.C. is not very effective for demagnetizing parts of large or even moderate size because of its inability to penetrate. The difficulty can be overcome by using a lower frequency A.C. Twenty-five cycles-per-second current is far more effective than

60 cycles. In some installations motor-generator sets delivering 25 cycle current have been used in order to achieve better demagntization and still retain the convenience of simple coil-type equipment.

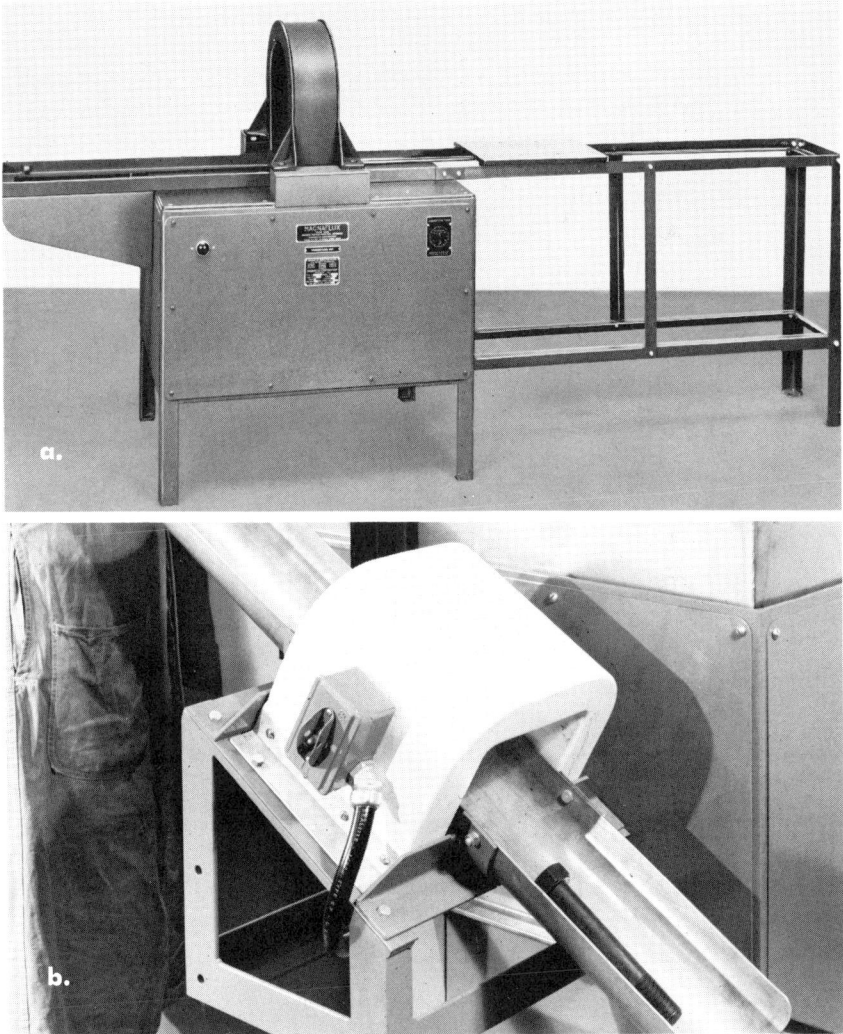

Fig. 144—Demagnetizing Units for Operation on 60 Cycle A.C. a) For intermittent operation. b) Feed-through type for continuous operation.

The decaying current method has been made simple to operate and is widely used. Built-in means for automatically reducing the alternating current to zero, by the use of step-down switches, variable transformers, or the newer saturable core reactors, makes this method easy to apply. When this method is used the current may be passed directly through the part instead of through the coil. This procedure is more effective on long circularly magnetized parts than the coil method, but does not overcome the lack of penetration due to skin effect, unless much lower frequencies than 60 cps. are used.

Another method of achieving "decaying current" demagnetizing is to draw A.C. from a motor-driven generator. By shutting off the motor, the voltage of the generator, and therefore the current in the demagnetizer, drops to zero as the generator dies down to a stop.

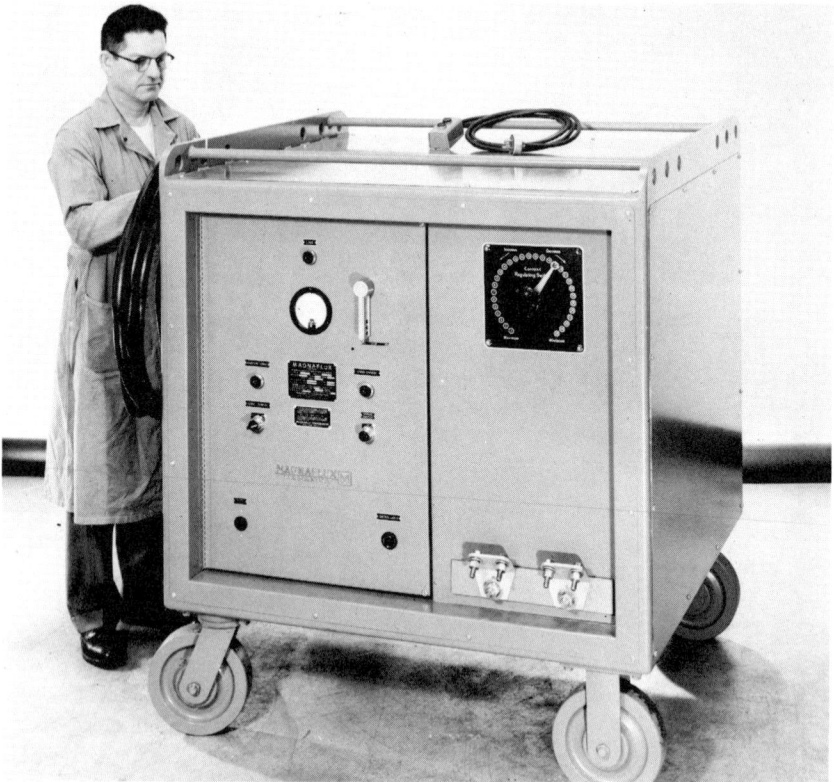

Fig. 145—Magnetizing Unit with Automatic Current Reduction for Demagnetizing, Using a 30-Point Step Switch.

9. DEMAGNETIZING WITH D.C. There are several methods of demagnetizing with direct current. Although more effective, they are essentially identical in principle to the A.C. methods just described. By using reversing and decreasing D.C., lower frequency reversals are possible, with resulting more complete penetration of even large cross sections. Mechanical switching makes possible automatic reversals at as low a frequency as desired. One reversal per second is a frequency commonly used. This method requires specially built equipment, but is most effective. It is one of the more successful means of removing circular fields, especially when the current is passed directly through the part, and will demagnetize even large sections. When using a coil in conjunction with this method, the part remains in the coil for the duration of the entire cycle.

A "trick" method for accomplishing demagnetization with D.C. involves the so-called "single shot" method. Reference to the hysteresis curve of Fig. 55, Chapter 5, shows that the residual field of point (b) corresponds to a coercive force of (c) on the minus H axis. By applying exactly the correct field to overcome this coercive force, the residual field in the part will be brought to the zero point on the loop. The method is not a very practical one however, because the amount of coercive force required is not readily obtainable for the varying kinds of steel encountered. The amount of field which must be applied to counteract exactly the coercive force must be arrived at experimentally by the cut-and-try method. Demagnetization by this means has been useful occasionally on large objects when other means have not been available.

10. DEMAGNETIZING WITH OSCILLATING CURRENTS. The use of oscillating circuits is an attractive means for securing a reversing decaying current for demagnetizing purposes. By connecting a large capacitance of the correct value across the demagnetizing solenoid, the solenoid becomes part of an oscillatory circuit. The solenoid is energized with D.C., and when the source of current is cut off, the resonant resistance-inductance-capacitance circuit oscillates at its own resonant frequency, and the current gradually dies down to zero. The system works well in special cases, but is not adaptable to general use. This is because the size and character of the part affects the value of the inductance component of the resonant circuit, and any considerable variation in this factor will prevent the circuit from oscillating. The system is in use in connection with D.C. yokes and a circuit designed for a specific part to be demagnetized (see also Section 11).

11. YOKE DEMAGNETIZATION. Yokes, either A.C. or D.C. provide a portable means for demagnetizing when other methods are impracticable for the circumstances involved. In some cases yokes are more effective than coil-type demagnetizers, because the field of the yoke can be concentrated into a relatively small area. Thus, even some parts with a high coercive force can be demagnetized in this concentrated field.

Fig. 146—Yoke for Demagnetization with A.C.

Yokes for demagnetizing are usually C-shaped. The space between the poles of the C should be such that the parts to be demagnetized will pass between them as snugly as possible. With A.C. flowing in the coil of the yoke, parts are passed between the poles and withdrawn. The method is very effective, but is limited to comparatively small parts. Sometimes the yoke is used on large parts for local demagnetization, by placing the poles of a U-shaped yoke on the surface, moving them around the area and then withdrawing the yoke while it is still energized. This type is sometimes called a "growler" because of the noise it makes when in contact with the part, due to the vibration set up by the alternating field. A modification of this method is a single pole A.C. electromagnet used in the same way. In general the single pole magnet is less effective than the yoke.

D.C. yokes are similar in appearance to A.C. yokes, but use low-frequency reversing D.C. instead of A.C. for demagnetizing purposes. This system is much more effective in penetrating larger cross-sections, and has the advantage, as has the A.C. type, of concentrating a strong field in a small area.

In some models the method of damped oscillations is incorporated in the design for the purpose of obtaining the reversing and diminishing field. By using a resonant "CIR" circuit (capacitance, inductance, resistance) the current oscillates and dies to zero with frequency and time dependent on characteristics of the circuit. This means of demagnetizing is, when used, designed for a specific part, since the part itself constitutes a portion of the magnetic circuit which is directly related to the inductance. Variations in inductance, which parts of different sizes, shapes and compositions would introduce into the circuit, could destroy the oscillating characteristics on which this system of demagnetization depends. This being the case, the usefulness of this system is limited to applications in which the parts to be demagnetized are closely similar in character and size.

D.C. yokes, properly designed for a specific job, probably provide the deepest penetration for demagnetizing of any of the possible methods.

12. DEMAGNETIZING WITH A.C. LOOPS. Portable units for maintenance inspection around a shop provide a flexible means for demagnetizing. A coil of any desired size and number of turns can be wound from heavy flexible cable—00 or 0000, or even larger—to suit the part. A.C. can be passed through the coil, which may be withdrawn while the current is flowing, or may be left in place while the current is reduced to zero by one of the devices that have been described.

13. SOME HELPFUL HINTS FOR DEMAGNETIZING. There are a few "tricks of the trade" that can be helpful for demagnetizing certain sizes and shapes of parts, or for achieving the minimum level of remanent field. Some of these are:

(a) When a short part is being demagnetized in an A.C. coil by the method of withdrawing the part along the line of the axis of the coil, it is helpful to rotate the part both around axes parallel to and transverse to the coil's axis. This should be done while the part is in the coil and during the entire time of withdrawal. The procedure is also effective in demagnetizing hollow or cylindrical parts.

(b) As a variant of the above procedure, ring-shaped parts may be rolled through the coil, which helps achieve a lower level of remanent magnetization.

(c) A short part with an L/D ratio of one or less can sometimes be better demagnetized by placing it between two "pole pieces" of soft iron of similar diameter but longer than the part. This combination is then passed through the coil as a unit. It has the effect of increasing the L/D ratio and facilitates the removal of the field in the part.

(d) For the demagnetization of ring-shaped parts an effective method is to pass a central conductor through the ring. The central co.iductor is energized with A.C., and the current caused to decay to zero by means of either a step-down switch or a stepless current control. The latter method of decay can be much more rapid (down to a few seconds) than the step-down switch, which requires about 30 seconds to complete its cycle.

(e) The method of (d) above can also be used with reversing, decaying or step-down D.C., instead of A.C.

(f) For long parts, an A.C. coil with current on may be moved along the length of the part, and the part then withdrawn from the coil's influence.

(g) As a variant of (f) move the coil on "high" current along the length of the part, then step-down by one of the current decay methods.

(h) For large hollow parts, a central conductor with high A.C. current is passed through and close to the wall of the part, and the part rotated 360°; then the A.C. is caused to decay to zero.

14. CHECKS FOR THE DEGREE OF DEMAGNETIZATION. Since demagnetizing methods vary widely in effectiveness as used on parts of different shapes and magnetic characteristics—hardness and coercive force—it is often important to check the success of the operation in some way. There is no effective way to check the degree of removal of circular magnetism without damaging the part by making a saw-cut into it. It is, however, relatively easy to check the effectiveness of demagnetization in the case of longitudinal fields, since external poles are always present. The simplest device is the small hand-held field meter described in Chapter 8, Section 5 (Fig. 87). In use, the meter is brought close to the location of suspected residual polarity with the meter in a position normal to the surface, and the dial and pointer farthest from it.

If there is no residual field the needle will remain stationary. If there is a residual field the needle will move in a plus or minus direction depending on the polarity of the field. The amount of movement indicates the strength of the field. Although the dial is not calibrated quantitatively, the field meters supplied by Magnaflux Corporation do give approximately quantitative readings. Full scale deflection (ten divisions) is equivalent to approximately 12 oersteds or gauss. At the low end of the scale, the value is about 1.5 oersteds or gauss per division.

Some process specifications use this device to specify the degree of demagnetization desired. These will often call for a maximum of less than two divisions—three oersteds—for critical parts; and less than five divisions for parts in which a somewhat higher residual field can be tolerated.

If a field meter is not available a simple substitute is a short length of tag wire, or any light iron wire. A piece of such wire, 4 or 5 inches long, can be balanced on the end of the finger, and when brought near a part having a local pole, the end of the wire will deflect toward the magnetized area. If the wire is hard and will itself retain a magnetic field (tag wire is usually hard from cold drawing) it should be demagnetized just before use, so that its own remanent field may not be the cause of deflection toward the part being tested. The method is not quantitative of course, but an experienced operator can use it to judge quite closely the amount of remanent field.

15. CHOICE OF DEMAGNETIZING METHODS. Following are some of the factors that must be considered in the selection of a demagnetizing method. The several methods are listed with their broad advantages and limitations.

(1) *A.C. Coil*

    (a) Suitable for relatively small and soft parts.

    (b) Handles large numbers of parts in high production.

    (c) Maximum part diameter for thorough demagnetization, 1 to 2 inches.

(2) *A.C. Through Current*

    (a) Works well on relatively large and hard parts.

    (b) Best suited for relatively low production.

(c) Works well with 30-point step-down, infinitely variable current reduction or reactor decay methods.

(d) Maximum part diameter for thorough demagnetization, up to about 3 inches.

(3) *Reversing D.C. Coil with 30-Point Step-Down, Infinitely Variable or Reactor Current Decay Methods.*

(a) Works well on small parts in standard coil.

(b) For larger parts use full length cable wrap.

(4) *Reversing D.C., Through Current—30-Point Step-Down, Infinitely Variable or Reactor Current Decay Methods.*

(a) Best for large hard parts.

(b) Best for diameters greater than 3 inches.

(c) Best for parts that are of difficult size, shape or composition.

(d) Works well with central conductor when applicable.

Table VII is a ready reference chart for these several demagnetizing methods, and the conditions with respect to size, hardness and production rate for which each is or is not suitable. Like the list above, this table is given as a guide only, since there are no clean-cut lines of division between, for example, "small", "medium" and "large" sized parts, etc. The classifications should be helpful, however, in selecting the most promising demagnetizing method to try, basing its success or failure on check tests with the field meter. For those circumstances for which the given method is listed as not applicable, the classification is based on either lack of effectiveness or impracticability.

16. DEMAGNETIZING WITH MAGNETIZING EQUIPMENT. In many cases demagnetization is a separate step performed on separate equipment after the magnetizing and inspection operation is completed. Often, however, it is convenient and possible to demagnetize on the same unit used for magnetizing. This is in general true of all A.C. equipment.

In its simplest form it would include provisions only for the coil method—withdrawal of the part from the coil's field. Equipment is also available which provides for decaying A.C., which makes possi-

TABLE VII

GUIDE TO DEMAGNETIZING METHODS

Method	Part Size			Hardness			Production Rate		
	Small	Med.	Large	Soft	Med.	Hard	Low	Med.	High
A.C. Coil 50/60 cycles	X	X	O	X	X	O	X	X	X
A.C. Through Current, 30 Pt. Step-Down	O	X	X	X	X	X	X	O	O
A.C. Through Current, Reactor Decay	O	X	X	X	X	X	X	X	O
D.C. Through Current, 30 Pt. Reversing Step-Down	O	X	X	X	X	X	X	O	O
D.C. Coil, 30 Pt. Reversing Step-Down	O	X	X	X	X	X	X	O	O
A.C. Yoke	X	Local only	O	X	X	O	X	O	O
Reversing D.C. Yoke	X	Local only	O	X	X	X	X	O	O

X signifies applicable
O signifies inapplicable

ble the use of the through-current method. In the case of D.C. equipment several forms of demagnetizing facilities are available:

(a) Simple A.C. coil method, the A.C. being provided to the coil especially for demagnetizing purposes.

(b) Decaying A.C. current method. In this case A.C. is connected to either the heads or the coil, as desired, for demagnetizing purposes.

(c) Reversing D.C. step-down methods, available to either the heads or coil.

When considering methods and equipment, the possibility of using one unit for both purposes should be looked into, as such an arrangement may constitute a saving of both time and effort for the demagnetizing operation. Often parts, which are moderately large and heavy, are inspected while still in place on the unit in the magnetizing position. It is then quick and convenient if, after inspection, the part can be demagnetized without removing it from this position.

17. EFFECT OF VIBRATION. In some magnetizing operations, especially in the case of fairly large complex welded structures, a stabilizing or "soaking in" effect can be produced on the magnetic field by hammering or otherwise vibrating the part during and after the application of the field. Complex assemblies of this sort are often very difficult to demagnetize due to the fact that different components are of varying size and section, and take individual directions in their long axes. Vibration during *demagnetization* of such parts seems also to be a help in removing stubborn points of residual fields. Yoke demagnetizers sometimes help remove local polarity from such assemblies, though more often they merely move the residual field from one location to another.

The earth's field plays a part in the difficulty of demagnetizing such structures. The part to be demagnetized should be placed so that its principal axis, or the axis of its principal, or its longest member, is in an east and west direction. A long part lying in a north and south direction can never be demagnetized below the level of the earth's field by any method. Rotating the part or structure on its east-west axis while demagnetizing often helps reduce the field in transverse members which are not lying east and west. Vibration of the structure during the demagnetization process is also helpful under these circumstances. Complete removal of all fields from such a complex structure is virtually impossible—but again it should be pointed out, complete removal is seldom necessary.

CHAPTER 18

EQUIPMENT FOR MAGNETIC PARTICLE TESTING

1. HISTORY. The many forms of automatic and special-purpose equipment for conducting magnetic particle testing which are being built today are the result of thirty-seven years of evolution and growth in technology and in the practical requirements of the method. In the days prior to the mid 1930's, when experience with the method was quite limited and its potential applications only partially visualized, the equipment that was used was small and quite simple. It was designed for all-purpose use—a unit should magnetize the greatest possible variety of parts.

This all-purpose philosophy dominated the design and development of larger and more versatile equipment until the early 1940's when the requirements created by World War II calling for many new applications at much higher production speeds, dictated the construction of many new equipment types and variations to fit special uses. The demand for speed to keep up with mass production rates caused a considerable number of automatic or semi-automatic special-purpose units to be built. Then the true age of automatic equipment began in the early 1950's and continues today, and automatic processing of parts, from miscellaneous small parts to massive steel billets and blooms, is now common practice in many industries.

It is not the intent to give in this Chapter a complete catalog of available equipment. The following sections are intended to give only a quick survey of the so-called "standard" or general-purpose types of magnetic particle testing equipment, and a few examples of some of the modern equipment now being built and used. The following Chapter, 19, will describe the evolution, use and scope of automatic and "special-purpose" equipment, the problems governing its design, and its advantages in many special applications.

2. NEED FOR EQUIPMENT. Magnetic particle testing is a process which does not depend for its successful application upon any particular piece of equipment as do the X-ray, ultrasonic and eddy current methods of nondestructive testing. The proper magnetization of most parts *can* be accomplished by very simple means, and at the outset the possible need for the large and complicated equipment being used today was not even visualized.

Magnetic particle testing equipment serves two basic purposes, which dictate the sizes, shapes and functions of modern units.

These are:

(1) to provide *convenient* means for accomplishing *proper* magnetization. *Proper* magnetization with respect to field strength and direction, is, as the previous chapters of this book have shown, of preeminent importance. *Convenient* means, providing sufficient power of the right sort, suitable contacts and coils, means for applying the magnetic particles and well-lighted space for careful examination of the processed part for indications, can only be achieved with equipment designed to meet these requirements for various types of parts and the conditions under which they must be tested.

(2) to make possible *rapid* testing of parts at whatever speeds are required, with assurance that the results will be reliable and reproducible. With suitable equipment all parts can be tested under identical conditions of contacts, current types and values, and techniques of particle application, when used in different locations and by different operators. Such control of the process becomes more and more important as the size of the defects sought becomes smaller and finer; or as it becomes desirable to find defects *over* a certain minimum size, ignoring the very fine ones.

3. SIMPLE EQUIPMENT. For occasional testing of small castings or machine parts for surface cracks, for which small, easily portable equipment is most convenient, and for the inspection of welds, magnetic yokes are often adequate and very easy to use. They are able to put a strong field into that portion of the part that lies between the poles of the yoke. Either dry powder or wet particles may be used. Yokes are available for either A.C. or D.C., and in one model permanent magnets are used. The latter permits inspections where no source of electric current is available, or where its use is not permissible because of the fire or explosion hazard in the area where the inspection must be made. For longitudinal magnetization of shafts and spindles and similar articles, the portable kits are furnished with a fixed coil covered with a heavy protective rubber coating. Figure 147 illustrates such a testing kit.

Fig. 147—Magnetizing Yoke and Coil in a Portable Kit.

4. LARGE PORTABLE EQUIPMENT. Larger portable equipment is made for use where larger power requirements or heavier duty

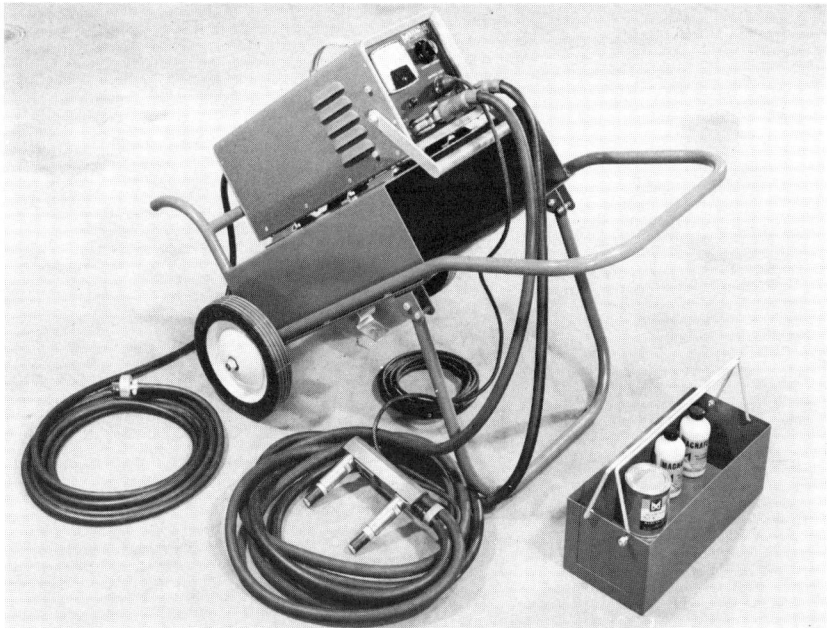

Fig. 148—Small Magnetizing Kit, Operating from 120 volts A.C.
Mounted on a Carry-all Truck.

cycles make the small kits inadequate. One of the smallest of this series is illustrated in Figure 148. It operates from 120 volts A.C., and delivers up to 700 amperes, either A.C. or half wave D.C. Dual prods for direct contact are provided, and flexible cable for making loops for longitudinal magnetization are included.

Large units with outputs of up to 6000 amperes, delivering either A.C. or half wave D.C. are used for testing of castings, forgings or weldments, where such heavy currents are required. The units are equipped with current control switches and with the 30-point step-down switch for demagnetization; or, more commonly today, with the saturable core reactor decay system. Figure 149 shows such a unit being used to test truck components as a part of a planned maintenance program. This unit is equipped with the saturable-core reactor which provides stepless current control, both for magnetizing and demagnetizing.

Fig. 149—A Portable Magnetizing Unit Being Used for
Inspecting Truck Components.

The most modern of the series of portable units is the power pack shown in Fig. 150. This type of unit delivers from 4,000 to 6,000 amperes output of magnetizing current, either A.C. or half wave, with infinitely variable current control. In addition it has a self-regulating current control feature. The current "Dial-Amp"

Fig. 150—A Modern Portable Magnetizing Unit.

system automatically delivers the selected amount of current to the external magnetizing circuit, and automatically compensates for variations in load impedance and line voltage. Thus, the pre-selected current is continually maintained throughout the magnetizing cycle. Further optional remote dial controls permit setting the desired amperage from a point 110 ft. or more away from the equipment.

5. STATIONARY MAGNETIZING EQUIPMENT. A large variety of stationary, bench-type units are available, with various characteristics to fit all different testing requirements. The smaller sizes are used for small parts easily transported and handled on the unit by hand. The larger ones are used for such heavy parts as long diesel engine crankshafts, where handling must be by crane. Such units are made to deliver A.C. or D.C. with various types of current control.

Figure 151 shows one of the smaller units (52 inches) of this type. This unit is for use with the wet method, either with colored

Fig. 151—A Modern Bench Type Magnetizing Unit.

magnetic particles or the fluorescent type. The portable black light is seen mounted at the left. The hood may be let down during black light inspection to secure the necessary exclusion of white light. Direct current up to 4000 amperes, derived from full wave rectified three phase A.C., is delivered to the adjustable contact heads. 100 inch units of this type will deliver up to 6000 amperes. A built-in coil is provided for longitudinal magnetization. This unit is equipped with the infinitely variable current control by means of a saturable core reactor, and also with the self-regulating current control.

Figure 152 shows the largest of the bench-type stationary units. It will handle any small or long heavy parts, such as Diesel crankshafts, up to 192 inches (16 feet) in length, and up to 28 inches in diameter or "swing". Alternating current up to 5,000 amperes is delivered for magnetizing purposes. The unit is for use with the wet method, and black light and hood are provided for use when fluores-

Fig. 152—Largest Bench-Type Magnetizing Unit for
Large Diesel Crankshafts.

cent particles are used. Demagnetizing can be accomplished by means of one of the A.C. decay systems.

6. LARGE HEAVY-DUTY D.C. EQUIPMENT. At the top of the list of heavy-duty stationary equipment are those direct current units designed for application of the "over-all" method of magnetizing, for the inspection of very large, complicated castings. In size these units are not large in comparison with the steel billet testing units (Chapter 19) but in current output they are by far the most powerful built. Rectified three phase A.C. is delivered with current values running as high as 20,000 amperes. Such high currents are needed to magnetize, at one time, an entire casting which may weigh many tons. Another feature of these units is that they deliver current, separately and in rapid succession, through three circuits, making possible the location of cracks in any direction in one operation. The system is known as multi-directional magnetization. By means of electromechanical switching, demagnetization can be accomplished by utilizing one circuit in conjunction with reversing D.C. and a 30-point step-down switch.

The cost of these units is easily justified by the saving in time and labor over the older point-by-point inspection with prod contacts. In addition, the system is much more reliable in the location

Fig. 153—20,000 Ampere D.C. Power Unit for Overall
Multi-Directional Magnetization.

of surface and near-surface defects than the prod method, and often shows indications of some serious defects which the latter method has missed. The unit is most often used with fluorescent magnetic particles by the expendable technique.

7. UNIT VARIATIONS. A great number of variations of these typical magnetizing units is available. These variations are in size, in current output and kinds of current, in the methods of current control, and in numerous types of fittings to expedite magnetization of odd-shaped parts. In addition there are many accessories, such as contact pads, automatic bath applicators, contact clamps, leech contacts, steady-rests for heavy shafts, prod contacts, special shaped coils, powder guns, etc.

8. DEMAGNETIZING EQUIPMENT. Demagnetizing methods and equipment have been described in the preceding chapter,. and in connection with the various magnetizing units in which demagnetizing features are embodied. When separate demagnetizing units are required—usually for the demagnetizing of large numbers of small parts—the A.C. coil is used. A large size of this type demagnetizer is shown in Figure 144, Chapter 17. Coils of this type are made in a number of sizes, since demagnetizing efficiency is higher if the parts nearly fill the coil opening. In the case of special units, special demagnetizers are often built into the unit to suit the size and operating cycles required for the particular application.

CHAPTER 19

AUTOMATIC AND SPECIAL MAGNETIC PARTICLE TESTING EQUIPMENT

1. INTRODUCTION. The need for equipment to do special testing tasks grew out of the application of magnetic particle testing methods to an increasing number and variety of parts, both as to size and to shape. As the use of the method expanded it quickly became apparent that the standard unit, intended for testing a large variety of parts, was not adequate to serve the testing needs which were being called for. Units to perform specific tests on a limited variety of parts were being demanded by industry.

This was a new philosophy, since previously it was felt by users of the process that to justify its existence equipment must be as versatile as possible. A single unit had been expected to accommodate almost any type, size or shape of part that was brought to it for testing.

In the present chapter it is the intention to give a comprehensive view of special and automatic units. Such units are being used in large numbers today in many applications. We will also analyze considerations involved in their design and use. Some of the modern special-purpose equipment will be described and discussed.

2. DEFINITIONS. Magnetic particle testing equipment available today may be considered to fall into two major groups—standard and special.

Standard units are those which are designed to handle a variety of sizes and shapes of parts, and are, as the name implies, made in quantity to a standard design. They may be stationary or portable, for use with either dry or wet methods, and are used with little or no modification for the testing of widely differing types of parts, in a large variety of industries.

Special units are those which have been especially designed to take care of some situation out of the normal, which standard units cannot, for one reason or another, handle. They may be special as to the method of magnetizing or particle application, or be designed to handle an unusual size, shape, or number of parts. They may or may not be automatic.

Special units can be further broken down into two groups:

(1) *Special-purpose*, which may be a) single-purpose or b) general-purpose.

(2) *Automatic.*

Special-purpose units are those which are built to do a single testing job. This special job may be a variation in magnetizing technique, in the way the magnetic particles are applied, or in the way parts are handled. They may be *single-purpose*, in which case they are to test a single type of part and no others, possibly by a special processing technique. Or they may be *general-purpose*, in which case they are designed to apply a special magnetizing or processing technique to a (perhaps limited) variety of parts.

Automatic units, on the other hand, are those in which part or all of the handling and processing steps are handled automatically. Either single- or general-purpose units may be partly or entirely automatic. Even standard units, by addition of standard accessories, may be made automatic in some of their functions. The principal purpose of automatic units is to speed up the entire handling-processing-inspection cycle. This is accomplished through automation of one or more of the important steps involved in any given testing operations.

3. HISTORY OF SPECIAL UNIT DEVELOPMENT. In the early days of the development of the magnetic particle method, magnetizing units were designed and built as needs for equipment developed. The first designs of what became standard units were made for some special testing purpose for which, at the time, there were no suitable existing designs. Thus we had small units and large units, vertical units and horizontal units, stationary units and portable units. Some were for A.C. and some for D.C. But each one was designed with the idea that it would handle as many applications as possible, similar to the initial one for which it was built.

The building of special-purpose and automatic equipment did not enter the picture until World War II created the need for faster inspection of mass-produced parts. A number of automatic units were built during the early 1940's for testing such items as projectiles, bolts and nuts, aircraft engine valve springs, etc. Special-purpose units for handling larger and heavier parts such as steel propeller blades, propeller hubs, engine cylinders and engine mounts, also came into being, some of them at least partly automatic.

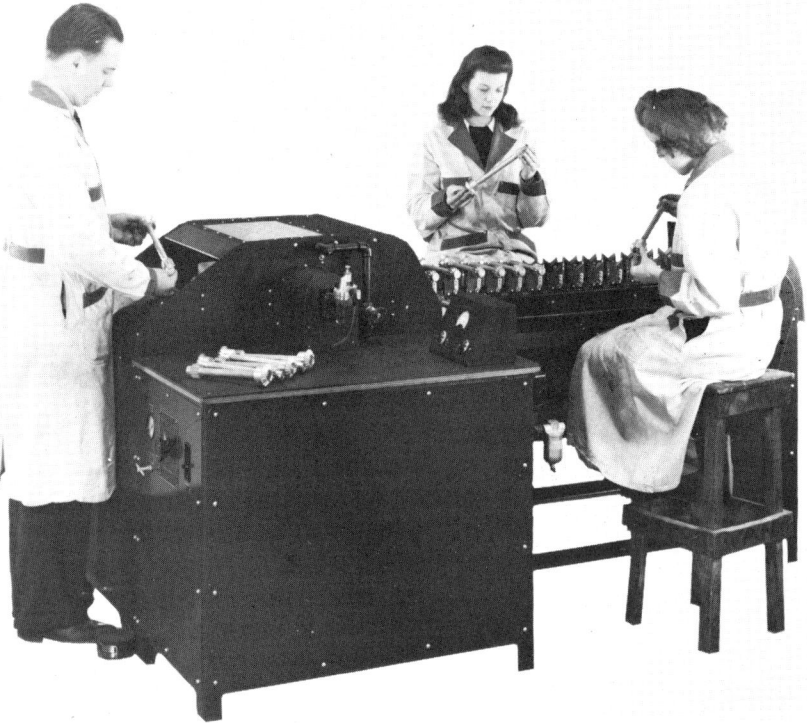

Fig. 154—Automatic Unit for Testing Bolts and Similarly Shaped Parts.

Following the war the demand for equipment shifted from war materials to industrial production for civilian use. The railroads undertook to test all axles of locomotives and passenger and freight cars, which parts had been severely taxed by war service. Figure 156 shows a special unit built in the late 1940's for railroad car axle testing.

When the diesel revolution came, additional special unit types were developed to handle critical parts of these locomotives. Crankshafts, connecting rods, pistons and various gears and castings were tested during manufacture and later at scheduled overhaul periods. Many of these parts were tested on special-purpose units.

Fig. 155—Special-Purpose Unit for Testing Steel Propeller Blades.

Fig. 156—Special-Purpose Unit for Railroad Car Axles.

Fig. 157 shows an installation of magnetic particle testing equipment for the inspection of diesel crankshafts.

During this period the automotive industry also became large users of automatic equipment to reduce costs and time of inspecting

Fig. 157—Magnetic Particle Equipment for Diesel Crankshaft Testing.

engine and steering gear parts. Manufacturers and users of oil field tools and equipment also made extensive use of the method. Figure 158 shows an automatic unit for testing couplings for oil well tubing and casing.

Fig. 158—Automatic Unit for Testing Oil Field Pipe Couplings.

The era of the jet engine, missiles and space vehicles has made it imperative that testing with magnetic particles (as well as all other methods of nondestructive testing) be increased in sensitivity and reliability beyond the levels which had been achieved during and just after the war. The special purpose and the automatic units are a large factor in accomplishing this end.

4. FACTORS DICTATING THE NEED FOR SPECIAL UNITS. The justification for a special unit in any given inspection situation is created by the presence of one or more of the following factors:

(a) The need to employ a processing technique not available with standard units.

(b) The size, shape or weight of the part or parts to be tested, when they are sucn as to be not readily processed on standard equipment.

(c) The necessity of insuring absolute reliability and reproducibility of results, at selected high *or low* sensitivity levels, which end can be achieved by automation of the processing cycle.

(d) The need to increase the output per man-hour of the testing unit, which can be accomplished by automation of the testing operation.

The individual importance of these factors will vary as the service and production requirements for the part to be tested will vary. When the safety of the public or operating personnel—as in aircraft operation—are dependent on the failure-free performance of the part, the need for absolute reliability (factor c) would be the controlling consideration. Or automation (factor d) may be needed in order to keep pace with production when 100% in-process inspection is called for.

One generalization that can be made is this; When "accept-reject" standards are to be used and the rate of production is high enough to use the full capacity of a standard unit, automation of at least the processing cycle is indicated. By this means the reproducibility of results will be achieved, which will permit the application of standards of acceptance and rejection without consideration of each case by the inspector. If, in addition, the rate of production is *very* high, further automation can be profitably employed. By controlling process variables by automation, the hazard of error due to the human factor is tremendously reduced.

The special unit need not always be complicated or expensive. Sometimes a special accessory can be built for a standard unit and do the job adequately. In other cases, of course, when all processing and handling is automatically controlled and sequenced, the unit is necessarily more complex and more costly.

For example, Fig. 159 shows a special unit for crankshaft inspection which is a standard unit equipped with special fixtures to process crankshafts automatically. This unit is quite inexpensive when compared to the one shown in Fig. 160. This is also a special unit for crankshafts, but is completely special, including automatic processing, conveyorized part handling and demagnetizing facilities. It is necessarily more complex and costly than the modified standard unit of Figure 159.

Fig. 159—Special Unit for Crankshaft Inspection of the Fixtured Standard Type.

5. SINGLE-PURPOSE AND GENERAL-PURPOSE UNITS. Whether or not they include any automatic features, special-purpose units may be built to meet two different objectives. If the testing job to be done is, for example, the inspection of crankshafts in a mass production operation, the unit is designed to handle crankshafts and nothing else. This is a *single-purpose* special unit. Other examples of this type of application are the testing of bearing balls, rollers and races; jet engine compressor blades; missile motor cases; railroad car axles; steel billets and blooms, etc.

In these cases, the nature of the equipment design is such that, to handle properly the parts for which it is intended makes the de-

Fig. 160—Crankshaft Testing Unit, Special Throughout.

sign *unsuitable* for other parts. The reason for unsuitability may be the processing technique or the method of part handling, or both.

A special processing technique may be developed for a specific application, resulting in a single-purpose special unit. However, if a wide number of essentially similar parts can be tested by this same technique, the special unit may be designed to handle the variety of parts involved. This, then, is a *general-purpose* special unit. An example is the equipment designed and built to test large castings by the overall method. (See Section 17, this Chapter.) The units are special, but they will test large castings of many sizes and shapes. Other examples of general-purpose special applications are the testing of forgings, small castings and miscellaneous machined parts such as might be produced in a job shop type of operation. For these applications the special unit design is usually dictated by the part-handling requirements—size, weight or production rate—rather than by processing technique.

When a special unit is indicated for a given testing job, there is usually little question as to whether it will be single- or general-

Fig. 161—Single-Purpose Unit for Testing Bearing Rollers.

purpose in design. The testing need itself dictates the choice. The need also indicates where the emphasis should be placed in the de-sign of the unit. The end to be achieved may be one or more of the following:

(a) Insurance of adequate controlled sensitivity.

(b) Uniformity of test application.

(c) Test production output, to keep pace with the production rate for the part.

(d) Avoidance of potential damage to parts during circular magnetization.

(e) Minimizing reliance on the human factor.

The final design of such a unit may well require a compromise among some of these objectives if more than one is involved. How-ever, such compromise must not go so far as to interfere with the satisfactory performance of the unit in fulfilling its primary reason for existence. Any special unit design *must be technically correct*

Fig. 162—General-Purpose Unit for Testing Automotive Steering Parts.

for the given application, and any design compromise must not violate this requirement. Figure 161 is an example of a single-purpose unit for testing critical rollers for bearings. Figure 162 shows a general-purpose type unit for testing of a variety of automotive steering parts.

6. AUTOMATIC EQUIPMENT. Use of automatic equipment is usually called for as a means of increasing the output of the testing operation. This end can be accomplished:

(a) By making the process cycle automatic. This relieves the operator of most duties and shortens processing time.

(b) By developing special processing techniques which may reduce processing and inspection time.

(c) By designing special means for handling the size, volume, weight and shape of the parts to be tested. In some applications automatic handling (as in the case of very heavy parts such as steel billets) is essential to permit effective and economically practical inspection of the part at all.

Some of the devices for increasing the speed of testing which are used in various operations are the following:

(a) Handling the part on a conveyor through one or more of the inspection steps involved in the entire "floor to floor" testing cycle.

(b) Automatic rotation or manipulation of the part for viewing by the inspector.

(c) Multidirectional (Duovec®) magnetization of the part to reduce the number of processing and inspection steps.

(d) Use of special fixtures to simplify the task of positioning parts accurately for the required magnetizing operation.

(e) Automatic segregation of the inspected parts, upon signal by the inspector, into such categories as "good", "repair" or "scrap".

7. ADVANTAGES OF AUTOMATIC EQUIPMENT. As has just been said, automatic equipment achieves a primary objective by increasing the per-man-hour output of the testing operation. However, there are other advantages which the use of automatic equipment accomplishes, which in some instances may be the major reason for the application.

The appearance of the indication of a given discontinuity in a part is affected by the following factors:

(a) Strength of the leakage field.

(b) Duration of the existence of the leakage field.

(c) The number of magnetic particles available in the near vicinity of the leakage field, and their freedom to move and form an indication.

(d) Sequencing of the magnetizing current shot and the application of the magnetic particles.

If the inspector is to judge the severity of a given discontinuity, he must do so on the basis of the appearance of the indication. Therefore it is important that the processing of all parts be alike and that no variation in the appearance of the indications be due to variations in processing, especially in respect to factors b, c and d, above.

Automatic equipment insures that the indications truly reflect the severity of the discontinuity in the part, in that factors b, c, and d are automatically maintained uniform, and are not dependent on the human factor.

One reason for favoring the residual wet method in automatic units on such materials which have sufficient retentivity, is that the two factors, b and c, are more easily controlled and d is not a factor at all. If the short-shot continuous method must be used because of the magnetic characteristics of the part, great care must be exercised in the design of the control of the current-bath sequence.

8. STEPS LEADING TO THE DESIGN OF AN AUTOMATIC OR SPECIAL-PURPOSE UNIT. The automatic or special-purpose unit *must* be designed to be technically correct in all aspects of the processing and handling of the parts to be tested. This is assumed by the user, and must be a first consideration of the designer. Therefore, in undertaking the design of a special unit to solve a given inspection problem, all the factors affecting the projected design must be carefully examined and analyzed in advance. An orderly progression through these factors and finding satisfactory solutions for each one in turn, can result in a special unit design which will produce the results called for.

9. THE FIRST STEP, ANALYSIS OF THE PROBLEM. The first step is a broad examination of the problem, to determine the overall conditions and requirements to be met and achieved. To do this it is necessary to:

(a) Understand thoroughly the purpose of the proposed testing program.

(b) Define the part or parts involved in terms of:
(1) Size.
(2) Shape.
(3) Weight.
(4) Surface finish.
(5) Magnetic characteristics.
(6) Heat treatment.
(7) Operations prior to and following the proposed testing.

(c) Define the service requirements for the part.

344

(d) Define locations and directions of suspected defects.

(e) Define the production requirements—that is, the required rate of testing—for the special unit, anticipating some downtime for the unit and some percentage of rejected parts.

With this information the designer is ready to proceed to a consideration of actual details of design.

10. THE SECOND STEP. CONSIDERATION OF METHOD FACTORS. As a second step the choices among the various method variations must be considered, and decisions made as to how the test is to be performed. These considerations can be broken down into separate items and examined separately.

A. *Method of Application of the Magnetic Particles.* This is a first consideration, and the choice lies between the wet and the dry methods.

The wet method is used almost exclusively in special units for numerous reasons. Some of these are:

(1) Ease of application.

(2) Speed of application.

(3) Assurance of complete coverage of the surface of the part.

(4) Greater sensitivity for most sought-for *surface* defects.

(5) Easier sequencing of particle application in an automatic processing cycle.

(6) Easier re-capture and return of excess particles to the circulating system for re-use.

All these considerations have to do with method reliability, and the economics of the process.

B. *Water vs Petroleum-Base Vehicle for the Wet Method Baths.* One of the reasons that water was first proposed as a medium for the wet method bath was to eliminate the fire-hazard assumed to be inherent with petroleum-base liquids. While the incidence of fire in equipment using petroleum liquids has been very low over the years, this *is* one of the considerations in making a choice.

For example, the inspection of billets and blooms is an area in which the fire hazard is very real if petroleum-based baths are used. In many cases bloom and billet ends are covered with a heavy non-conductive scale. Special magnetizing circuitry has been devised to burn through this scale to make contact, by deliberate arcing at the contact points. This arcing could easily ignite a flammable bath, and because of this water is used exclusively in these units.

But there are other reasons for the use of water as a bath, which involve the economics of the process, especially where large volumes of bath are required, as in billet testing. In a typical installation of this sort, 50 to 80 gallons of bath must be added per eight hour day to replace that lost by "carry-off". Water is essentially "free", whereas an oil medium may cost 40¢ per gallon.

On the other side of the economic ledger, however, is the fact that water baths, in a recovery system, tend to foam. This foaming is caused by additives to improve the ability of the bath to wet completely the surfaces of the part to be tested. In some cases it is necessary to use other additions to control this foaming tendency. These additions increase the cost of particle concentrates for use with water. However, costs of the particle concentrates for water and oil are comparable, so there is still the advantage of the difference between the cost of oil and water.

At first thought, the expendable technique would appear to be wasteful and costly. However, for certain applications it can well be a less costly technique when the bath medium is water. For example, when applied to the testing of large castings using the overall method, its use obviates the need for a large bath reservoir and a filtering and recirculating recovery system. Water also presents less of a disposal problem than does a petroleum based bath.

C. *Color of Particles Used.* The choice here is based on the particular color of magnetic particle which provides the best average contrast with the surface of the parts. Because of their greater visibility, fluorescent magnetic particles are far more widely used in special units than are the visible colors, wet or dry. Indications delineated by fluorescent particles are much more clearly and quickly seen than are those indicated by daylight visible particles. However, there are some few cases where daylight-visible particles offer an advantage. This is particularly the case when the sought-for defects give poorly defined or diffuse fields. This would occur at subsurface defects or at

346

fairly shallow and open surface defects. In such cases daylight examination not only provides better observation of the indication, but also permits a more thorough evaluation of the defect than can be made in a darkened room.

D. *Type of Magnetizing Current.* The choice of currents usually lies among four types—alternating current, rectified 3 phase A.C. direct current, half wave rectified single phase A.C., and full wave rectified single phase A.C.

Since most automatic and special-purpose equipment involves the detection of surface defects, any of the four types of current are suitable. The choice may therefore be based on considerations like the following:

(a) Consistency with the type of magnetizing current used in prior or subsequent testing of the same part.

(b) Consistency with the type of magnetizing current used in testing similar parts.

(c) Standardization with the type of magnetizing current preferred in a particular company or plant.

(d) Limiting the current drawn by the magnetizing circuit from the power lines. A.C. and half wave D.C. operate from a single phase A.C. line input. These draw more current per leg and tend to unbalance the shop distribution system. D.C. from rectified 3 phase A.C. does not present this difficulty.

(e) D.C. is easier to control for conveyorized coil-magnetization of parts. The timing of magnetizing current cut-off is less critical with D.C. than with A.C. due to the demagnetizing effect of the latter.

(f) When a "fast break" is required, full wave rectified A.C. is the most effective type of circuit to use.

In the case of multi-directional magnetization, different types of magnetizing current may be used in the different circuits, for the same part. A.C. or half wave D.C. for the head shot is often combined with single phase *full* wave rectified D.C. in the coil shot. The reason for this is the critical requirement for "fast break" for the coil shot is better achieved with D.C.

347

E. Direction of Magnetizing Current. This choice is determined by the most likely direction of the sought-for defects. There are three choices:

(a) Circular magnetization.

(b) Longitudinal magnetization.

(c) Multi-directional magnetization.

If both circular and longitudinal magnetization are called for, the multidirectional system should be considered and may often be applicable. If usable, it will shorten total inspection time. It often may not be useable however, because such factors as shape, surface finish and magnetic characteristics, individually or in combination, may rule it out.

The manner of applying the magnetic fields selected is also subject to numerous considerations dictated by sensitivity requirements and part characteristics.

F. Continuous vs. Residual Methods. Since the residual method application is easier to control, it is preferred where the test sensitivity requirements and the part characteristics permit. There are some cases where a combination of continuous and residual methods is employed. This involves the continuous application of magnetic particles before, during and after exposure of the part to the magnetizing force. Such a combination favors simplicity in the design of equipment. The sensitivity level of this technique is likely somewhere between that produced by the continuous and residual methods separately. However, its effectiveness in any particular case should be carefully established, as it may be affected by such factors as surface finish, manner of applying the bath and the magnetic properties of the part.

11. THIRD STEP. FINAL DESIGN SPECIFICATION. Having reached decisions with respect to each of the above outlined considerations, the designer is ready to make up his final design specification. It frequently happens, however, that some of the optimum desirable features are in conflict and compromises must be made.

In making such compromises care must be taken that they are not made in the area of technical accuracy or in insuring that the prime objective of the special unit is accomplished. For example, where the ultimate in test sensitivity is required, as for aircraft parts,

348

the unit design must be first one that provides the most effective and sensitive technique for the defects sought. It must also insure that this technique is properly applied. Rate of output of the unit would in this case be of secondary importance. Alternatively, if the application is an in-process testing to up-grade the quality of parts later to be subjected to final inspection, a compromise in sensitivity may be permissible in favor of maintaining high production output.

12. EXAMPLES OF SPECIAL UNIT APPLICATIONS. From the preceding discussion it might be assumed that special equipment is primarily designed for testing parts that might be easily handled manually, in order to insure better quality or increase production. It is no doubt true that the majority of parts tested on special equipment do fall into this category. However, there are many parts outside this group that involve such large sizes, heavy weights, complex configurations or unique test requirements, that they could not be tested at all without special equipment.

In the remaining sections of this chapter we will describe five cases in which the use of special units was the only way to apply magnetic particle testing on any practical or economically feasible basis.

13. TESTING BEARING BALLS AND RACES. For many years hardened steel bearing balls resisted all efforts to test them with magnetic particles to locate cracks and other metallic flaws. Devising of a special magnetizing technique, and the subsequent design of an automatic unit for applying it, finally solved the problem.

The usual two-step head shot and coil shot technique did not satisfactorily disclose all defects in whatever direction, for two reasons:

(a) A ball's L/D ratio is such that a coil shot is ineffective.

(b) Without marking each ball there is no way to orient the ball properly for the three head shots that would be necessary to disclose all defects.

A method was finally worked out wherein the ball was treated as a cube, and separately processed and inspected, using circular magnetization, through its X, Y and Z axes, respectively. Controlled rotation of the ball between each of the three processing and inspection positions assured proper orientation and therefore a re-

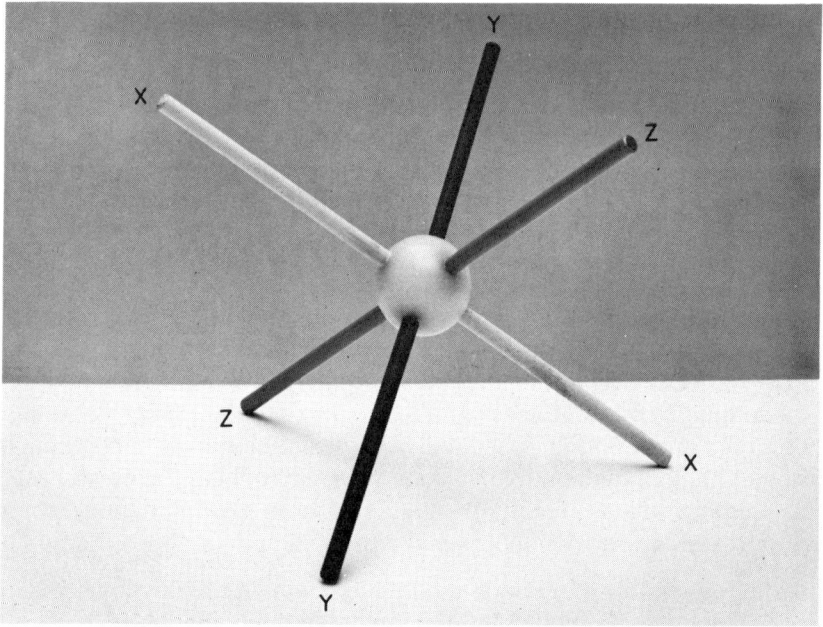

Fig. 163—Model Showing the X, Y and Z axes of a Bearing Ball.

liable overall test. Subsequent designs permitted magnetization without any electrical contact with the ball (using a variation of the induced current method) thus eliminating any possible danger of damage to the ball due to arcing at the points of magnetizing current contact. Figure 164 illustrates this unit.

Another interesting example is the testing of bearing races. Although this ring-shaped part had long been tested on standard equipment, it was a slow process, three inspections being required. Also, the need to make contact on the hardened, highly polished surface of the race, resulted in a certain amount of burning, which in some instances ruined the race.

Defects lying on any of the surfaces in a direction transverse to the circumference of the ring can be easily located by circular magnetization using a central conductor. Circumferential defects, parallel to the ball path, are a different matter entirely. Field in the proper direction can be induced by a head shot across the diameter of the ring. But in the area of the contact points the current splits and follows around to the other contact in two portions. At the points of separation of the current the field is distorted so that

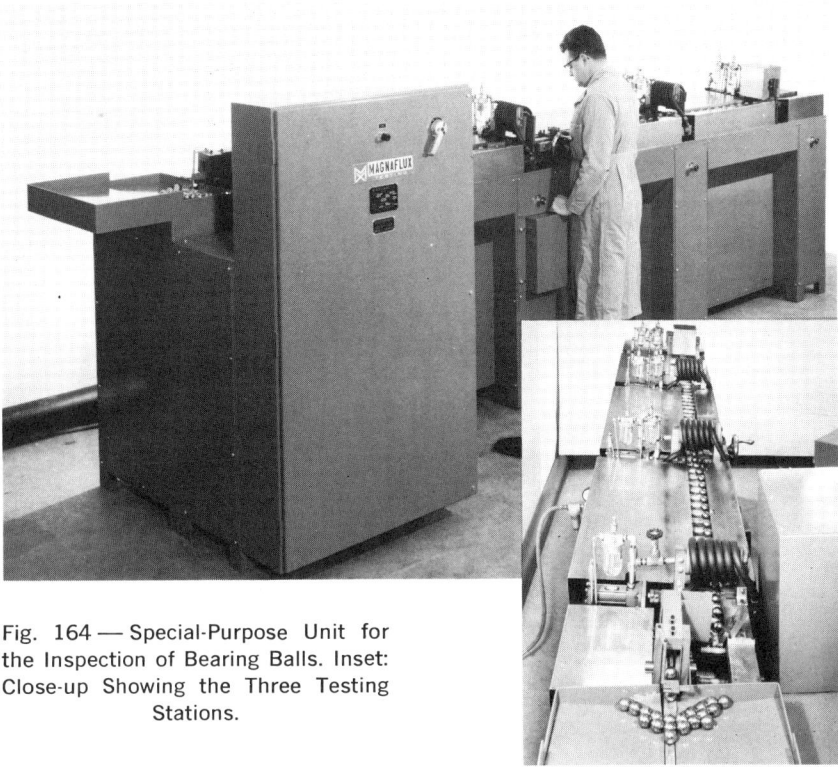

Fig. 164 — Special-Purpose Unit for the Inspection of Bearing Balls. Inset: Close-up Showing the Three Testing Stations.

reliable inspection of those areas cannot be had. Consequently the ring must be turned 90° and given another shot, and reinspected. In addition to the time consumed this method has two major objections:

(a) The possibility of damage due to arcing or contact resistance heating at points of contact.

(b) The possibility of distortion of the race due to excessive clamping pressure.

Figure 165 is a diagram of the current and field in a bearing race when being magnetized circularly by giving it a head shot.

The contact damage problem was solved by the use of the induced current method using an automatic special-purpose unit. Figure 166 shows the current and field distribution when the race is magnetized by the induced current method. This technique provides a complete and reliable test without making contact on the race, and

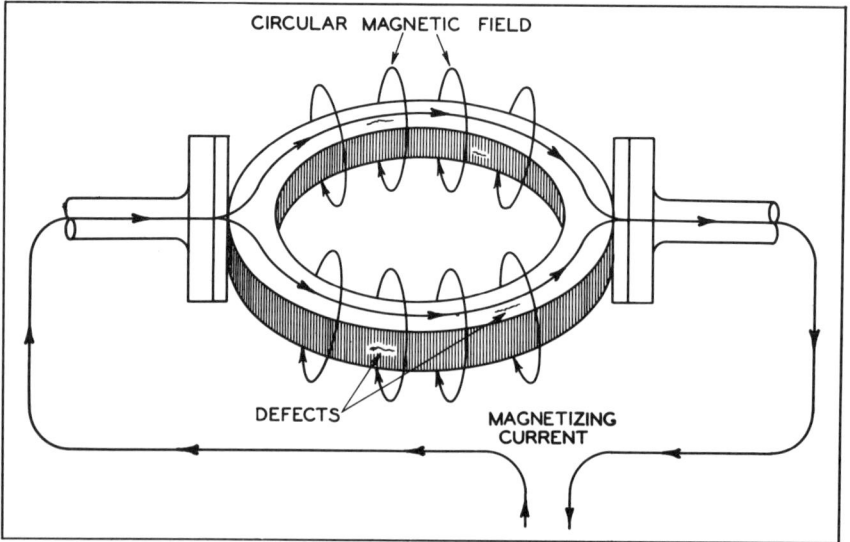

Fig. 165—Field and Current Distribution in a Bearing Race
Being Magnetized with a Head Shot.

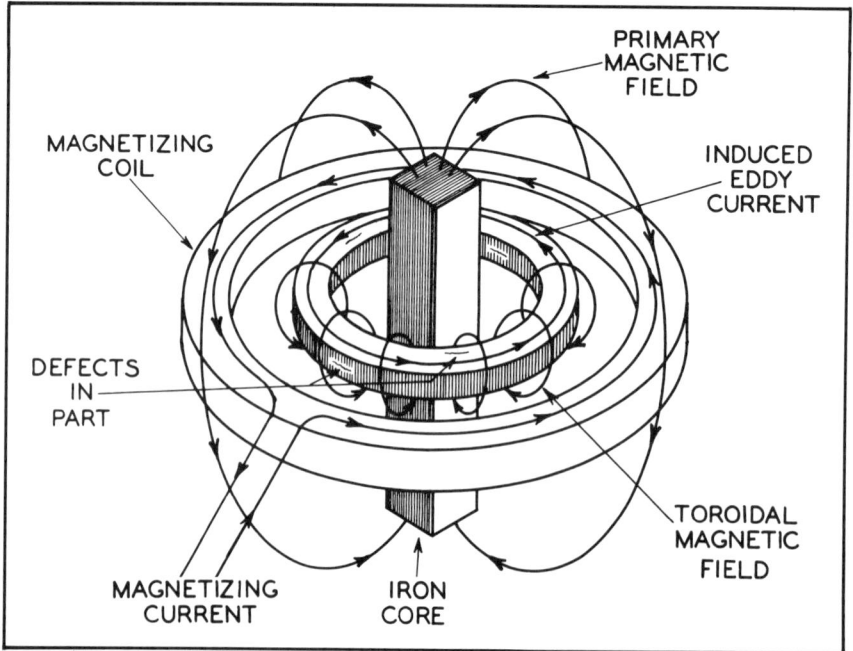

Fig. 166—Current and Field Distribution in a Bearing Race
Being Magnetized by the Induced Current Method.

Fig. 167—Special-Purpose Unit for the Inspection of Bearing Races by the Central Conductor and the Induced Current Method.

requires only a single processing. Figure 167 shows this unit. It provides for magnetization by central conductor, inspection, magnetization by induced current, inspection and, finally, demagnetization. A similar technique has also been applied to certain critical jet engine parts, both during manufacture and at overhaul.

14. BILLET TESTING. Billet conditioning in the past has been conducted by visual examination (which usually required pickling) for seams, which were then chipped, ground or flame scarfed out. The system was far from satisfactory, because the inspectors missed many seams entirely, and removed much good metal where no seams or only very shallow seams existed. Scale removal, by pickling or grit blasting, improves this situation somewhat, but is a costly procedure. The economics of billet testing and conditioning is a complex subject, but it is easy to appreciate the advantage of eliminating the pickling and of being able to remove all those seams, and only those seams, which would be detrimental to the finished product.

Magnetic particle testing systems have, in the past ten years, been developed which accurately locate significant seams and indicate their approximate depth. Properly designed equipment, circular magnetization, the continuous method and water-based fluorescent magnetic particles provide the means for doing this. Billets may be classified without the need for scale removal, and segregated, usually into three groups according to their degree of seaminess. These groups are usually selected within the following ranges:

(a) *Prime*. No seams deeper than a pre-determined level, usually 0.020 to 0.030 inch. These billets are shipped or further processed "as is". The very shallow seams will "scale out" or "wash out" on reheating.

(b) *Condition*. Seams deeper than the above level, but shallower than a level established as too deep for economical removal, call for surface conditioning. Seams of this group may be from 0.050 to 0.070 in maximum depth.

(c) *Down-grade*. Billets with seams deeper than 0.070 are diverted and used for lower grade products, or scrapped.

15. SEAM DEPTH DISCRIMINATION. Billet testing with magnetic particles requires that the inspectors judge the seam depth so as to mark the billets properly and classify them as to quality. This is made possible by controling the entire processing cycle, so that the fluorescent brilliance of the seam indication reflects its approximate depth.

With equipment properly designed, operated and maintained, it is easily possible to discriminate with a high degree of accuracy between the indications formed by seams varying in depth by approximately a two to one ratio. This does not mean that magnetic particle testing can be employed for seam depth *determination*. In testing billets we are looking at indications of seams only on the basis of *relative* depth. For example, are they shallower than 0.025 inch; over 0.025 inch but under 0.070 inch; or over 0.070 inch? We do not care what the *actual* seam depth is—we care only about which of the three classes it falls into. Extreme accuracy at the dividing line between the classes is not required.

Control of the bath concentration and the magnetizing current level can weaken indications of shallower seams so that they are scarcely seen. The real determination to be made, therefore, is whether a seam is under or over the 0.070 inch level used in the example above. The method does result in classification of billets

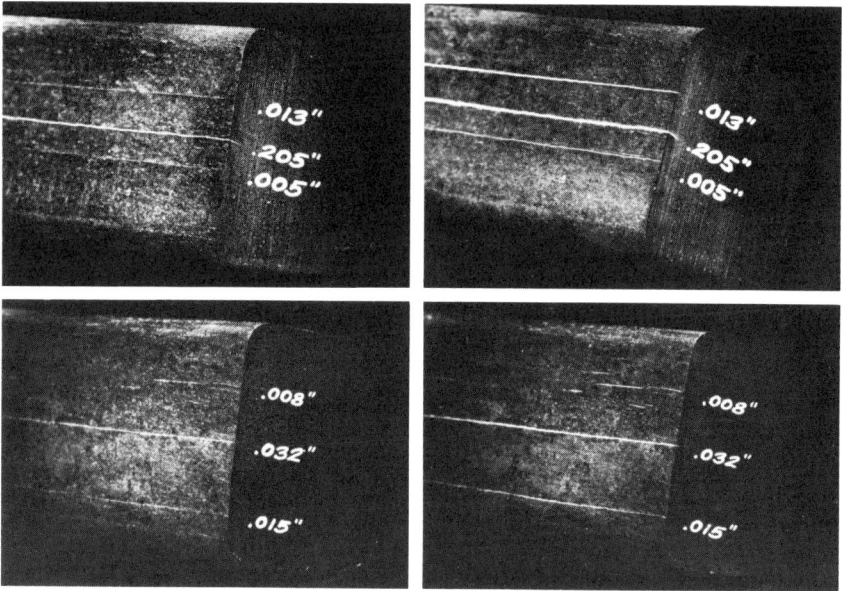

Fig. 168—Fluorescent Magnetic Particle Indications of Seams of Various Depths, Shown by the Residual and the Continuous Methods.

by a satisfactory separation of the billets into the three levels of quality.

Seam depth discrimination is made feasible by the following:

(a) Specially developed magnetic particles, deliberately insensitive to weak leakage fields at very fine seams.

(b) Specially developed low foaming wetting agent, permitting practical use of water as the suspending medium, eliminating the problem of excessive foaming of the bath.

(c) Specially designed electrical magnetizing circuitry to provide a high enough voltage to arc through the scale on the billet ends, then quickly drop back to that required for the desired pre-set magnetizing-amperage level.

This arcing device is made necessary because three phase full wave D.C. is used for billet testing, often at current densities as low as 200 amperes per inch of cross-section. Thus the magnetizing voltage required to produce this current is frequently too low to break through the mill scale on hot-sawed or hot-sheared billet ends in order to make contact.

(d) Precise sequencing of bath application cut-off and magnetizing current cut-off. Both are "on" for long time periods—several seconds—compared to the fractional seconds usually associated with wet method particle testing.

(e) Effective separation of mill scale from the bath suspension, to preserve bath quality.

(f) Suitable black light type and placement, to provide adequate black light intensity on the billet surface, without restricting the inspector's vision or freedom to mark seams.

In short, the method is de-sensitized, and the sensitivity level is then controlled along with all the other method variables, so that reliability and repeatability of results are assured.

16. DESIGN CONSIDERATIONS, OTHER THAN METHOD FACTORS, FOR BILLET TESTING. In addition to the testing method, design considerations involving billet handling and mill operating conditions, create problems not encountered with the average special unit. For example:

(a) Ambient temperatures in the billet testing and conditioning areas may range from $-20°$ to $120°$ F. This calls for special attention to operator and inspector comfort (and therefore effectiveness) and avoiding of freezing of the water bath recirculating system, and of the bath on the surface of the billet during testing.

(b) The design must anticipate the billets being bowed or twisted, for they often are.

(c) Billet transfer from station to station must be rapid and positive, yet cushioned to avoid excessive punishment to the equipment.

(d) Delivery of a processed billet to, and discharge from the viewing station must be accomplished rapidly so as to allow maximum viewing time with maximum safety of the inspectors.

(e) Since two or more inspectors view and judge parts of the same billet, each may judge his section to be better or worse than the sections of the other. Therefore, electrical circuitry is incorporated to permit each inspector to judge the billet as he sees *his* section. Classification of the billet is then determined automatically based upon the poorest grade any one inspector chooses.

(f) Billet transfer between processing and handling during the viewing cycle must be such that indications are not smeared, nor false indications transferred onto the billet's surface.

(Courtesy Youngstown Sheet & Tube Co.)
Fig. 169—A Typical Large Billet Testing Unit. Inset: Close-up of Inspection Station Showing Billet on Chain Sling Billet Turner.

17. INSPECTION OF LARGE CASTINGS. The large castings which justify the use of special equipment are those intended for critical service conditions, and therefore require exceptionally thorough testing. They may weigh from a few hundred pounds up to several tons. Conventionally such castings, too large to be handled on a standard unit, are tested using prods and expendable dry powder. Although this technique can be very effective there are several drawbacks to its use in production. Some of these are:

(a) Great care must be taken that prod contacts are made in a regular pattern to make sure that the entire area to be inspected has been properly covered.

(b) It is time-consuming.

(c) Some casting designs do not permit proper prod magnetization in critical areas due to difficulty in making contact at the right points because of the shape of the casting.

(d) Arc burns from prod contacts are sometimes damaging.

(e) Usually prod inspection is a two man operation.

The special equipment most frequently used for large casting inspection is a unit composed of a high amperage, high duty cycle, 3-phase full wave D.C. multi-directional power pack; a system for spray-gun application of expendable fluorescent magnetic particles; and the necessary cables, contact clamps and black lights. The entire casting may be magnetized at one time. This method is called the "overall" magnetizing method.

Fig. 170—Heavy Duty Equipment for Applying the Overall Magnetization Method to a Complex Casting.

The magnetizing current is switched rapidly and sequentially into two or three separate circuits. Heavy cables are wrapped around, threaded through or clamped to the casting and a separate set of coils or contacts is used for each high amperage circuit.

Fluorescent magnetic particle bath is used, and is flowed over the casting while current is "on," covering all areas to be inspected. This technique overcomes, at least in large measure, the five drawbacks to the prod and dry powder technique listed above. The advantages achieved through the use of the overall method are so obvious that it is easy to see the justification for the cost of the installation. Processing and inspection times have often been reduced to one quarter or less of those required for the prod magnetization and powder method. In one application the time was cut to 10%.

The overall method can also be applied through the use of special magnetizing fixtures or handling devices. Where the inspection operation is such that the range of casting sizes and shapes is not too great, a special magnetizing unit and handling fixture using the overall method is often employed. Figures 170 and 171 each illustrate such an installation.

Fig. 171—Overall System Applied to a Miscellaneous Assortment of Castings.

18. INSPECTION OF WELDED STEEL MISSILE MOTOR CASES. Missile motor case inspection poses a unique challenge to the design engineer, because often it is necessary to devise unconventional magnetizing methods to secure the necessary sensitivity for this most critical inspection. This is an excellent example to illustrate the great adaptability of magnetic particle testing. In reading this book the impression may have been given that hard and fast rules govern the proper use of the method, but this is not necessarily the case, as demonstrated by the solution of the missile motor case problem.

The highest attainable strength-to-weight ratios are sought for all the materials and components used in the design and construction of missiles. Designers are forced to use lower safety factors than they would for more ordinary products. To achieve this result they rely mainly on three things:

(a) Use of the best and most advanced materials for their purpose.

(b) Closely controlled manufacturing methods.

(c) Use of all available methods of nondestructive testing. Magnetic particle testing is only one of the methods involved.

Motor cases are pressure vessels, subject to extremely high pressures when fired. They are also as thin-walled as a reasonably minimum safety factor will permit. Therefore the most minute defect is cause for concern. And since they are stressed both longitudinally and circumferentially, defects in any direction are significant.

Some of the major design considerations for equipment to be used for testing missile motor cases are the following:

(a) Fluorescent magnetic particles should be used, for maximum sensitivity to surface cracks.

(b) 100% surface inspection should be provided for—both inside and outside—for defects which may occur in any direction.

(c) Magnetization should be accomplished without current contacts being made directly to the surface of the motor case. Arc burns from electrical contacts would themselves constitute defects, and cannot be tolerated.

(d) Automatic control of the entire processing cycle is desirable to insure reliability and reproduceability of test results.

(e) Handling and processing equipment design should be such that it insures against any form of damage to the motor case during the entire testing process.

A large number of units have been built for motor case testing, including some for the Titan III. These sections were 10 feet in diameter and up to 11 feet in length. Standard central conductor and coil magnetizing techniques can sometimes be used for the smaller diameter cases of simple design, but for the larger sizes and more complex configurations special techniques must be employed.

Ordinary coil and central conductor techniques are not practicable for several reasons:

(a) The motor case dimensions are such that unreasonably high magnetizing currents would be required.

(b) Its configuration is such that vital areas would be shielded from the magnetizing force, or the magnetic field would be distorted, resulting in a loss of reliability, to an unknown degree, for the location of *all* defects. Field direction and distribution in a complex object such as a missile motor case is impossible to predict with certainty. (See Chapter 10.)

(c) Longitudinal magnetization with a coil, again because of the size and shape of the motor case, would be ineffective. In addition it would impose almost insurmountable problems in the design of practical handling equipment.

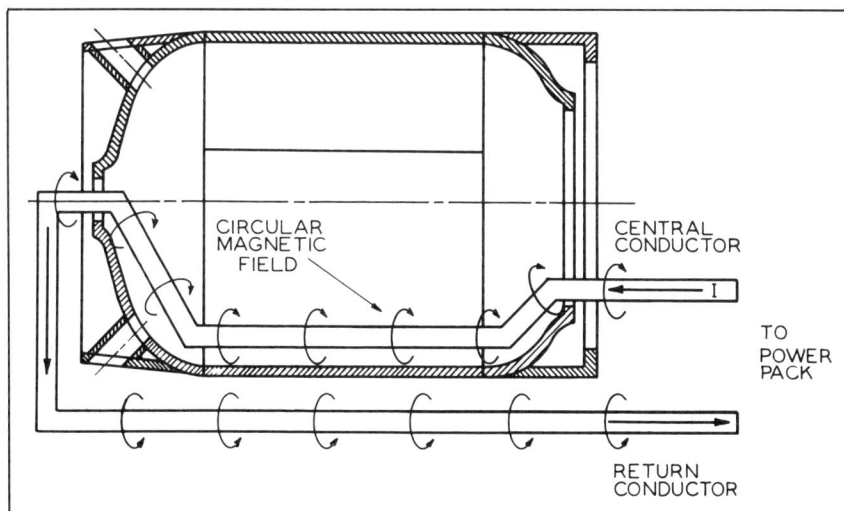

Fig. 172—Schematic Drawing of Conductor Arrangement for Circular Magnetization of Cylindrical Missile Motor Cases.

Figure 172 illustrates, schematically, a Polaris motor case and the means used to provide effective circular magnetization. This "central conductor", which is actually a shaped single turn coil, closely conforms to the inside profile of the case. Due to its close positioning to the wall of the case, much stronger fields are generated in its immediate vicinity than could be secured at reasonable current levels by a truly central conductor. Furthermore, the return conductor, also closely conforming to the outside surface, reinforces the magnetic field. Complete circular magnetization of the entire cylinder is accomplished by rotating the case one full revolution while 3-phase, full wave D.C. is flowing in the conductors.

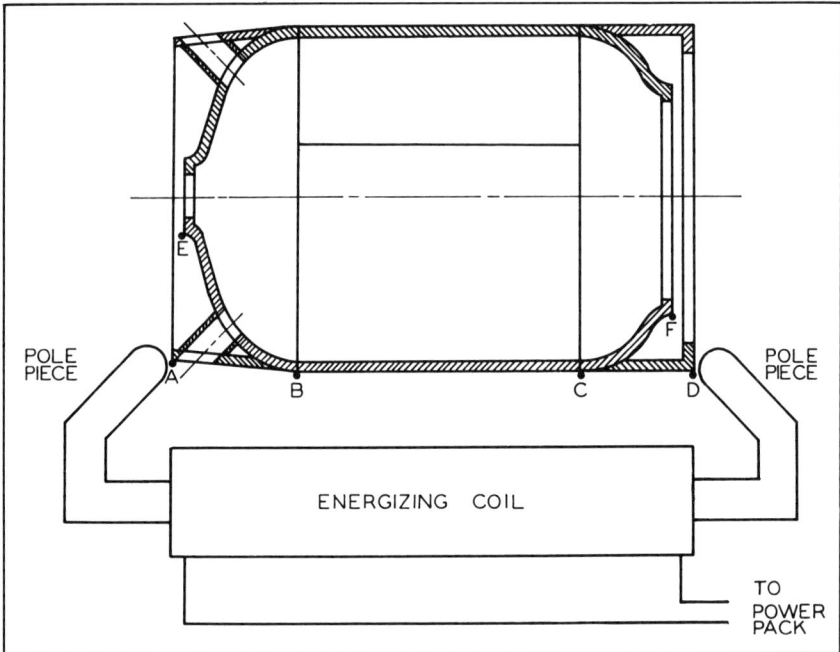

Fig. 173—Schematic Drawing of Conductor Arrangement for Longitudinal Magnetization of Cylindrical Missile Motor Cases. (See Fig. 209, Chapter 23.)

Figure 173 illustrates schematically the same motor case with a yoke-like induced field fixture for providing the longitudinal magnetic field. The longitudinal field is induced in the A-B-C-D sections (see drawing) and complete magnetization is accomplished by rotating the case a full revolution while 3-phase full wave D.C. is flowing in the fixture. Repositioning the pole pieces to points E and F and repeating the process provides effective magnetization in the E-B and C-F sections. Before this technique can be applied to any specific motor case, proof of its effectiveness must be obtained by experimental means. If the case were rolled out flat with the magnetizing fixture in place, it can readily be seen that the flux lines do not follow a straight line from pole to pole of the yoke. They spread out in an eliptical pattern (see Fig. 72, Chapter 7) resulting in a weakened field as it progresses from the poles to the midpoint between them. If the case in question were longer, the entire length should be magnetized in separate stages through sections A-B, B-C and C-D, etc., rather than singly from A to D.

Bath application is critical for both circular and longitudinal magnetizations. It must be applied in the area close to the magnetizing sources, where the field is strongest. The flow of bath must cover the surface thoroughly, but must be gentle, without much velocity, so that patterns are not washed away by the flow of the bath.

Port areas, such as thrust-reverser stacks, call for additional special consideration. Circumferential defects in the weld zones, as well as in the parent metal itself, must be looked for. The previously applied circular and longitudinal magnetizations are not entirely reliable in these areas because of field distortion around the port openings.

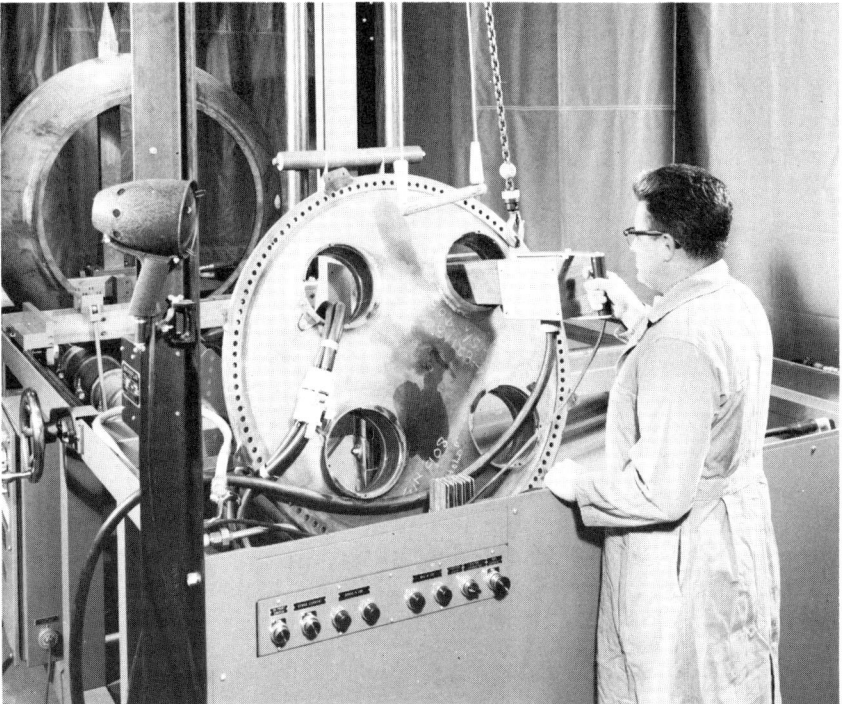

Fig. 174—Induced Current Fixture for Missile Motor Case Ports.

Figure 174 shows how this phase of the problem was handled with a semi-portable A.C. induced-current fixture. The core of the fixture is inserted into the opening of the port and the current turned on to produce a toroidal magnetic field around the walls of the cylindrical section. While the current is on, the bath of magnetic particles is applied from a hand hose. After turning off the magnetizing current (to avoid demagnetizing action by the A.C. if the fixture is withdrawn while the current is flowing) the fixture is withdrawn and the inner and outer surfaces examined for defects.

19. MULTIPLE TEST SYSTEMS. From the foregoing discussions it should not be concluded that special magnetic particle testing equipment is any kind of panacea, or that magnetic particle testing is the single nondestructive test for any and all defects associated with ferrous parts. This is certainly not the case. Although magnetic particle testing methods hold an outstanding place in the testing of parts made from magnetic materials, radiographic methods, eddy-currents and ultrasonics also have a definite place in this same area. There are numerous instances where two or more tests are applied to the same part. For example:

(a) Magnetic particles and radiography for welded pipe, steel castings and forgings, and missile motor cases.

(b) Magnetic particles and ultrasonics for welded pipe, steel castings, general weld inspection, forgings, and billets and blooms.

(c) Magnetic particles and eddy currents for welded pipe and pipe couplings.

These multiple test applications have not usually been conducted simultaneously or by use of a single station of equipment. They were applied separately with totally separated facilities. This is not surprising, since the introduction of the various nondestructive testing methods have followed each other over a long period of years. Manufacturers "made do" with available test methods until other methods were developed and offered. In recent years, however, planned multi-method nondestructive testing installations have been placed into service for such parts as:

(a) Welded pipe—magnetic particles plus radiography.

(b) Welded pipe—magnetic particles plus eddy currents.

(c) Welded pipe—eddy currents plus ultrasonics.

(d) Billets and blooms—magnetic particles plus ultrasonics.

The trend toward multi-method nondestructive testing systems will undoubtedly continue, because there are very definite economic as well as technical advantages to be gained by combining more than one test into a single handling and testing installation.

20. FUTURE TRENDS. In the course of this chapter we have discussed special equipment—automatic and special-purpose, single-purpose and multi-purpose—for testing parts varying as widely as from one quarter inch bearing balls to ten foot diameter bearing races; 30 foot long gun tubes, ten foot diameter missile motor cases, five ton blooms and 30 ton steel castings. The trend toward special equipment of all types and sizes has continued at an accelerated rate during recent years. The reasons for this trend are numerous. The wide demand for higher quality by users of materials, brought about by modern design trends, along with the demand of manufacturers for increased production rates, has forced the move toward special equipment. The above demands are, of course, the result of technological advances in materials and fabricating techniques spurred by the need for higher strength-to-weight ratios.

On the side of the nondestructive testing industry, a large factor in the increasing use of special equipment is the availability of improved special unit designs in such areas as mechanical handling, magnetizing systems, wet method bath application methods and electrical control circuits. A further aid to the design of special applications is the development of greatly improved magnetic particles, to make the results obtainable from the use of special methods much more effective.

Finally, as always, the demonstration by successful installations of the advantages of such systems causes more and more users to be attracted to them. There seems to be little doubt that the trend toward special equipment and methods will continue, since the underlying reasons for it are not likely to lessen in number or value.

CHAPTER 20

DETECTABLE DEFECTS

1. THE MAGNETIC PARTICLE TESTING FUNCTION. A great deal has been said in the preceding chapters about defects and discontinuities, and how they may be detected with the magnetic particle testing method. It has been emphasized that all discontinuities are not defects, and that a given discontinuity may or may not be a defect, depending on its location with respect to operating stresses in the part, and whether or not it will interfere with the performance of the part in its intended service.

It is the function of magnetic particle testing to show the presence of discontinuities, but it is not part of its function to determine whether or not such discontinuities constitute defects. But since it cannot be known in advance how serious discontinuities may be, or whether they will be harmful or not, the function of the method is to find *all* discontinuities which its proper application is capable of indicating. How harmful to the service life of the part such discontinuities may be is a matter to be decided by human judgement, by those who know the service requirements of the part, and can appraise the magnitude of the effect on these requirements which a given discontinuity may have. The problems of interpretation and evaluation of indications and discontinuities will be discussed in a following chapter (22).

The usefulness of magnetic particle testing in the search for defects, and the controlling factor in the decision whether to use the method or not, depends, therefore, on exactly what discontinuities the method is *capable* of finding. In various places in the preceding text, the proper means and procedures to use in magnetizing a part so as to provide the optimum conditions for finding the greatest number of defects have been fully discussed. Rules have been given also regarding what magnetic particles to choose and how to apply them in order to give the best possible indications. The many variations of techniques have been discussed at length with respect to their effect on the defect-detecting abilities of the method. It would seem worthwhile, before ending these discussions, to look at the problem from the other side of the fence, as it were—

from the point of view of the discontinuity itself. What are the characteristics of the *defect itself* which allow or prevent its detection by one or more of the procedures which have been worked out?

2. DEFECTS CLASSIFIED. In Chapter 3 defects were classified, and described with respect to when and how they were caused during the production, or the service life of the part. From the point of view of detectability, however, we are not concerned with the *origin* of the discontinuity. Of importance only is its size, shape, orientation and location, with respect to its ability to produce leakage fields. The only classification which is pertinent to this analysis, therefore, is that which separates discontinuities into two groups: 1) those that constitute a break in the surface of the part, and 2) those that lie entirely below the surface. It is this consideration more than any other that determines the selection of specific techniques for magnetizing and applying particles.

3. IMPORTANT CHARACTERISTICS OF DISCONTINUITIES. In considering discontinuities broadly, and their characteristics which have a bearing on their detection with magnetic particles (as well as their possible role as defects) a number of points are of importance, as listed below:

A. Discontinuities *open to the surface*. See Fig. 175.

(1) Depth, D. The distance from the surface to the farthest point of the defect, measured in a direction normal to the surface.

(2) Length, L. The longest dimension measured at the surface, in a direction parallel to the surface.

(3) Width, W. The longest dimension measured at the surface, in a direction parallel to the surface, and at 90° to length.

(4) Shape. Sharpness at the bottom—V or U.

(5) Angle of penetration with respect to the surface.

(6) Orientation of principal dimension

(a) with respect to the longitudinal axis of the part.

(b) with respect to the transverse axis of the part.

(c) with respect to the surface.

(7) Frequency. The number per unit of surface area.

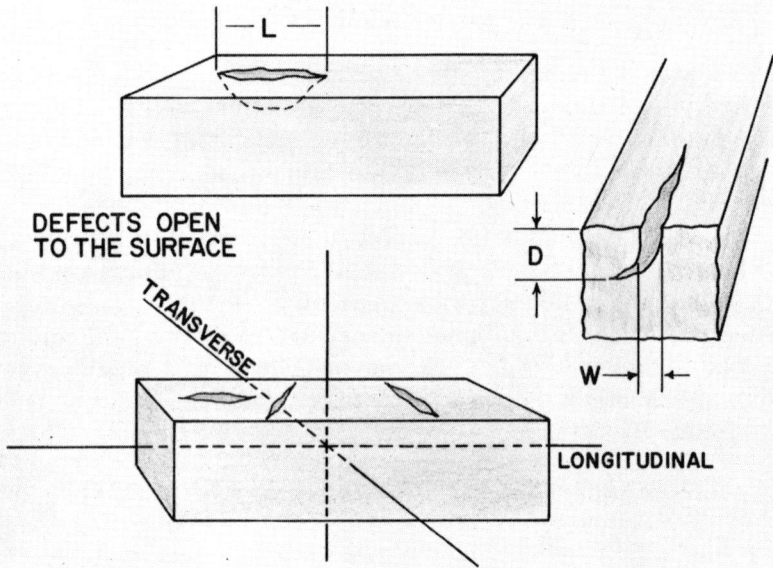

Fig. 175—Characteristics of Surface Defects.

(8) Interrelationship. Grouping, alignment, etc.

(9) Relationship of all characteristics to service stresses in the part, and to critical stress locations.

(10) Stress-raising effect from all considerations.

B. Discontinuities lying *wholly below the surface.* See Fig. 176.

(1) Length, L. Longest principal dimension, measured at the surface, and in a direction parallel to the surface.

(2) Width, W. Longest dimension, measured at the surface and in a direction parallel to the surface, and at 90° to Length.

(3) Height, H. Dimension normal to the surface.

(4) Depth, D. Distance from the surface to the nearest part of the discontinuity, measured at 90° to the surface. Note that the dimension, D, has a different meaning as between surface and sub-surface discontinuities.

(5) Shape. Globular, angular, flat, sharp-cornered, etc.

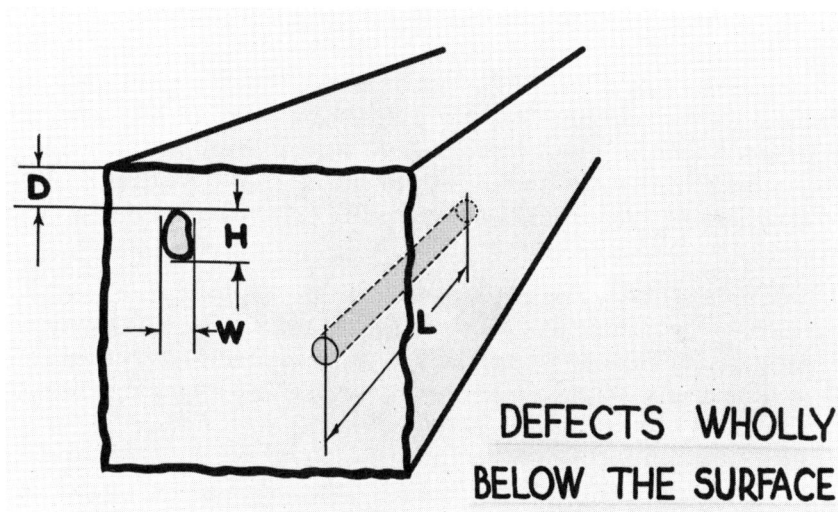

Fig. 176—Characteristics of Defects Lying Wholly Below the Surface.

(6) Orientation of planes of principal dimensions
    (a) with respect to the surface,
    (b) with respect to the longitudinal axis of the part,
    (c) with respect to the transverse axis of the part.

(7) Frequency. Number per unit area of cross-section.

(8) Interrelationship—grouping.

(9) Relationship of all characteristics to service stresses in the part and to critical stress locations.

(10) Stress-raising effect from all considerations.

Note: In the above listings, the term "surface" in all cases refers to the *surface on which the inspection is being made.*

From the standpoint of detectability, only items one through six have a bearing. The entire list, one through ten, bears on the question of the discontinuity as a potential defect.

4. SURFACE CRACKS. Surface cracks and other surface discontinuities make up by far the largest and most important group which magnetic particle testing is used to locate. This is true for two principal reasons. First, the surface crack is the type most effectively located with magnetic particles; and, second, surface cracks as a class are much more important and dangerous to the service life of a part than are defects lying wholly below the surface. They are therefore more frequently the object of the inspection. Fibre stresses are usually highest at the surface of a part, and any break in the surface constitutes a point of still higher concentration of stress. A surface crack by its nature is very sharp at the bottom, and is the most severe kind of stress-raiser. For the latter reason surface discontinuities are looked for with extreme care, if expected stresses are even moderately high.

The surface discontinuities looked for with magnetic particle testing include all fatigue and service cracks, and such serious sources of potiential failure as seams, laps, quenching and grinding cracks as well as many surface ruptures occurring in castings, forgings and weldments.

5. DETECTION OF SURFACE CRACKS. It has been stated repeatedly that the magnetic particle testing method is *the most sensitive and reliable* method for locating surface cracks in ferromagnetic material. That this is true is due largely to the fact that in the great majority of cases, no extremely critical conditions or techniques are required for the detection of surface discontinuities by this method. Magnetizing and particle application methods may be critical in certain special instances (See Chapter 19), but in the case of most applications the requirements are relatively easily met, because leakage fields tend to be strong and are highly localized. A few simple but important principles are involved which must be observed, but there is usually a fairly wide latitude in the selection of procedures and materials when surface cracks *only* are being sought. If the discontinuities are of a size or character to be in the threshold area of detectability special techniques may be necessary. It is, however, relatively easy to define the characteristics of a surface discontinuity that make it favorable for detection.

The material must, of course, be ferromagnetic, and should have a maximum material permeability of not less than 500. (See Chapter 9, Sections 9, 10 and 11.) Five hundred is the material permeability of nickel, and surface cracks are quite easily found in this metal. Cobalt and some types of Monel metal which have a material permeability of less than 500 are still sufficiently magnetic that surface cracks may be located, using high-sensitivity techniques— i.e., strong direct currents with circular magnetization, and wet magnetic particles.

For successful detection in any given case it must be possible to set up a field of sufficient strength and in a generally favorable direction to produce strong leakage fields. This is especially true if the discontinuities are small and fine. Assuming that this has been done, the most favorable characteristics of a discontinuity itself for its detection are

(a) that its depth be at right angles to the surface,

(b) that its width at the surface be small, so that the air gap it creates is small,

(c) that its length at the surface be large with respect to its width,

(d) that it be comparatively deep in proportion to its surface opening.

The field set up in the part then, should be in the direction at right angles to the length of the defect.

Incipient fatigue cracks and fine grinding checks often have a depth of less than 0.001" and a surface opening of perhaps one tenth that or less. Such cracks are readily located using some form of the wet method. The *depth* of the crack has the least effect on its detectability, except that, up to a certain limit, the deeper the crack the stronger will be the indication for a given level of magnetization. This is because the leakage flux is stronger as the crack becomes deeper, due to the greater distortion of the field in the part. However, this effect is not particularly noticeable beyond perhaps one quarter of an inch in depth. If the crack is not close-lipped but is wide open at the surface (has a large dimension W) the reluctance of the resulting longer air gap reduces the strength of the leakage field.

It is almost self-evident then, that detectability involves a relationship between surface opening and depth. A surface scratch, which may be as wide open at the surface as its depth, does not usually give a magnetic particle pattern, although in some cases it may do so at high levels of magnetization. Because of many other variables that may enter, it is not possible to set up any exact values for this relationship, but in general a surface discontinuity which is at least five times as deep as its opening at the surface, will be detectable.

There is some evidence of a limitation at the other extreme, namely, a crack of some depth but with its surface opening so small that no indication is produced. If the faces of a crack are tightly forced together by compressive stresses, the almost complete absence of air gap may produce so little leakage field that no particle indication is formed. Shallow cracks produced in grinding or heat treating and subsequently put into strong compression by thermal or other stresses have been reported, which gave no indications with magnetic particles. Sometimes, with careful, maximum sensitivity techniques, faint indications have been produced in such cases. The operator should be alert to the possibility of this occurrence when dealing with a part the surface of which may have residual *compressive* stresses from any cause.

One other condition sometimes approaches the limit of detectability and should be mentioned. This is the case of the lap, produced in forging or rolling which, though open to the surface, emerges at an acute angle. (See Fig. 32, Chapter 3.) Here the leakage field produced may be quite weak, because, due to the small angle of emergence, and the relatively high reluctance of the actual air gap which results, very little leakage flux takes the path out through the surface lip of the lap to jump this high reluctance gap. When laps are being sought — usually always when inspecting newly forged parts — high sensitivity methods, generally with the use of fluorescent particles, are desirable. Figure 177 shows the faint

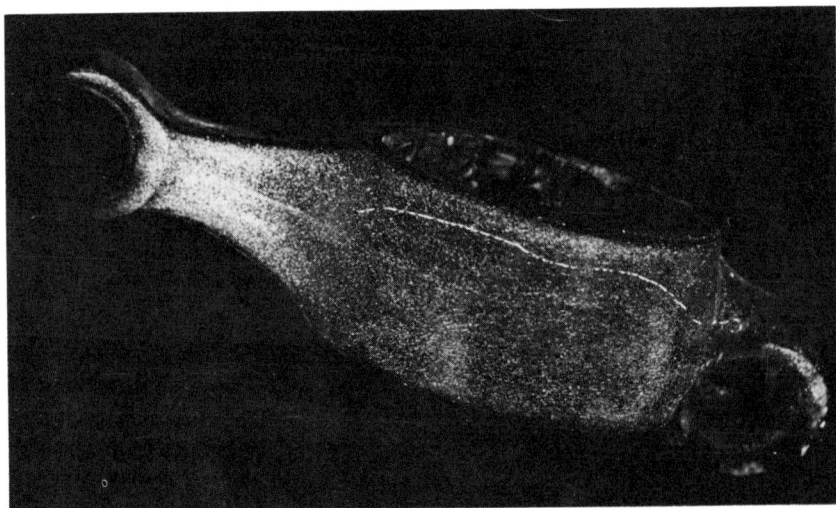

Fig. 177—Fluorescent Magnetic Particle Indication of a Forging Lap.

372

indication of a forging lap, produced with fluorescent particles, which sectioning showed to extend, at an angle, quite deeply into the body of the part.

6. DISCONTINUITIES LYING WHOLLY BELOW THE SURFACE. The magnetic particle method is capable of finding many defects which do not break the surface of the part in which they occur. This is an important ability, since there are circumstances when radiography and ultrasound, methods whose primary field is locating such defects, cannot be used. These two methods are inherently better adapted to the location of interior discontinuities than magnetic particles, but sometimes the shape of the part, location of the defect, or the cost or availability of the methods and the equipment needed, makes the magnetic particle method the best one to use. As a group those discontinuities which lie wholly below the surface are less dangerous from the point of view of potential failure than are surface cracks. This is because they are usually (though not always) more or less rounded in shape and, lying below the surface, are in an area of fibre stress below the maximum. They are, therefore, less severe stress-raisers than even a very small surface crack.

The detection of such discontinuities with magnetic particles is nonetheless often important, and much work has been done to determine the optimum conditions for success in this area.

7. DETECTION OF DEFECTS LYING WHOLLY BELOW THE SURFACE. Definition of the limiting conditions that determine whether or not a discontinuity below the surface is likely to be found with magnetic particles, is not nearly so simple as is the case with surface cracks. A large number of variables are factors, any one of which may be determining in a given case.

The question, often asked, "How deep below the surface can a defect be detected with magnetic particles?" has no answer in specific terms. But some of the factors and variables that affect the detectability of deep-lying discontinuities *can* be defined and understood, so that an operator can be aware of what the problem really is.

8. TWO GROUPS OF SUB-SURFACE DISCONTINUITIES. The sub-surface discontinuities which magnetic particles will locate may be put into two groups. The first of these comprises those small voids or non-metallic inclusions which lie close to and often *just* under the surface of the part. Non-metallic inclusions are present in all steel products to a greater or lesser degree. They may occur as scattered individual entities, or they may be aligned as long stringers. In the inspection of machined, highly finished parts, where the wet continuous D.C. method using high magnetizing currents is a common technique, these stringers are often found. They are seldom significant unless they occur in excessive numbers or lie in a transverse direction in an area of high stress. Since they are usually very small they are not found unless they lie very close to the surface.

Because of the nature of these discontinuities they produce highly localized but rather weak leakage fields. Contrary to the rule earlier laid down, that the dry method excels for finding deep lying defects, the wet method is the best one to use for this inclusion type of discontinuity if its detection is important. This is because these fine, non-metallic stringers are not really "deep-lying", for the reason that, though sub-surface, they must be very close to the surface to be found at all, by any method.

9. DEEP-LYING DEFECTS. The second and much more important group of sub-surface discontinuities are those larger and more serious conditions which may be quite deep in heavy sections— perhaps one quarter inch to two inches or more below the surface. These may be, in weldments, lack of penetration, sub-surface lack of fusion, or cracks in the beads below the last in heavy welds. Such deep discontinuities in castings may be internal shrinks, slag inclusions, or gas pockets. Although it is important to know and define limiting conditions for the detection of such discontinuities, to do so presents a real problem. But because of the often-expressed interest that exists in the problem of locating deep-lying discontinuities, it is worth while to consider it in some detail.

10. DEFINITION OF TERMS. In Section 3 of this chapter (refer to Fig. 176), the meanings of the terms "Depth", "Height", "Length" and "Width" as applying to discontinuities lying wholly below the surface were defined. The important point to remember here is that these dimensions are always related to that surface of the part *on which the magnetic particle inspection is being made.* In some instances, as in a rectangular cross-section, the discontinuity being sought may be contiguous to two or three surfaces of the body in which it is located. Detection may be more successful on one of these surfaces than on the others because of shape or orientation of the defect.

In the case of a tube, for example, the defect may be "deep" with respect to the outer surface, but may be quite close to the surface of the inner wall of the tube.

11. CONCEPT OF DEPTH. The principal reason why one cannot set a limit in terms of inches or fractions to the depth to which magnetic particle testing can reach in locating internal disconti-nuities, is that size and shape of the discontinuity itself in relation to the size of the cross-section in which it occurs, is of controlling importance. This idea of relative dimensions can be easily expressed qualitatively by saying that "the deeper the discontinuity lies in a section, the larger it must be to be detected with magnetic particles."

This is, however, not a very satisfactory way to dismiss the prob-lem, and the concept of depth can be better visualized by taking a concrete example. Suppose we consider a round bar having a radius of ⅜ inch and that it has a defect having a height, H, of 0.025 inch, lying at a depth, D, of 0.025 inch below the cylindrical

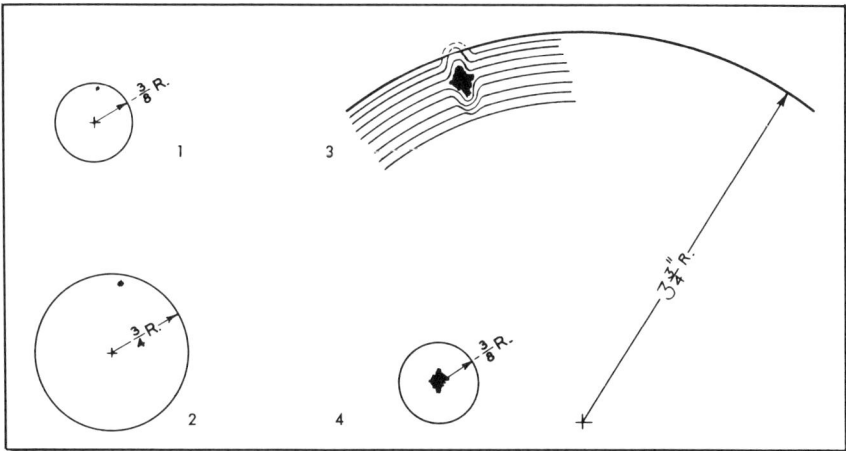

Fig. 178—Concept of Depth of a Deep-Lying Defect.

surface. Let us further assume that the width, W, is 0.02 inch and the length, L, is 0.025 inch. In another bar suppose we consider that all these dimensions are doubled, and in a third bar that they are all multiplied by ten. The latter two bars would have radii, respectively, of ¾ inch, and 3¾ inches.

The percentage relations of these dimensions in each of the three cases are the same, and with steel of ordinarily favorable magnetic properties, and by using proper flux density and suitable magnetic particles, we would expect no difficulty in getting good indications of any of these three defects—in fact, they would all show about equally well. However, if we take the discontinuity of Bar #3 (H = 0.25 inch) and put it into Bar #1 (Diameter = ¾ inch) at the same depth below the surface at which it was located in Bar #3 (0.25 inch), it will be found to lie at the center of Bar #1. In this location, in spite of its size, we would not expect to get an indication, at least with circular magnetization. There would be no appreciable field distortion. Nor would the discontinuity of Bar #1, if transferred to Bar #3, without change of size, but at the same depth as the discontinuity of Bar #3, be expected to give a good (or any) indication.

Stated another way, the "depth" of a defect, as it affects detectability with magnetic particles, is a relative term, and cannot be considered alone. On the contrary, depth must be considered in relation to the size of the defect, and the total dimensions of the cross-section. It is relatively easier to find a defect of a given size and shape at one half inch below the surface if it occurs in a four inch section, than if it occurs in a section only 1 inch thick. In the latter case, it would be at the center of the section. (See also Section 18, this Chapter).

12. SPREAD OF EMERGENT FIELD. Though we have said that the defects in bars 1, 2, and 3 discussed above would be found with equal facility, although they have widely differing depth, D, dimensions, the appearance of the indications would be quite different. The width of the particle pattern on Bar #3 would be probably as much as ¼ inch, whereas the pattern on Bar #1 would be compact and only some 20 thousandths of an inch wide. The reason for this difference lies in the angle through which the leakage field spreads as it leaves the surface. This is a function of the actual depth, D. The more metal between the defect and the surface, the more "spread" of the leakage

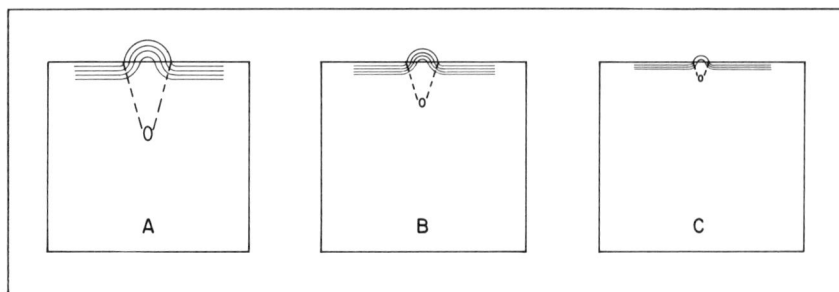

Fig. 179—"Spread" of the Emergent Field at a Defect.

field can occur. Of course, the *width* of the defect also enters into the result, especially when the defect lies quite close to the surface. The dimension of depth is, however, the dominant factor in determining the amount of spread.

Thus the width of the particle pattern and the compactness and amount of build-up over a discontinuity lying wholly below the surface, when produced under controlled conditions, is known to furnish a good indication of the actual depth, D—that is, the distance from the test surface to the nearest edge or surface of the defect.

13. EFFECT OF WIDTH. Width is defined as the longest dimension of the discontinuity measured in a direction parallel to the surface and at 90° to length. Consider a discontinuity such as a flake in a forging, for instance, which is in general circular in outline, lens-shaped, and about the size of a half dollar. Assume that the flake is oriented 90° to the surface, and at right angles to the flux lines. It makes little difference whether the width (or thickness, measured in a direction parallel to the field) of the defect is 0.04 inch or 0.40 inch. Here the height and length (or in this case the diameter, since we have assumed a circular shape) are the dimensions of prime importance.

If, however, the opposite faces of the discontinuity were in contact, or separated by only, say, one ten thousandth of an inch, the defect would be *much* less detectable than if this dimension were the 0.04 inch mentioned above. There must be an appreciable high-reluctance gap to set up an obstruction in the path of the field, else no leakage flux will be crowded out to the surface.

14. EFFECT OF HEIGHT AND LENGTH. These dimensions, in the case of the forging flake discussed above, are measured in a plane at right angles to the flux lines, when we are considering a discontinuity from the point of view of detectability. Their magnitude determines the size of the obstruction presented to the lines of force, and consequently also the amount of crowding of flux to the surface of the part. If the lens-shaped flake with a diameter of about one inch (that is, the length and height equal, and both equal to one inch) were altered in shape so as to shrink either its vertical or horizontal diameter to say one eighth or one sixteenth of an inch, the chance of detecting it would shrink to the vanishing point. The width, measured in a direction parallel to the flux would enter into the result not at all.

If, instead of a lens-shaped flake we assume a discontinuity having a sort of tubular configuration such as illustrated in Fig. 176, having a longest dimension, L, parallel to the flux lines, but presenting a substantial circular face at right angles to the flux path, the effect would be quite different. The distortion of the field would be less sharp, and although there would be crowding between the discontinuity and the surface, the leakage field, if any, would be spread out along the distance corresponding to L, or longer, and would not be expected to give a very readable indication.

In considering the detectability of a discontinuity lying below the surface, we are concerned primarily with the projected area which is presented as an obstruction to the lines of force, and the sharpness of the distortion of field produced. It is helpful to think of the field in the specimen as flowing through it like a stream of water. The

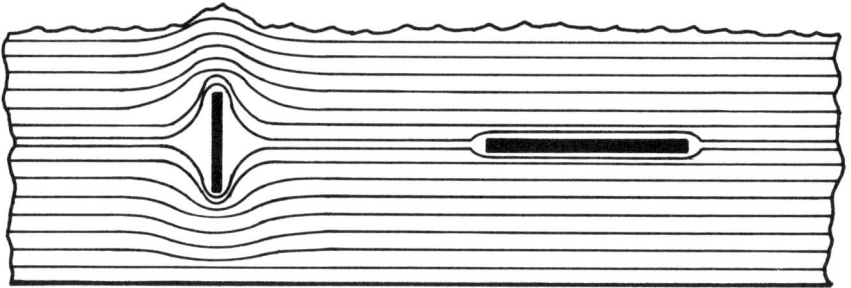

**TURBULANCE IN A STREAM
MAGNETIC FLUX OR WATER**
Fig. 180—Effect of Interruption to the Flow of Water (or Magnetic Flux) Due to Shape and Orientation.

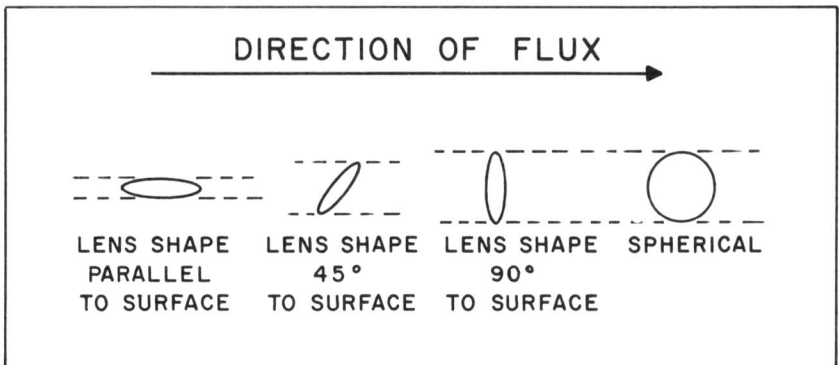

Fig. 181—Effect of Shape and Orientation on Detectability.

378

top of a tin can, on edge below the surface and at right angles to the surface and the flow, would cause a sharp disturbance in the movement of the water; but a straight, round stick placed at a similar depth and parallel to the direction of flow, would have very little effect.

15. EFFECT OF SHAPE. We have been considering a lens-shaped defect and the area it presents to obstruct the lines of force. A spherical gas or slag inclusion of the same diameter as the lens- or coin-shaped discontinuity would present the same projected area, but would be *much* less likely to be detected. In such a case the flux lines would "streamline" around the sphere, and the disturbance created in the field would be much less sharp.

16. EFFECT OF ORIENTATION. Another most important factor in detectability is the orientation of the defect. The lens-shaped obstruction discussed above was considered to be at $90°$ to the direction of the flux. If, however, the same defect is inclined, either vertically or horizontally, to an angle of only $60°$ or $70°$ instead of $90°$, there would be a noticeable difference in the amount of leakage field, and therefore in the strength of the indication. This difference would be due not only to the fact that the projected area of the discontinuity would be reduced, but also to a "streamlining" effect as in the case of the sphere.

17. MOST FAVORABLE DEFECT FOR DETECTION. We find emerging from this analysis a better concept of the kind of defect we are most likely to be able to find with magnetic particles when it lies wholly in the interior of the part. The most favorable discontinuity is one which has a height and length of the same order of magnitude, both being distinctly greater than the width or thickness. The most favorable orientation of this discontinuity is to have it lie so that the plane of its general direction is at right angles to the direction of the field.

The least favorable shape for detection is a flat discontinuity lying parallel to the surface instead of $90°$; and of discontinuities presenting considerable projected area to the flux path, the spherical shape is least favorable.

To this summary should be added the thought—already expressed —that the nearer the discontinuity lies to the surface, whatever its shape, the smaller it can be and still be located; or, conversely, the deeper it lies, the larger it must be to be detected.

18. EFFECT OF METHOD OF MAGNETIZATION. Other factors than the size, shape and orientation of defects play a part in whether they will be detected or not. The direction and strength of the magnetic field is obviously an important consideration. Direct current—or better, half wave current—should be used for deep-lying defects, and the dry powder almost always gives the best indications.

For *really* deep-lying defects, such as are sought in thick weldments, the prod method of magnetization is superior to any other. D.C. yokes are effective if the discontinuity is fairly close to the surface, but when depth is of the order of one quarter inch to two inches or more, prod magnetization is by far the best. The overall method now being widely used instead of prods, on large castings particularly, is superior to prods for finding all *surface* discontinuities, but less effective than prods for finding defects lying wholly below the surface.

With prod magnetization it is possible to produce indications, on the surface to which the prods are applied, of discontinuities lying at the center of the section, or even of those lying near to or even on the opposite surface. This is commonly done in locating lack of penetration in a single V butt weld. Spacing of prods is important, and this has been discussed in detail in Chapter 10, Section 6 and following sections.

19. EFFECT OF PERMEABILITY. It is more or less self-evident that the permeability of the part should be high if deep-lying defects are to be located in a given specimen. A maximum material permeability of at least 500 has been mentioned in connection with the detection of surface cracks, but the value should be much higher for deep-lying defects. Since the need to find internal voids is most urgent in the case of weldments it is fortuitous that the steel involved is very often low carbon steel in the form of plate or shapes, which has a very high material permeability. The growing use of harder steels for pressure vessels and piping is unfavorable to the location of deep defects in such materials. The effect of reduced permeability was shown by the tests on the tool steel ring, unhardened and after hardening. See Figs. 115 and 118, Chapter 12.

20. OTHER FACTORS. Other factors which may play an important part in the detectability of any given sub-surface discontinuity, are the condition of the surface and the method of applying the powder. Obviously the surface should be clean and dry; but beyond this, the smoother the surface is, the more easily can the powder move to arrange itself into a pattern when leakage fields are weak and diffuse. The surface on which the inspection is made should be horizontal for best results, so that gravity does not tend to cause the powder to fall away. Weld beads are often machined off to provide a smooth surface for this inspection.

The powder should be applied as a light cloud, which is especially well accomplished by the air-operated powder gun. Patterns can be observed as the powder drifts to the surface. Such patterns often disappear as more powder is applied. When the excess of powder is blown off, even with a very gentle air stream, such patterns are often also blown away, leaving only those held by stronger and more concentrated leakage fields. Thus, the limit of detectability may often be dependent on the skill of the operator.

21. THE OPERATOR. The detection of deep-lying defects requires a greater degree of skill on the part of the operator than any other field in which magnetic particle testing is used. It is probably true, that out of several inspectors who might be given the same part containing a small, deep sub-surface defect, not all would find the indication. It is for this reason that operators who are to engage in the inspection of weldments with the magnetic particle method, are given qualification tests in order to assure that they are competent in the application of the method for this purpose.

It is extremely desirable that new operators being trained for this purpose be given the opportunity to see the chipping-out and re-welding of some of these deep-lying defects, so that they can see exactly what caused the indications they produced on the surface of the weldment. In this way they are able to confirm their work and build up experience which makes them better able to recognize significant patterns when they occur.

CHAPTER 21

NON-RELEVANT INDICATIONS

1. DEFINITION. Reference was made in Chapter 3, Section 4, to indications caused by magnetic or metallic discontinuities which are present by design, or by conditions which have no bearing on the strength or service usefulness of a part. Such patterns have no relation to potential defects, whether they are caused by design or accident.

Before an operator has progressed very far in his experience with magnetic particle testing he will have become aware of this type of indication. They can sometimes be very puzzling, since in many cases investigation to account for them at first reveals no apparent reason for their occurrence.

Such indications are true particle patterns, actually formed and held in place by magnetic leakage fields; but the leakage fields responsible are not caused by conditions that are relevant to the strength or useability of the part. The name "non-relevant" has been given to this type of pattern. Obviously the magnetic particle operator must be acquainted with these non-relevant indications and be able, when they occur, to recognize them for what they are.

2. FALSE INDICATIONS. The term "false" has sometimes been applied to all non-relevant indications, but the name is not a good one, since such indications are in nearly all cases magnetic in origin. There is perhaps one truly false indication, and that is the case of particles held mechanically or by gravity in surface irregularities, with no relation whatever to leakage fields. Similarly, when using the wet method, a "drainage line" of particles will often form. In such cases it is only necessary to shake or blow, or to rinse off the particles to prove that they are not magnetically held.

3. EXTERNAL POLES. Particles will adhere to local poles at sharp corners, sharp ridges or surface irregularities, but these effects are not usually very confusing. Such patterns are most likely to be encountered when longitudinal magnetization is being used. Their occurrence is often an indication that too strong a magnetizing force has been applied.

4. ALL-OVER PATTERNS. Sometimes surface patterns or all-over patterns are produced when magnetizing circularly. These are the indications of flux lines of the *external* field produced by the magnetizing current. The lines of the pattern will always appear at right angles to the direction of current flow, and are usually produced only when too much current is being used.

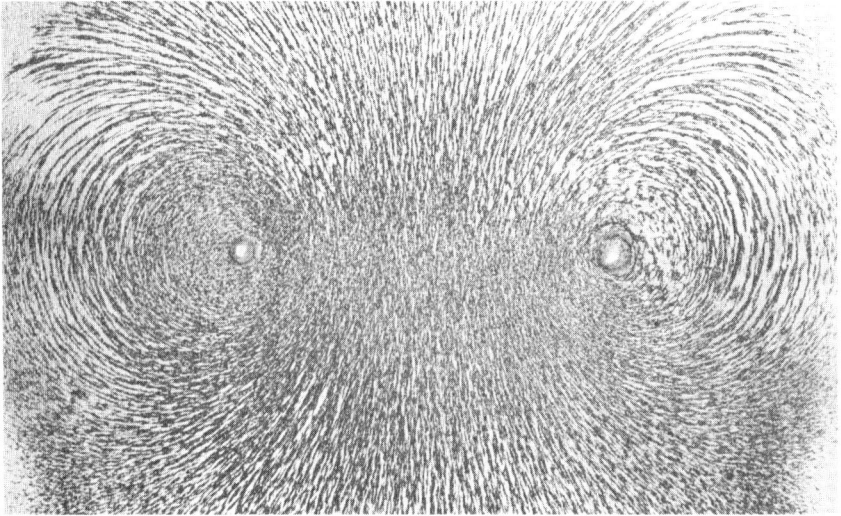

Fig. 182—Magnetograph Showing the External Field Pattern Produced When Prods Are Used to Magnetize Steel Plate.

5. EDGE OF SCALE. A rather obviously non-relevant indication is that which can appear at the edge of a patch of tightly adherent mill scale. It is usually a patch of very thick scale that produces such an indication. Since the mill scale is magnetic but with a very much lower permeability than the steel from which it was formed, this indication is really the result of a magnetic discontinuity. Though obvious, it is surprising how frequently a careful examination is required to convince an operator that the indication is not a genuine metallic surface discontinuity. Visual examination is in most cases sufficient, but occasionally a more careful cleaning of the surface is found to be necessary. Such heavy scale patches are today not very common on rolled or cast products.

The extent to which scale patches, if they occur, may tend to confuse the test, depends on the kind of defects being sought. For example, if looking for seams in hot-rolled bars or billets, an irregular patch of scale could not be confused with the straight-line indication of a seam. Furthermore, the irregular scale pattern is not in any way similar to any defect likely to be found in bars or billets.

6. CONSTRICTION IN THE METAL PATH. One of the most common non-relevant indications is that caused by a constriction in the metal path through which the flux must pass. Such constrictions are caused by the shape and construction of the part. The crowding out of flux to produce a leakage field is exactly the same as that which produces indications over a sub-surface defect, only in the case of these non-relevant indications no questions of defects is involved. Figure 183 shows a gear and splined shaft magnetized circularly by means of a central conductor, with particle patterns on the outside of the shaft showing the inside splines. The field is crowded out by the reduced thickness of the metal at the base of the splines. Patterns also appear on the gear between certain of the holes in

Fig. 183—Gear and Shaft Showing Non-relevant Indications
Due to Internal Splines.

the web, and at the base of certain gear teeth that lie adjacent to these holes.

Such patterns can usually be recognized by the following characteristics:

(1) Given the same magnetizing technique they will appear on *all* parts and in the same location. They will all have substantially the same appearance—a condition almost never met where real sub-surface defects are involved.

(2) They can always be related to some feature of construction or cross-section which accounts for such a leakage field.

(3) They seldom look, to an experienced operator, like indications of any sort of relevant discontinuity.

If there is any question of the presence of a real defect at such locations, the part should be demagnetized and then remagnetized at a lower level, reducing the magnetizing force until the non-relevant pattern disappears. There will still be field enough to give an indication if a real defect exists in such areas.

7. SHARP FILLETS AND THREAD ROOTS. Similar in character to the non-relevant indications just described, are those due to leakage fields at sharp fillets or roots of threads. Here the flux lines tend to bridge the air gap at such points rather than to follow the metal path exactly. Confusion is particularly likely to occur in locating actual cracks at the roots of threads and in sharp fillets—places where cracks are very likely to occur.

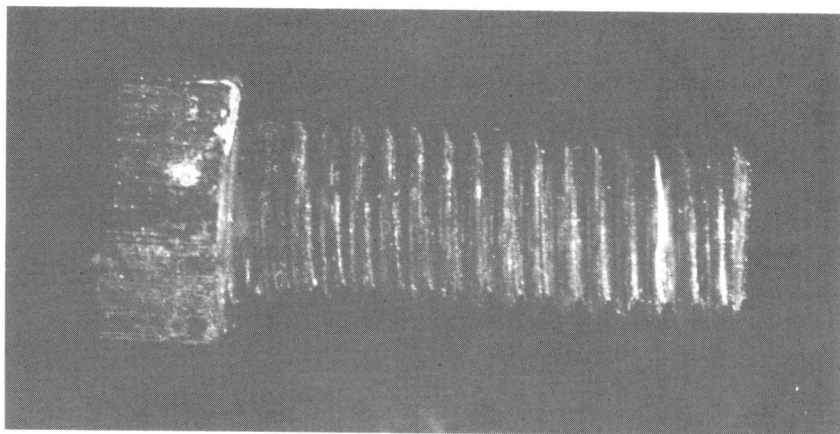

Fig. 184—Fluorescent Indications of a Crack at the Root of a Bolt Thread.

To make certain that indications of actual cracks are not over-looked at such locations because of the interference of the non-relevant pattern, the level of magnetization should be adjusted below that which produces the non-relevant leakage field, at which level indications of true cracks will still be strong. The fluorescent particles are especially good for tests at such locations since they emphasize the contrast between the particle build-up at a real crack and at the non-relevant field.

8. MAGNETIC WRITING. If two pieces of steel come into contact with each other when either or both are in a permanently mag-netized condition, but not magnetized to the same degree, local poles are apt to be formed at the areas of contact. As these areas are usually the result of one piece being rubbed against the other, they resemble lines, and the local changes in flux density or polarity also resemble lines. If powder or a liquid suspension of magnetic par-ticles be applied, the particles tend to adhere along these lines. Their

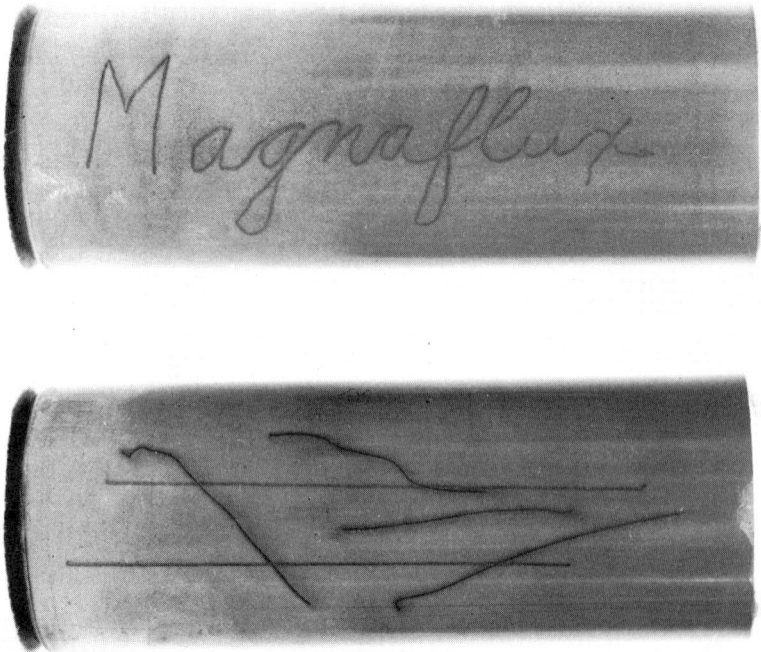

Fig. 185—Examples of Magnetic Writing.

appearance occasionally resembles a scrawl and is called magnetic writing, whether done intentionally or not. Figure 185 shows two examples of magnetic writing; one intentional and the other accidental.

Magnetic writing seldom gives any trouble in interpretation because its appearance is characteristic, the particles being loosely held and peculiarly fuzzy in outline. They seldom resemble the pattern formed by an actual discontinuity. If the object is thoroughly demagnetized and remagnetized as usual, and the inspection medium again applied, magnetic writing will not again appear, whereas a real flaw will repeat its indication after any number of such cycles. A requirement for the occurrence of magnetic writing is that the steel in which it occurs must have a sufficiently high retentivity, so that it will retain an extremely local increase or decrease in magnetization. The phenomenon is more apt to occur in circularly magnetized parts than in those longitudinally magnetized. To prevent it, such parts should not be allowed to come into contact with ferromagnetic objects after being magnetized.

9. EXTERNAL MAGNETIC FIELDS. Non-relevant indications may appear on hardened steel parts due to residual local poles, which may be caused by proximity of magnetic fields from power lines, magnetic chucks or permanent magnets of one kind or another. They are similar in character to magnetic writing (though not usually in appearance) and are removed by demagnetization. In Fig. 186 is shown a roller bearing assembly, having a quenching crack at point A. The patterns at points B are local poles left by the poles of a magnetic chuck. Demagnetizing followed by remagnetizing will erase the magnetic chuck indications, but the indication of the crack will, of course, reappear.

10. COLD WORKING. The type of plastic deformation called cold working produces a hardening in steel with a consequent change in permeability. When the cold working is very local, the abrupt permeability change is often sufficient to cause a particle pattern. The indication produced is at times similar in appearance to magnetic writing. On demagnetizing and remagnetizing, however, the indication from cold work reappears, whereas that due to magnetic writing does not. If the object is sectioned and etched, no discontinuities are found to account for the pattern, but examination under the microscope will usually show the typical grain distortion representing cold work. Cold work indications can be removed only

Fig. 186—Roller Bearing Assembly with Non-Relevant Magnetic
Particle Patterns Produced by a Magnetic Chuck.

by heating the part, usually to 800°F. or above, and allowing the part to cool slowly.

One example of such cold work indications was found in the case of some new helical springs which gave a pattern resembling discontinuous seams. The indication on each turn of the helix was immediately opposite the indications on adjacent turns. This occurrence was a most unlikely coincidence if the indications were caused by seams. Further investigation revealed that the springs were being compressed by complete closure with a strong blow during the acceptance tests, and cold working was occurring at the contact areas between turns.

Another example was found in the case of new aircraft bolts. These had a circumferential or transverse indication which, in some instances, had the appearance of a fatigue crack—obviously impossible in new bolts which had neither been in service nor strained in any way. Examination under the microscope of additional specimens indicated cold working, which was confirmed, being found to have been caused by calipers or snap gauges used during the final grinding operation. The layer of cold work was extremely shallow, since the bolts were hardened, but the effect undoubtedly was responsible for the non-relevant indications.

Figure 187 shows another example of cold work patterns. This is a piston pin on which a heavy roughing cut was taken, possibly with a dull tool. The metal was cold worked under the point of the tool to such a depth that subsequent finishing cuts and grinding did not completely remove it. The spiral pattern was brought out when the pin was longitudinally magnetized, and suggested the cause in machining.

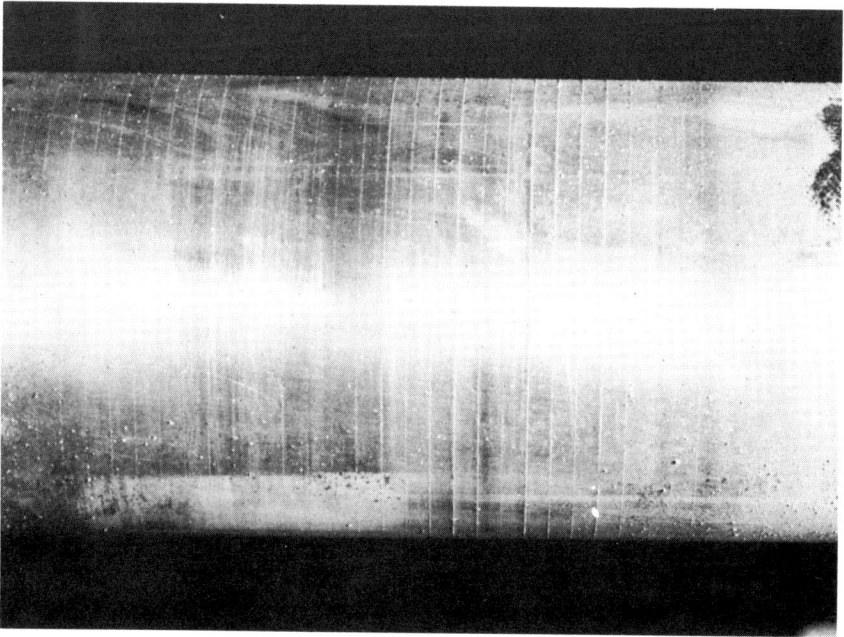

Fig. 187—Magnetic Particle Pattern Produced by Cold Working Due to Machining.

11. LUDERS LINES. Planes of slippage, called Luder Lines, in materials which have been stressed beyond the yield point, are occasionally the cause of non-relevant magnetic particle indications.

Bent plates for assembly for tank members which have been formed cold by means of rolls, if magnetized before stress-relieving by passing current through the plate with widely spaced prods, will show indications of Luders Lines when dry powder is dusted over the surface. They appear as an incomplete criss-cross pattern of lines running at 45° to the axis of curvature.

12. GRAIN BOUNDARIES. When grain size is very large, the macrostructure showing grain outlines may be found to be shown by a magnetic particle pattern, even though no metallic discontinuity exists. The pattern is due to sharply different permeability as between the grain and the boundary material.

13. BOUNDARY ZONES IN WELDS. In weld inspection an indication is often obtained at the boundary between the fused metal and the base metal. Other indications in the form of lines may appear at the edges of decarburized zones. These occurrences actually indicate an abrupt change in permeability in the path of the magnetic flux, but are not *necessarily* indicative of an objectionable condition. Many sound welds will yield a powder line at the junction of base and weld metal. If in doubt, metallurgical evaluation should be obtained.

14. FLOW LINES. Many steels, particularly in forgings, will show a pattern as shown in Fig. 20, Chapter 3. This magnetic particle pattern is similar to the grain flow in forgings brought out by deep etching, and is, in fact, produced by these grain-flow lines. The indications are brought out only by high sensitivity methods and at levels of magnetization higher than ordinarily used for inspection. The occurrence of such a pattern is not indicative of defective steel.

Dendritic segregation and segregation of carbides and other metallic constituents may also cause indications. Obtaining such indications usually depends on some increase in sensitivity over normal inspection practice.

15. BRAZED JOINTS. When two pieces of ferromagnetic material are joined by brazing, the film of brass forms a *magnetic* discontinuity, even though the joint may be perfectly sound structurally. A magnetic particle pattern will be produced outlining the joint. Since the braze metal is not ferromagnetic, local poles at the edges of the magnetic material will be formed just as though there were no metallic junction at all. This pattern has no significance whatever from the point of view of the strength of the joint which must be evaluated by other tests. See Fig. 19, Chapter 3.

16. JOINT BETWEEN DISSIMILAR MAGNETIC MATERIALS. Sometimes a piece of hard steel is butt-welded to a softer piece by any of several methods. At the point of welding there will be a sharp change of permeability, the soft steel having a high permeability and the hard steel a much lower one. If a magnetic field be set up so as to flow across this joint there will be a concentrated leakage field and consequently a magnetic particle pattern. This pattern, however, does not give any information regarding the soundness of the welded joint.

Figure 18, Chapter 3 shows an example of this condition. A piece of soft steel rod is butt-welded to the end of a similar rod of hardened steel. The strong magnetic particle indication is caused by the sharp change in permeability at the weld. Under such circumstances it is difficult to get a true magnetic particle indication of an actually unsound joint, even at low levels of magnetization.

17. FORCED FITS. One other example of a non-relevant indication should be mentioned. When two members of an assembly are very *tightly* fitted together, as in a pressed fit between a shaft and pinion gear, a magnetic particle pattern of this joint may be formed. If the fit is tight *enough* no indication may be produced, since the air gap between the two members may be almost zero. If an indication appears it is never misleading, unless the joint is hidden by paint or rust and the operator does not know that the joint exists. Usually polishing with fine emery cloth will reveal the line between the two members of the assembly. See Fig. 17, Chapter 3.

CHAPTER 22

INTERPRETATION, EVALUATION AND RECORDING OF RESULTS

1. ESSENTIAL STEPS. In any nondestructive testing process there are three essential steps. These are:

(a) Production of the indication, whether on film, a meter, oscilloscope, tape or other indicating device, or directly on the surface of the part as in the case of penetrant and magnetic particle testing.

(b) Interpretation of the indication as to what condition is present to have caused it.

(c) Evaluation of the condition indicated, as to its effect and seriousness from the standpoint of useability of the part.

2. DEFINITIONS.

A. *Production of the Indication.* The first step obviously is to apply the nondestructive test properly so as to produce indications of the discontinuities or other flaws which the method is designed to detect and which it is desired to locate in the particular situation. The previous chapters of this book have been devoted to this step for the magnetic particle method. The various conditions and requirements for producing magnetic particle indications under all sorts of circumstances in all kinds of ferromagnetic materials have been thoroughly discussed.

The first step of the testing process has therefore been achieved when the test has been properly and intelligently applied. Those parts on which no indications have appeared may then be presumed free of flaws, and can be passed directly to the next process in their manufacture. Those parts on which indications *have* appeared require further consideration before they can be used, repaired or scrapped.

B. *Interpretation of the Indications.* The indications of the presence of flaws which the various methods of nondestructive testing produce are, literally, just that. They are indications that something in the part is not normal. But they do not, in themselves, tell exactly what the condition is which has produced them. In magnetic particle testing, every pattern is produced by a magnetic disturbance setting up a leakage field, but further knowledge and information is needed to enable us to say whether the pattern, and therefore the disturbance, is really significant or not.

Given, therefore, an indication of a magnetic disturbance evidenced by a magnetic particle pattern, someone must decide what the condition in the part is that has caused the pattern—in other words, he must *interpret* the indication in terms of its cause.

C. *Evaluation.* Before the part can be disposed of, however, a further and final decision remains to be made. Once it is determined what the condition in the part is, then the condition must be *evaluated* in terms of its effect on the useability of the part for its intended service. After this decision the part may be accepted as satisfactory or rejected as scrap—or perhaps salvaged by reworking it in some manner. A further cost-saving analysis can often determine what control of processes can do to prevent future similar occurrences.

3. THE PROBLEM OF INTERPRETING. We have just said that interpretation consists of deciding what is causing the particle pattern. Obviously, if our evaluation of the condition is to be intelligent, it is not sufficient to say that there is a surface crack or that there is a discontinuity under the surface. The exact nature of the condition, if known, will determine whether rejection is warranted or whether salvage is possible. Such knowledge may, further, lead to the correction of processes to eliminate recurrence of the condition in the future.

The magnetic particle pattern does not, in itself, tell us what condition is causing it. Some of the typical characteristics of patterns produced by surface and sub-surface discontinuities have been shown in foregoing chapters, and a great amount of information is always obtainable from an observation of the indication itself. But to translate such observation into a thoroughly accurate identification of the defect is not always easy. This is the problem of interpretation.

4. OUTSIDE KNOWLEDGE REQUIRED. The fully competent non-destructive testing engineer should be prepared to interpret indications wherever they are encountered. It is true that in many applications special-purpose units or controlled tests are set up to find indications of specific defects, which, when found, need no further interpretation. But to make a decision as to the identity of a defect without such built-in knowledge calls for considerable experience and general information regarding the part. This information is not always available to an otherwise perfectly competent inspector. To be able to produce indications properly does not in itself at all qualify an inspector to interpret his results dependably.

In order to do this, at least some, but preferably all, of the following points and sources of information should be possessed by, or be available to the inspector:

(a) A knowledge of the material from which the part is made —whether high or low carbon steel, and what alloys are present and in what amount. To make use of this knowledge he must already know something of steels and steel-making, and the character of the defects likely to occur in various types of steel.

(b) A knowledge of the processing history of the part—whether made from rolled stock, or from a forging or casting; also what machining operations—grinding, lapping, etc. have been applied to it; also the heat treatment it has received —whether hardened, carburized, nitrided, etc.; and he should know enough about such processes to be familiar with the defects each may introduce into the part.

(c) A knowledge of how metals fail and what conditions are likely to lead or contribute to such failure.

(d) Past experience with similar parts. Such experience often indicates what to be on the lookout for and helps to identify an indication when it is found.

(e) Facilities for making destructive tests on specimens containing indications of which the cause is not clear. Cutting up the part and examining the defect in section often makes possible identification of similar indications when they are encountered.

(f) A general knowledge of metallurgy is extremely helpful, if not essential.

An experienced inspector, well equipped with knowledge such as outlined above, will know at the outset that certain defects may be present and that others cannot be. His problem of interpretation is at once narrowed to perhaps a very few possibilities, since the history of the part rules out the rest. Often he is actually looking for a specific kind of defect, as, for instance, flakes in a large forging. On the other hand, in any given plant such as a foundry or forge shop the types of defects that can be present are usually clearly understood, so that here again the problem is simplified.

5. THE OPERATOR. It is very evident from the above considerations that the "operator", to whom frequent reference has been made in these pages, or the inspector or other person conducting the tests, is the key to the attainment of the greatest possible value from the use of magnetic particle testing. This is indeed equally true of any method of inspection where readings must be taken on meters or instruments, or photographic negatives must be examined. The necessity for interpreting and evaluating the observations made by any nondestructive testing method necessarily introduces the human element into the inspection results. Even in the case of electronic methods where the equipment is pre-calibrated and set to give automatic indication of a rejectable discontinuity, setting the limits of acceptability involves human judgment, as does also monitoring the equipment to insure that it continues to operate properly.

For the above reasons, the better equipped the inspector is with a background of knowledge and experience, the better will be the results from the test. In a large plant, it is not always possible to find well qualified personnel to operate the magnetic particle units, and if the training and experience of an inspector is not adequate, he should not be charged with the responsibility of interpreting or evaluating his results. It is not the intention to imply that all indications must be referred to some higher authority in the event that the inspector is not qualified to interpret what he finds. Where many similar parts are being examined, rules for acceptance and rejection can be laid down, since in such cases the types of defects likely to be found can be predicted. But the indications that do not fit such specifications must be individually interpreted.

In most cases the operator of the test equipment is a member of the plant inspection staff, in which case the assistance of the

Metallurgical Department and the Testing Laboratory should be easily available to him to assist in the interpretation of indications. In some plants the responsibility for interpretation may be taken entirely out of the hands of the inspector who makes the test, and be delegated to the metallurgist, chief inspector or a qualified supervisor—except as pre-determined rules have been prepared for the inspector's use.

The test operator should always possess good vision, since he must, in the first place, *see* the indications. As far as his mental equipment goes, he must be conscientious and have a temperament which can be relied upon not to become careless under possible monotony. These qualities are usually found to a sufficient extent in any good inspector in a plant making quality steel or quality steel parts. Such a man will make a good magnetic particle inspector and, if sufficiently intelligent and alert, he should soon acquire the experience to interpret most indications. But he should not be entrusted with the responsibility of working out new magnetizing procedures for new and widely differing classes of parts made of different kinds of steel unless he has or acquires additional background along the lines of the subject matter of this book. To do this he must possess considerable native intelligence, initiative and a strong sense of responsibility.

If the inspector himself is expected to pass on the acceptance and rejection of costly or important parts, then the best man available should be used for the work.

6. SOURCE OF KNOWLEDGE AND EXPERIENCE. In order to acquire the knowledge and experience which is required for adequate interpretation of indications, various sources of information are available to the inspector. Assuming that he is competent and well informed as to the magnetizing procedures and techniques involved in producing indications, he should study the operations and processes within the plant to become familiar with the defects typical of those operations. He should, further, read texts on such subjects as steel-making, metallurgy, fatigue of metals and how metals fail in service. The plant metallurgist may have, or at least can recommend, suitable books for this purpose. The necessary knowledge, if not already in the inspector's experience, is not quickly or easily acquired; and yet, the knowledge of what to look for and where to look for it is indispensible to the intelligent application of the test and the interpretation of indications.

But further than this, accurate interpretation depends equally on actual experience with typical defects, and in order to gain such experience he must actually identify a sufficient number of cases so as to be familiar with the appearance of the indications the various discontinuities are likely to produce.

7. SUPPLEMENTAL TESTS. The absolute identification of a given discontinuity involves the use of supplemental tests, and these tests often require the actual cutting up of a specimen, or otherwise probing into it to see what the discontinuity really is that lies at or below the surface to produce the indication.

Many of these tests are simple and can be performed with a minimum amount of laboratory or testing equipment. Others require specialized equipment and experience involving the use of testing laboratories and the methods of metallographical investigation.

8. SIMPLE TESTS. To wipe off the magnetic particles forming an indication is almost instinctive, and sometimes the defect, if a surface crack, is quite readily seen, once the exact location has been revealed by the indication. A low power hand glass is a most convenient pocket tool to aid the eye in such a first check. If the surface is rough, or is covered with a light film of rust, polishing the area where the indication appeared with fine emery cloth usually renders the defect more visible for study.

Another simple check is to wipe off the indication and again apply the powder or liquid suspension of particles, to see whether the indication will be reproduced by the residual field in the specimen. Since the residual method is always less sensitive than the continuous, it is obvious that if the original indication was produced by the continuous method, the manner in which it reappears by the residual method gives at once some indication of its severity and extent. As a matter of fact, this criterion of reproduction by the residual field has in some cases been made the basis of acceptance or rejection of parts. There are, however, too many factors involved for it to be safe to conclude that if the indication does not reappear, no defect is present.

Often times it is worth while, in order to confirm the indication, to demagnetize the part and repeat the test from the start to make sure that the indication really does come back in the same form as originally.

9. BINOCULAR MICROSCOPE. This instrument is an invaluable aid in examining surface discontinuities. It is easy to use, moderate in cost, and does not require a skilled technician. Magnifications from 10 to 32 diameters are available with this instrument, and are adequate to resolve even exceedingly fine cracks and other surface conditions.

A good light to illuminate the surface just under the lens is essential, and usually cleaning of the surface and polishing with very fine emery cloth is helpful. Sometimes light etching brings out the crack more distinctly.

If none of these surface examinations reveals a discontinuity which will account for the magnetic particle indication, and the indication persists after the surface is smooth and clean, the presumption is that the discontinuity is below the surface. Confirmation of this may be had from the appearance of the powder pattern itself, under a glass or a microscope. The indication of a sub-surface defect will appear diffuse rather than compact, and is likely to be a bit fuzzy at the edges.

10. FILING. Use of a file is perhaps the easiest way to determine the depth of a surface crack or seam. Either a flat file used cornerwise or a triangular file will readily cut to the bottom of a crack,

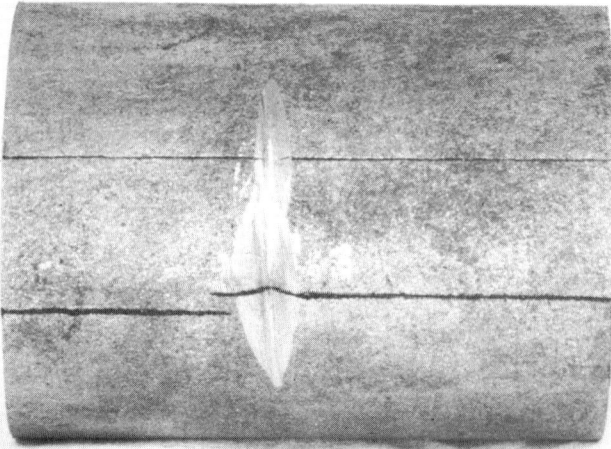

Fig. 188—File Cut Applied to Cracks in a Bar, with
Magnetic Particles Re-applied.

or else determine quickly that the crack is so deep that the specimen is unuseable. Rechecking at the filed point with magnetic particles is a means of knowing whether the file cut has gone to the bottom of the defect or not. Such a test is shown in Fig. 188. One of the two cracks is obviously much deeper than the other. The amount of particle surface-build-up had already indicated this fact before the file test was applied.

11. GRINDING. A similar test for determining the depth and extent of surface discontinuities is to grind the surface of the part

Fig. 189—a) A Small Forging, Ground at a Lap with Defect not Completely Removed. b) Exploring Crack Indications on Line Pipe with Small Hand Grinder.

at the indication. A permanently mounted grinding wheel, or portable wheel mounted on a flexible shaft may be used, depending on whether the part is small or large.

Figure 189a shows a small forging which has been ground at a lap. It is common practice at many plants to investigate all indications of this nature by such grinding. Shallow laps are usually ground out entirely at the time, so that the objectionable condition is at once removed from the part.

Another very useful tool for investigating defects is the small hand grinder shown in Fig. 189b. This tool is provided with a number of small grinding heads or burrs, which permit grinding at the indication with a minimum removal of metal beyond that strictly local to the crack.

Such grinding methods permit exploring a crack or other defect over a greater depth or area than can be done by filing.

12. CHIPPING. A still more rapid method for exploring cracks or other defects in large objects is by chipping. This is usually done with an air chisel, though sometimes a hand chisel and hammer is more conveniently used.

Chipping methods are most commonly used on welds, castings and large forgings—not only that the inspector may determine the actual depth of the defect, but also to remove the defect in preparation for re-welding or repair if this is permissible.

13. CHIPPING FOR REPAIR AND SALVAGE. Grinding and chipping methods, therefore, often serve the dual purpose of determining the extent of the defect, and at the same time actually removing the defect from the part. As stated above, forgings are commonly ground to remove shallow discontinuities entirely, if they do not penetrate below the dimensional tolerance of the part. Rechecking with magnetic particles should be done to insure complete removal of such defects.

In chipping operations it is useful to observe the manner in which the chip comes away from the chisel as the cut proceeds either along or across the crack. If the crack extends completely through the chip being removed, the chip will separate into two parts. When the cut extends below the bottom of the crack the chip will not separate. A recheck with magnetic particles is still, however, easily applied and is a safeguard to make sure that the bottom of the defect has been removed before welding repair is begun. In most cases there is sufficient residual magnetism in the material to bring out the indication when particles are reapplied, and remagnetizing is then not necessary.

This removal of the defect can also be done very rapidly by flame gouging as in the case of conditioning steel billets and blooms. By suitable manipulation of the burning torch, metal may be removed very rapidly in local areas at the defect. By watching the surface of the metal under the flame before actual burning commences, it is often possible to observe the presence of a surface crack, and thus see when it is all removed. However, rechecking with magnetic particles is easy to do and is a better and safer precaution to make sure that the part is ready for repair or further processing.

14. DESTRUCTIVE METHODS—FRACTURING. In order to derive the maximum possible information regarding a questioned indication a number of tests which involve complete destruction of the part are most useful. Such tests cannot be regularly applied in the course of routine inspections, but they are invaluable to the inspector who is attempting to improve his knowledge. By observing the nature of the defects which occur in the products for which he is responsible, he thus increases his ability to make reliable interpretations in the future.

Such tests are also frequently used after a part has been rejected because of an indication, as a sort of autopsy to confirm, and thus improve, the judgement of the inspector who made the rejection. This again leads to increased knowledge and confidence to aid future decisions.

One of the easiest of these destructive confirming tests is to break the part or attempt to break it, through the discontinuity. The part may be placed in a vise in such a manner that the metal on one side of the crack is held firmly, leaving the rest of the piece projecting from the vise so that, when struck heavy blows with a hammer it will tend to break at the crack. Such procedure works best on hard parts, but it can often also be applied on relatively soft materials by using a very heavy hammer blow.

Another way of completing the fracture through a surface crack if the section is too large for the hammer and vise method, is to put the part in a large testing machine and apply a load in such a fashion that the part tends to break at the crack. Such methods are particularly effective for studying the origin of a fatigue crack, or in distinguishing between a heat treating and a grinding crack. The faces of a heat treating crack are likely to be oxidized whereas a grinding crack will not show such oxidation. Sometimes it is important to make sure that the fracturing operation has not extended the depth of the original crack. This can be done by heating the specimen containing the crack to a sufficiently high temperature (500° F.) that the surface becomes blued. When

Fig. 190—Fracture Through a Fatigue Crack that has been Blued by Heating.
a) The Bar Showing the Fatigue Crack. b) The Fracture.

fractured at the crack, the original crack face will be blue, the rest of the fracture will not be. Fig. 190a shows a fatigue test bar with a number of small fatigue cracks. Fractured after blueing, the extent of the original crack is clearly shown. Note the circular propagation of the crack from the point of origin.

Another trick that can be applied to facilitate this fracturing test, particularly on small parts, is to chill the part in dry ice and then immediately fracture. At these very cold temperatures, well below zero degrees F., most steels, even those which may be soft at ordinary temperatures, will become brittle enough to break easily.

This would not be true of austentic steels, which do not become brittle at even very low temperatures.

15. SECTIONING BY SAWING. Very often it is desirable to examine a crack in cross-section, and various methods for sectioning a part are used. If the material to be sectioned is soft enough, the easiest thing to do is to saw across the crack. The depth and direction of the crack is shown on the section. This may be done either with a hand or power hack saw, or with a slotting saw on a milling machine, or with an abrasive cutting wheel.

Sometimes when large forgings, castings or other bulky parts are being investigated, a complete section is difficult and costly to make, and it is simpler to remove a piece of metal by one means or another so as to include a cross-section of the crack for its entire depth. Slotting saws, chisels, drills and various ingenious devices are sometimes necessary to get such a piece out of the larger part.

One relatively easy technique is to use a core drill. This is a hollow cutter which drills a core, usually about an inch or thereabouts in diameter, to as great a depth as may be required to reach the bottom of the crack. This method of cutting out a section is called "trepanning", and is often used in the investigation of welds in heavy plate.

When it is necessary to make a cross-sectional cut on very hard materials the usual sawing methods are either too slow or cannot be used at all because the saws are less hard than the part itself. In this case sectioning is done by means of a high speed abrasive wheel or disc. In using this method great care must be taken to avoid heating of the part during the cutting operation. Heating at the cut is very rapid, and even slight local temperature increases may extend the depth of the crack being investigated beyond its original extent. Also, such heat may change the structure locally, which may be a matter of importance. Frequently removing the specimen and cooling it under water is necessary. Much better is to flow a stream of cold water over the saw and specimen at the cut. Some wheels actually operate under water, which is undoubtedly the best technique.

16. EXAMINATION OF THE CUT SURFACE. Sometimes these cut cross-sections can be examined without further treatment, either directly with a hand glass or a binocular microscope, or by the application of magnetic particles after a suitable magnetization. More often, however, it is necessary to smooth the surface, usually done by draw-filing followed by some degree of polishing with emery cloth. Filing and polishing should be in a direction parallel to the crack to avoid dragging metal across the discontinuity and thus obscuring it from observation. Following such polishing, magnetic particles may be applied or the surface examined under the binocular microscope, with or without a light acid etch.

17. ETCHING. Light etching with dilute nitric or hydrochloric acid is often useful as a method for studying a defect either on the original surface of the part or on a cut cross-section. Deeper etching brings out flow lines, which may be important (especially on forgings), and also reveals much information regarding a defect. Figure 191 shows how deep etching on the section of an automotive

Fig. 191—Deep-Etched Section of an Automotive Steering Arm Showing Forging Fold and Flow Lines in a High Stress Area.

steering arm shows by the flow lines that the indication found with magnetic particles (arrow) was caused by a fold at the upset boss on the original forging, and not by any subsequent machining or heat treating operation.

When differences in grain structure are also of interest, the macrostructure may be brought out by etching the section with a solution of ammonium persulphate. This technique is sometimes helpful in studying the cross-section of a weld, and reveals the pattern made by the weld metal and the heat-affected zone in the base metal.

18. ETCHING CRACKS. When etching is used—especially deep etching—there is a danger that must be borne in mind. Hardened, quenched articles which have not been thoroughly stress-relieved by tempering or drawing, but which are not cracked, may contain residual stresses to such an extent that when the surface fibres are attacked by etching reagents, these stresses may cause cracks to appear in the surface as the stresses relieve themselves. Such cracking, often spoken of as etch-cracking, appears on the etched surface and is often indistinguishable from cracks which may have been present before etching. Figure 192 shows magnetic particle indica-

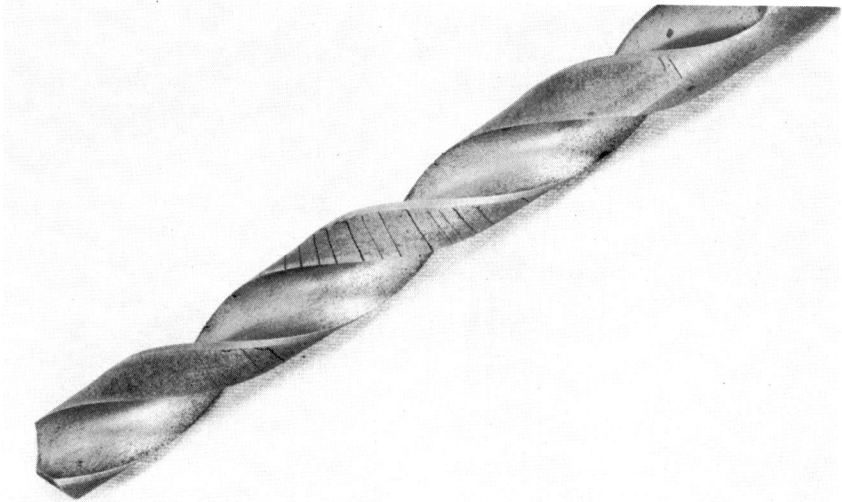

Fig. 192—Etching Cracks on a Twist Drill.

tions of cracks on the etched surface of a hardened twist drill, which were not present before etching.

Failure to recognize this occurrence has led to false conclusions regarding the presence of cracks in such articles. If, therefore, the surface of a highly hardened part is to be etched as a means of confirming the presence of cracks or of investigating cracks revealed by magnetic particle testing, the specimen should first be stress-relieved by heating to temperatures which range from 700° F. for hardened carbon steels up to as high as 1400° F. for some alloys, holding for a short time and then cooling slowly.

19. MICROSCOPIC EXAMINATION. A great deal of information as to the character and origin of a crack or other defect can be obtained by examining sections at high magnifications by metallographic methods. Such examinations are usually beyond the skill of the inspector and are usually in the province of the metallurgical laboratory. The sections, after careful preparation, are examined, either unetched, or etched to bring out the grain structure, at magnifications which may range from 50 to 500 diameters or even higher.

20. INSPECTION LIGHTING. It should be obvious that satisfactory results from an inspection operation cannot be expected if the inspector is handicapped by poor lighting and cannot plainly see the indications he may be required to judge. Still, it is surprising to note how many plants allow their inspectors to work under lighting conditions which are unsatisfactory, despite the fact that results of the inspection may be of great importance, economically, to company operations.

Inspectors should not be required to judge indications without fully adequate light. Among the requirements of good inspection lighting are, first, that it should be uniform from one inspection to another, as for instance, between the inspection unit and the quality control laboratory. Daylight varies from hour to hour, so that satisfactory lighting can only be obtained throughout the day by artificial light. Artificial light for magnetic particle purposes should be white where possible for visible magnetic particles. If colored, the color should be such that the maximum color-contrast of the magnetic particles will be brought out by the lighting. For example, when the lighting is by incandescent lights, the red magnetic particles are more visible than the gray or black.

Another important requirement is that highlights should be minimized as much as possible. Daylight, when strong, fulfills this requirement probably better than any other, but unfortunately it is not consistently available in proper intensity. It is extremely difficult to prevent highlights entirely, particularly on parts having small diameters and curved surfaces, when artificial light is being used. A properly designed lighting system, however, can do a very satisfactory job. When the indications being examined are made with fluorescent particles the same requirements apply, but are more easily met, since the light source (black light) is usually moveable, and outside sources of light which would interfere, can be minimized.

21. RECORDS. Permanent records of the appearance of indications can be of great value for a number of purposes. Records showing the typical appearance of acceptable or rejectable indications of discontinuities are useful for the guidance of inspectors in the testing of large numbers of similar parts. A record of indications of discontinuities which are subsequently investigated by sectioning or other means is an essential part of such a report. Sometimes a part is put back into service containing a known discontinuity which experience has shown will grow slowly. Comparison of the indication obtained at the next inspection with the records of the previous ones is a positive means for checking such growth rate. And records may be useful in statistical studies of the occurrence of different types of discontinuities.

22. FIXING AN INDICATION. Sometimes it is desired merely to fix an indication on the surface of a part to protect it from smearing or other damage during handling. This can be done easily by the use of a clear lacquer. The indication is developed on the part and a light coating of clear lacquer applied. Spray cans of lacquer make this a simple process.

If the indication is produced with dry powder, surplus powder should be carefully removed so as to get the indication itself in clear outline. A light lacquer spray can then be gently applied.

If the indication is produced by the wet method it is necessary to allow the bath liquid to dry before applying the lacquer.

23. LIFTING AN INDICATION. Very often it is desired to transfer an indication from the surface of the part to some permanent record place, such as a report or a notebook. There are two good and simple ways to do this.

(1) *Use of Transparent Plastic Tape ("Scotch Tape")*. The three-quarter inch width of clear colorless tape is preferred. If the indication is dry powder it is necessary only to lay a strip of the tape smoothly down onto the indication, and press it firmly along the line of the indication. The powder build-up should not be allowed to be heavy, as better delineation is obtained if the powder does not spread out when the tape is pressed into place.

The tape can be immediately stripped off from the surface, and the indication will come away also, stuck to the tape. The tape can then be put onto the page of a notebook or a report. Duplicates, if wanted, can be made by repeating the process with a fresh powder indication, or by photographic printing of the transparent tape on sensitized photographic paper. Also, the tape can be mounted on glass and used as a slide for projection on a screen if desired.

If the indication is produced by the wet method it is essential that no bath liquid remain in the particle build-up. If this is not fully dry, the indication will "squash out" like an ink-spot test and be of no value as a record. Two or three hours may be required for the indication to dry thoroughly. The drying may be hastened by carefully applying a small amount of a volatile solvent such as naphtha or carbon tetrachloride if the bath liquid was oil; or alcohol can be used if the bath liquid was water. Application and removal of the tape is then the same as for dry powder indications.

A few articles which may be useful when applying tape are a small pair of curve-bladed scissors for shaping the tape; broad-end tweezers for handling the tape; and a wooden stick or lead pencil for introducing the tape onto interior surfaces. These are, however, not essential.

It is frequently desirable to record not only the appearance of the indications, but also their locations on a part. In this case the tape should be cut long enough to cover not only the indication, but also extend to a corner, hole, keyway or other change of section which may be used as a "bench mark". If some of the inspection bath or dry powder is allowed to fall on this base point the tape will pick it up and thus reference the location of the indication on the part.

(2) *Use of Stripable Lacquer.* A clear lacquer is now available in spray cans which, when dry, can be stripped from the surface of a part, giving a lacquer strip similar in character to the transparent plastic tape. This material is very easy to use and gives a clearer film than does the plastic tape. This is a desirable characteristic when the record is to be duplicated by photographic printing or used as a slide for projection.

Dry powder records should be made from clean surfaces free of powder background. The lacquer is sprayed along the indication with as little scatter as possible, then built up in sufficient amount to give a film thick enough to permit stripping off. Stripping can be done as soon as the lacquer is dry. The strip of removed lacquer can be trimmed to suitable length and width with scissors or knife.

With wet method indications the bath liquid should be removed by a solvent or allowed to dry as with the tape transfers. The stripping lacquer is then applied as above.

The records made with stripping lacquer are likely to be better than those made with tape because the film is more transparent, and because the indication has not been spread even a little bit by pressure, as it may have been in the tape process. Also the particles are incorporated into the lacquer film and not just stuck in the sticky surface, as in the case of the tape. This makes the record more durable if it is handled very much, or for long-time records.

24. PHOTOGRAPHY. Photography is one of the best ways to make permanent records of the appearance of indications, but takes more time and equipment than the tape or lacquer techniques. Also, if precise delineation of the contour of indications is important, the tape or lacquer records lifted directly from the surface of the part can be more exact. A photograph, on the other hand, shows the indication in its natural environment on the part in which the discontinuity occurs, and has this advantage over the lifted records.

Black and white photographs of parts containing indications sometimes require some ingenuity in "posing" the specimen and in securing lighting that sets off the indication so that it gives a clear picture and be a faithful reproduction of the indication on the resulting photograph.

Use of the Polaroid camera and film process is a most convenient and quick method of making photographic records. The immediate availability of the picture makes it possible to make corrections in lighting or positioning quickly. Several successive shots can be made if duplicates are wanted.

25. BLACK LIGHT PHOTOGRAPHY. When the fluorescent particle method has been used, photography of the results is the only way to get an accurate record of the appearance of the indications. Photography under black light is a bit "tricky", but techniques which are fairly easily followed have been worked out, and with some experience and care, good photographs of fluorescent indications can be taken to make a striking record of their appearance, in either black and white or color.

In addition to the technical skill required of the photographer, he is also under the obligation to see that the photograph neither exaggerates the size and brilliance of the indication, nor minimizes it. This requires good objective judgment, since photography can in this instance easily give a false value to the appearance of the indication, as compared to what the eye actually sees.

In spite of the difficulties, photographs are often desirable and the following outline, based on experience, will make the task easier for the photographer who has not previously attempted to take pictures under black light. Figure 193 shows diagramatically a suggested set-up for this purpose. A K2 or G filter over the lens of

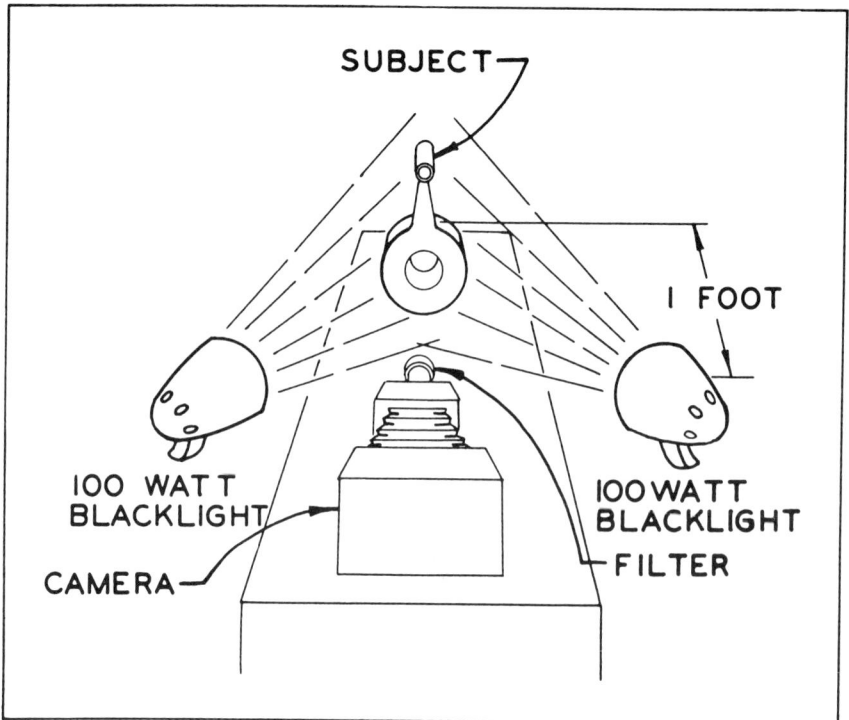

SUBJECT

1 FOOT

100 WATT
BLACKLIGHT

CAMERA

100 WATT
BLACKLIGHT

FILTER

Fig. 193—Diagram for Set-Up for Black Light Photography.

the camera is essential to filter out the *black* light which would affect the film. A fast panchromatic film is preferred. The parts should be cleaned of all fluorescent background, and set up in the dark room with the two 100 watt black spot lights placed so as to bring out the brilliance of the fluorescent indications with as little reflective high-lighting of the part as possible. A light colored non-fluorescent background is usually desirable so that the black outline of the part shows in silhouette against it.

The exposure time varies greatly with the brilliance of the indication. With a G filter exposures vary from 20 minutes at $f/32$ for heavy bright indications up to an hour at $f/22$ for fine indications of low intensity. If the thinner K2 filter is used, the exposure time is cut to about one half and the definition of the part is improved, but undesirable highlights from reflected black light may come through stronger.

To increase the definition of the part as a whole and separate it from the background, white light may be used for a short time during the exposure, but the white light should be placed so that it does not illuminate the indication areas either directly or by highlight.

Since there is no practical means for pre-calibrating exposure, one or more test negatives should be made for each set-up. A Polaroid camera is very useful for making these test negatives. The negative after normal development should show the indications solid black, but not spread wider than they appeared on the specimen. The rest of the negative should be thin, but with parts clearly defined. It should be checked especially for highlights interfering with or resembling the indications. Such highlights can often be moved, weakened or diffused, or eliminated entirely by re-positioning the lights.

Printing can be handled normally, usually on medium contrast paper. The object is to produce a quite dark impression of the part as it would appear under black light, with clear white indications in the picture where fluorescent indications appear on the part. If color transparencies are to be reproduced later, the "color correction" techniques should be used, to hold the high contrast within reproducible bounds.

26. Polaroid Film Technique. Use of Polaroid film has considerable advantage in photographing fluorescent indications, since trial shots can be made and corrections applied in much less time than with ordinary film. There is much less experience behind the technique for the use of the Polaroid camera, but the following is a suggested procedure which has been successfully used.*

In a dark room two 100 watt black lights are positioned on either side of the fluorescent indication at 24 inches distance and 45° angle from it. A 4″ x 5″ camera with a G lens filter is used. Focusing is by ground glass camera back. Positioning of the camera is such as to secure a full size reproduction of the indication, although this may be altered depending on the size of the indication and the part. A 4″ x 5″ Polaroid film holder with Type 57, 3000 speed film is used. Exposure time is one to three minutes at f/22 lens opening.

The photograph is then copied using 55 P/N Polaroid film to secure a negative for making multiple contact prints.

27. COLOR PHOTOGRAPHY. Photography with color film is sometimes attractive, but exposure time and lighting are even more difficult to work out than with black and white film. The photographer must first have mastered the technique of black and white photography, after which the challenge of fluorescent pictures in color may be sufficient for him to make the attempt. It is not, however, recommended as a method of *recording* indications, unless fairly repetitive tests are involved, to permit standardized techniques.

28. SUMMARY. After reading the foregoing pages of this chapter the reader may have received the impression, perhaps a discouraging one, that the interpretation of magnetic particle indications is such a difficult and complicated matter as to be beyond the ability of the average inspector to master. It has certainly not been the intention to exaggerate the difficulties of interpretation, but rather only to emphasize the absolute necessity of his acquiring some experience before too much reliance can be placed on his judgment, and to point out the kind of experience required and the sources from which it can be obtained.

That satisfactory interpretation, for all practical purposes, is not too difficult of attainment is borne out by the extensive use of the magnetic particle method after many years of evolution and development. It should also be repeated that the problems of interpretation and evaluation exist to as great or a greater extent

*Reported by S. W. Gearhart, Birdsboro Corporation, Birdsboro, Penn. 1963.

in every other method of nondestructive testing, partly because service requirements are not always well defined.

29. EVALUATION. We have stated that "evaluation" of a defect means the determination, involving a decision by some person or persons who are charged with that responsibility, as to whether the condition found by magnetic particles in a given part is cause for rejection, or whether the part may be either used as is or salvaged.

The decision is an entirely separate one from that involved in interpretation, since it is usually only where pre-determined standards have been well established that parts are automatically accepted or rejected on the basis of the mere presence of a certain type of discontinuity. Furthermore a great many factors enter into such a decision which are in no way connected with the matter of locating, and then determining the nature of, a given discontinuity.

30. THE PROBLEM OF EVALUATION. The problem of evaluation is one of whether a given discontinuity is of such a magnitude, or shape, or is so located that it will endanger the satisfactory performance of the material or part in the service for which it has been manufactured or designed. Some considerations entering into the answer to this question are the following:

(a) What type of actual service stresses is the part to be called upon to withstand, and what is their magnitude, direction and duration—that is, will the stress be steady, or pulsating or reversing? Have these been positively and experimentally determined, with reasonable accuracy?

(b) Where does the defect lie with respect to these stresses—is it in an area of high stress or low stress, and is it parallel to or at some angle to the maximum stress direction?

(c) What is the nature of the defect and how severe a stress-raiser is it? Also, how is it oriented with respect to other stress-raisers that may be present due to design or construction of the part?

(d) What is the importance of the part to the entire structure or assembly, and how serious would the result be if the part failed?

(e) What is the history of experience with similar parts in similar service—have they a record of frequent or occasional failure, or do they never break? And, have design or fabrication techniques changed, to improve or deteriorate this history record?

It is scarcely within the scope of this book to discuss at great length these many and complex factors, since they vary in nature and importance with each application and design. It is rather the purpose here merely to point out some of the more obvious considerations as they may influence the end result of inspection with magnetic particles.

31. DESIGN. It was one of A. V. de Forest's favorite comments that many more failures of structures and machines are due to lack of care in design than are due to defects in the materials of which the members are made. This "lack of care" does not refer to errors in estimating expected loads or in selection of materials which are in themselves unsuitable for such service.

What is meant is failure to evaluate with sufficient exactness the probable distribution of stress in the part or assembly; or rather, failure to recognize the stress-raising effect of certain design features, and to allow for or avoid the stress concentrations which are thus produced.

Modern methods of stress analysis make possible today's design by locating serious stress concentrations in advance; and design can now take the necessary steps to reduce such concentrations to values below the danger point for the materials in question.

32. STRESS-RAISERS. Even if we assume that all possible use has been made of such stress analysis tools, the designer still can only design for calculated or field-test-measured stresses, or make reasonable allowance for foreseeable overloads. Stresses in service are sometimes raised far above those for which the part is designed, by conditions outside the designer's control or his ability to foresee and provide for.

Some of these conditions are listed below:

(a) The accidental presence of notches such as cracks, scratches, etc., which act as stress-raisers.

(b) Residual stresses, remaining from fabrication, heat treating or assembly processes.

(c) Unknown distribution of stresses due to shape or size, for the pre-analysis of which available methods may have been inadequate or impractical.

(d) A loading applied for any reason, in service, different from that designed for.

(e) Vibration.

(f) Overstress due to accident.

33. STRENGTH OF MATERIALS. The use of a "factor of safety" in design is a device for allowing for unknown or unpredictable stress concentrations or operating conditions. When weight is not a primary consideration the factor of safety may be high—the part or structure may be made five or ten times as heavy as the best calculations the designer has been able to make indicates that it needs to be. A large factor of safety increases costs, however, and still does not necessarily make sufficient allowance for stress-raisers. Fatigue cracks continue to occur in massive forgings and castings, due to locally high stresses resulting from the presence of stress-raisers.

Even with the extensive knowledge now available of the behavior of materials, and with the information obtainable by modern methods of stress analysis, a designer *must* at some point assume a value for the strength and other properties of the materials he decides to use. He must—or should—take into consideration the very different behavior of these materials under different types of loading—static, dynamic or impact. And for services where weight *is* a primary consideration, as in the aircraft and missile fields, he designs as closely as he dares to this assumed strength, cutting his factor of safety as low as possible. In the design of missiles, utilization of the strength of materials well up toward the ultimate strength is not uncommon. It is therefore of the utmost importance that the strength of the material *actually entering into the structure* be as near as possible to the *strength capability he has assumed,* and that *accidental* stress-raisers be not present.

Magnetic particle testing, along with other nondestructive testing methods, is one of the links in this design chain, since it provides a means for insurance against stress-raising or other weakening conditions for which the designer has not or could not allow. However, intelligent evaluation of a condition revealed by magnetic particles must be preceded by knowledge and understanding of the *extent* to which the designer has *counted on freedom of the finished part from such a condition.*

34. OVER-INSPECTION. The purpose of magnetic particle testing as applied to materials and parts is, therefore, to help make sure that the finished articles are as good as they need to be to do the job for which they are designed; but as there is in general no need for them to be any *better* than such considerations demand, intelligent evaluation must be on guard against over-inspection.

A condition which is considered objectionable in one part for a particular service may not be at all objectionable in another part intended for a different service. Examples could be cited in which parts containing defects of a certain size and shape have been placed in service without any detrimental results, while defects of a similar size and shape in other parts, more highly stressed perhaps, or in more critical locations in the same part, would have been considered damaging enough to have warranted rejection. Evaluation, if intelligent, must therefore come down to a consideration of each individual case in the light of all possible knowledge as to the design of the part and the service for which it is built.

Avoidance of *over*-emphasis on magnetic particle indications is, therefore, important—and such over-emphasis is a definite tendency where complete information and sufficient experience are lacking. To "play safe", inexperience will cause the scrapping of many good parts, since to have the courage to accept parts containing discontinuities requires knowledge and the intelligent acceptance of the responsibility involved.

35. GENERAL EVALUATION RULES. Everything that has been said in this discussion thus far has emphasized the fact that general rules for evaluation of conditions revealed by magnetic particles cannot be laid down. And general rules are not necessary when the evaluator is equipped with adequate knowledge and experience. Nevertheless, inspectors are sometimes—all too frequently, in fact —called upon to make decisions regarding the seriousness of a defect when they lack such adequate information.

As a guide for inspectors, a few basic considerations—not rules— may be stated which, even though more or less obvious after a thoughtful analysis of the subject, may still be of some help when set forth as they are below:

(a) A defect of any kind lying *at the surface* is more likely to be harmful than a defect of the same size and shape which lies wholly below the surface. The deeper it lies below the surface the less harmful it is.

(b) Any defect having a principal dimension or a principal plane which lies at right angles or at a considerable angle to the direction of principal tension stress, whether the defect is surface or sub-surface, is more likely to be harmful than a defect of the same size, location and shape lying parallel to the stress.

(c) Any defect which occurs in an area of high tension stress must be more carefully considered than a defect of the same size and shape in an area where the tension stress is low.

(d) Defects which are sharp at the bottom, such as grinding cracks, for example, are severe stress-raisers and are, therefore more likely to be harmful in any location than rounded defects such as scratches.

(e) Any defect which occurs in a location close to a keyway or fillet or other design stress-raiser is likely to increase the effect of the latter and must be considered to be more harmful than a defect of the same size and shape which occurs away from such a location.

36. PROCESS SPECIFICATIONS. The problem of evaluation has been solved in another way in certain large plants where many parts of the same design are manufactured, and where acceptance and rejection must be expedited in the interest of production output. Qualified engineers have taken each part of a given design, considered the defects which might be encountered, and set up standards *for that particular part* which can take into consideration once and for all *for that part* all engineering and service requirements. Such specifications should include an exact statement of how the part is to be magnetized, the sequence of magnetizations if more than one is required, the amount and kind of current to use, and any other necessary points to insure that the inspection is carried out in identical manner on all parts of the type, no matter by whom or on what kind of a unit. Rejectable and acceptable conditions can then be spelled out and acceptance or rejection becomes "automatic." But if the design of the part is changed, resulting perhaps in a different stress pattern in the part, the test specification must be revised.

37. SUMMARY. This discussion of the problems attending the evaluation of the results obtained by use of magnetic particle testing has sought above all to present in proper perspective the part which the method should play in the business of producing better, safer and more reliable machines and structures. The concept that a "defect is only a defect when it affects the useability of the particular part in which it occurs," is fundamental.

This concept implies a further principle which should also be appreciated, namely, that this test method (as indeed any other) should be used only when the information it is able to give is significant and critical from the point of view of determining the suitability of a particular part for its intended service.

One additional thought is also worth keeping well in mind, and that is that the most valuable use of magnetic particle testing, as well as any other nondestructive testing method, both in the inspection of new materials and of materials that have been in service, lies not in the rejection of unsuitable parts, but in helping to locate the reason for the occurrence of the defects which it detects, to the end that processes can be corrected or designs improved so that future occurrence of similar defects can be avoided.

CHAPTER 23

INDUSTRIAL APPLICATIONS

1. INDUSTRIAL USES OF MAGNETIC PARTICLE TESTING. Magnetic particle testing, with increasing usage over 35 years, has become a standard testing practice in most sections of industry where iron or steel is made, fabricated or used for important end use. Its applications are so numerous and varied that it would be difficult to give examples of all of its industrial uses.

It is the intention in the following sections to give a brief outline of the several principal purposes for which the method is being used, with a few examples of interesting or unusual applications.

2. CLASSIFICATION OF MAGNETIC PARTICLE TESTING APPLICATIONS. The following is a list of the principal industrial use-areas for the method:

(a) Final Inspection.

(b) Receiving Inspection.

(c) In-Process Test and Quality Control.

(d) Maintenance and Overhaul in the Transportation Industries.

(e) Plant and Machinery Maintenance.

(f) Testing of Large Objects and Components.

3. MAGNETIC PARTICLE TESTING FOR FINAL INSPECTION. Historically, when the method was new, final inspection was the first area in which it was used. The test was applied to finished articles before they were shipped to the customer, to make sure that the product shipped was not defective. Customer dissatisfaction, costly returns and adjustments, handling of replacements and rework were all reduced by this means. At the same time reputations for quality products could be built up. One manufacturer of tool steel was so sure of the quality of his product after installing the method that he publicly advertised that he would replace without cost any bar or billet in which a seam was found. All this was—at that time —sufficient incentive to add the relatively inexpensive testing procedure of magnetic particle nondestructive testing. Final inspection was a necessity anyhow, and this new procedure was a refined tool.

Today the emphasis has shifted toward in-process testing, so that defective material and parts can be culled as early as possible, and before final inspection. In many instances final inspection is, however, still a must. This is especially true with parts on which a final operation may be the source of surface defects. The largest group of products in this class is precision hardened and ground machine parts, such as gears, splined shafts, crankshafts, and innumerable jet engine, aircraft and automotive parts as well as many other articles of this type.

One good example of final inspection with magnetic particles is the case of hand tools such as hammers, axes, chisels, cutting and planing blades, etc. Heat treatment—hardening and tempering— followed by grinding of edged tools, is the final operation in their manufacture. Presence of grinding or quenching cracks is highly objectionable for safety for the user, or early failure of the tool. Final inspection, usually with fluorescent particles, can guarantee absence of cracks in tools shipped to customers.

In the automotive and aircraft manufacturing fields, many hardened and ground precision parts are used. These include gears, valves, steering knuckles, spindles, axles, crank- and cam-shafts and many others. Such parts are commonly given a final inspection

Fig. 194—Semi-automatic Unit for the Final Inspection of Automotive Steering Knuckles.

before assembly. Figure 194 shows the loading end of a semi-automatic unit used for 100% final inspection of automotive steering knuckles. The parts are being magnetized circularly between the heads of the unit shown. A stainless steel (non-magnetic) conveyor carries the magnetized parts under a curtain of fluorescent particles in water suspension, and then to a hooded inspection booth where they are examined under black light. Subsequently the parts go through a second station, are magnetized longitudinally and bath reapplied, and again inspected. The parts are hard and have good retentivity, lending themselves well to this version of the residual wet method.

4. RECEIVING INSPECTION. In plants where many finished or semi-finished parts are purchased from other manufacturers, the function of receiving inspection is to make sure that such parts, when they are received from the vendor, are suitable for further processing or use. When large numbers of a given kind are involved in a lot, the inspection tests are often made on samples taken from the lot according to statistically prescribed rules. If the sample lot shows defects beyond a set tolerance limit, the lot may be inspected 100%. Extremely critical parts are inspected 100% in any case. Magnetic particle tests are in many cases part of the receiving inspection program, whether or not the specifications under which the parts are purchased have required the manufacturer from whom the parts were bought to make such tests.

Receiving inspection, however, encompasses much more than the check-testing of parts received from outside sources. Testing of raw material when received, before any work is done on it, is a very profitable way in which to weed out any initially defective material. Such inspection of incoming raw material is really the starting point of a program of in-process testing. Any work done on material or parts already defective is a costly waste of time and effort. Magnetic particle testing is used extensively on incoming rod, bar, forging blanks, rough castings, etc.

As an example, in the case of coil spring manufacture, such as valve springs, bars or wire, rod-ends are tested for seams with magnetic particles prior to drawing the spring wire. Seams are particularly objectionable in wire for this purpose, since they lead to early fatigue of coil springs. Cold-heading stock is another example where seams lead to defective bolts, since the heads will split on up-setting. In this case the rod is received in coils, and these are tested by examining a foot of rod cut from each end of a coil. The pieces are magnetized circularly and checked for seams using fluorescent particles. Fig. 195 is a view of an automatic unit for this purpose. The pieces of rod are placed on the fixtures by hand and carried by conveyor through the magnetizing station where the current is passed through them and they are flooded with the fluorescent particle bath in water suspension. The conveyor then takes them to the black light inspection booth.

Figure 196 shows another installation used for the inspection of incoming castings and forgings, using the fluorescent magnetic particle method on a semi-automatic unit.

5. IN-PROCESS INSPECTION. In the last ten years or more industry has fully realized that one of the most valuable uses of nondestructive testing is the detection of flaws in products at various stages in the production process. It is easy to see that a part, cracked or otherwise damaged in some operation during its manufacture, should be withdrawn from the production line at once, before further expensive work is done on it. The basic rule is, *find and remove a defective part as soon as it becomes defective.* To accomplish this may require two, three or even more inspections (at least on a sampling basis) before a part is finished. The cost is often justified by the increased control of the process, and the savings in labor and machine time that would have been wasted on the defective piece.

A typical example is in the manufacture of hardened and ground precision gears. In a modern plant the in-process testing starts with the raw material, and blanks are tested with magnetic particles for seams or other surface defects. The production sequence is arranged to facilitate the testing program. On a forged blank, for instance, the first operation is to machine the hole for the shaft. At this point internal defects such as pipe or forging bursts, associated with the center of the piece, are opened up and can be detected. After hardening, another magnetic particle test can be made for quenching cracks.

The final operation is grinding of the teeth and surfaces of the gear to close tolerances. In the case of grinding cracks, the testing is done on a basis of keeping the process in control, and portable magnetic particle units are moved from grinder to grinder to check the work as it is being done. Each unit may take care of up to 30

Fig. 195—Special Automatic Unit for Testing Rod Ends.

Fig. 196—Semi-automatic Unit for the Inspection of Small Castings and Forgings.

grinders. Tests are concentrated on new set-ups, to make sure that the feed and coolants are correct, that the proper grit wheel is being used, and that the set-up is producing no cracks. When the time for dressing the wheel has come on a machine, the testing unit is again on the job to make sure that "loaded" wheels are not causing cracks at that stage. Such a program reduces losses from grinding cracks to a very low percentage. Figure 197 shows some cracked gears, with indications produced with fluorescent magnetic particles.

Fig. 197—Fluorescent Magnetic Particle Indications of Grinding Cracks.

Figure 198 shows an induction hardening operation being carried out on conveyor wheels. After high-frequency induction heating, they are quenched in water. Cracks sometimes result from this severe treatment and magnetic particle testing is applied 100% as the next operation. The insert shows cracks indicated by fluorescent magnetic particles on a similarly hardened part.

At the upper end of the size range, Fig. 199 shows the 100% production inspection of seamless tubes, ranging from 4 to 10 inches in diameter and up to 40 feet long. The automatic unit magnetizes the tube by passing direct current through it while fluorescent magnetic particle bath—a water suspension—is applied. Inspection under black light reveals surface seams, tears and other damaging defects. This is really a final inspection, coming at the end of the

Fig. 198—Quenching of Conveyor Wheels, Followed by Magnetic Particle Testing for Quenching Cracks.

production process, but prior to end trim, threading or cutting into shorter lengths, for pipe couplings for example.

6. MAINTENANCE AND OVERHAUL IN THE TRANSPORTATION INDUSTRIES. The transportation industries—railroad, truck and aircraft—were early users of magnetic particle testing. Planned overhaul schedules at which critical parts are tested for fatigue cracks have reduced equipment failures in service to an extremely low value. When first installed on a western mountain railroad—thirty years ago, before the age of diesel power—many severe road failures with costly lóss of equipment and freight were occurring. During the first few months of magnetic particle testing, 45% of all rods and moving parts of locomotives were found cracked. After two years of a planned overhaul program, during which, by means of magnetic particle testing, all cracked parts were repaired or replaced, road failures due to fatigue cracking of parts became almost unknown.

The lesson was well learned by the railroads, and today every overhaul shop for diesel locomotives is equipped with magnetic particle testing equipment. Figure 200 shows a unit for the inspection of diesel locomotive chankshafts, as well as for many other miscellaneous parts. Provision is made for both circular and longtitudinal magnetization, and for the use of fluorescent particles. Figure 201 shows typical indications of fatigue cracks found in traction gears.

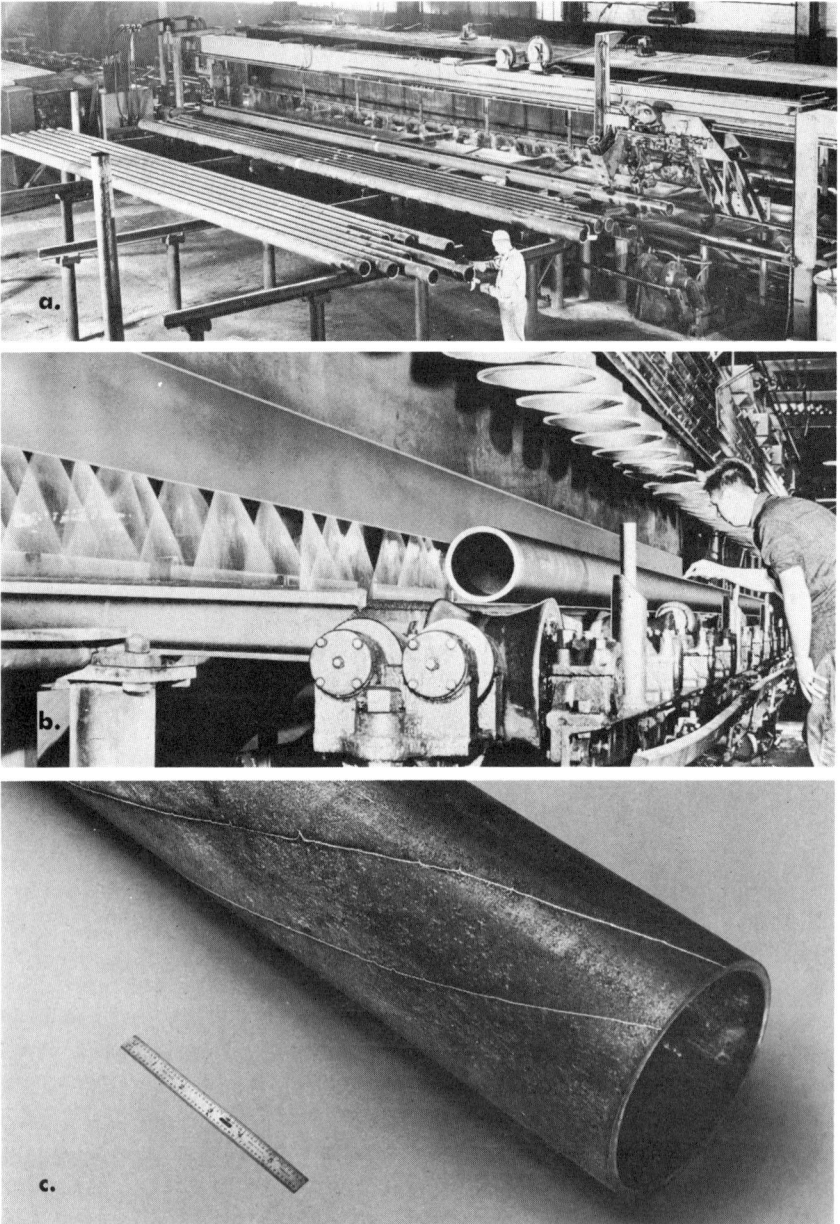

Fig. 199—Magnetic Particle Testing of Seamless Tubing on an Automatic Unit.
a) Loading Side. b) Inspection Station. c) Fluorescent Particle Indications
of Spiral Seams.

INDUSTRIAL APPLICATIONS

Fig. 200—Sixteen Foot Long Magnetizing Unit for Diesel Crankshafts, Installed in Overhaul Shop.

Fig. 201—Typical Fatigue Cracks in Gears.

The safety of aircraft operation today has become something generally taken for granted, and airline aircraft accidents are practically never traceable to failure of structural or engine components. This situation was not achieved, and is not maintained, without the most intense and painstaking tests at intervals dictated by experience. Every available method of nondestructive testing is employed in the elaborate facilities for engine and aircraft overhaul which the airlines maintain. The intervals between overhauls is measured in hours of engine operation, and have been lengthened as designs and testing procedures have been improved. Spot tests at shorter intervals, without dismantling the aircraft, are made on critical components such as landing gear parts.

Magnetic particle testing and fluorescent penetrants play a large part in this program. Figure 202 shows the magnetizing operation on a compressor disc of a jet engine, and Fig. 203 the examination

Fig. 202—Special Magnetizing Unit for Jet Engine Compressor Discs, Employing the Induced Current Method.

of a compressor rotor assembly for transverse defects in the blades. Both operations are employing magnetizing units designed especially for these parts.

Figure 204 shows the use of a special portable magnetizing unit in making tests on aircraft parts in the field. It is shown being used to examine a helicopter landing gear, with the dry powder method.

Fig. 203—Special Magnetizing Unit for Aircraft Gas Turbine
Compressor Blading—Assembled.

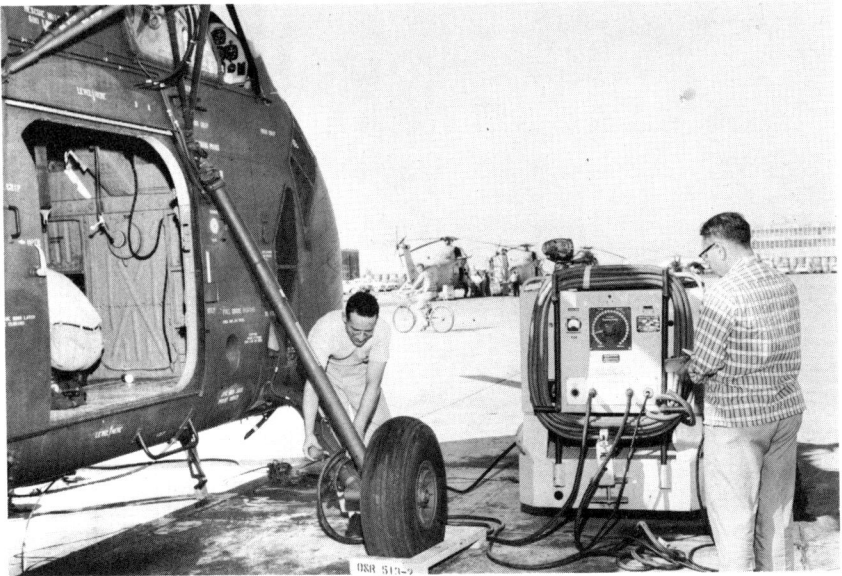

(Courtesy Jacksonville Naval Air Station)
Fig. 204—Testing Landing Gear Components of a Helicopter on
the Field, Using a Special Portable Unit.

429

Automotive overhaul maintenance programs have been particularly effective for keeping truck and bus fleets and off-the-road equipment in safe operating condition. Many operators of large fleets of vehicles have installed shops for tear-down, inspection and repair and re-assembly of engines and vehicle components, follow-

Fig. 205—Truck Engine Connecting Rods and Caps With Fatigue Cracks.

Fig. 206—Testing Truck Front Wheel Spindle with Portable Unit.

ing the plan used by the airlines and railroads. The result has been a large reduction in service failures of equipment, and the costly accidents and delays resulting from them. Figure 205 shows two truck engine connecting rods with fatigue cracks well advanced in both rods and caps. Figure 206 shows the coil of a portable magnetizing unit being used to test the front wheel spindle of a truck without removing it from the assembly.

7. PLANT AND MACHINERY MAINTENANCE. Planned inspection programs are also widely and profitably used in keeping heavy equipment in operation without breakdowns while in use, which can be very expensive and often unsafe. Punch press crankshafts and frames, because of the sudden and severe stress applications, are particularly vulnerable to fatigue failures. With small portable magnetizing units, checks can be made of known or suspected trouble spots, and fatigue cracks found early in their progress, without dismantling the press. When the cracks are found in time, repair by welding is in many cases easily accomplished.

The plant safety program also makes good use of portable magnetizing units. Crane hooks become work-hardened on the inside surface of the hook where concentrated lifting loads are applied. Fatigue cracks develop in this area and often propagate quite rapidly. Testing all crane hooks in a plant at intervals of a few weeks or months, depending on the service they are subjected to, is a safety requirement in many plants. The lifting fingers of fork trucks are also very vulnerable to fatigue failure. Figure 207 shows the magnetic particle test being applied, and the fatigue cracks found at the inside bend radius of the lifting member.

The magnetic blading, shaft and case of steam turbines have long been profitably examined for incipient failure at planned down-times. Magnetic particle testing with portable equipment has played an important part in detecting fatigue cracking in power-generating steam turbines over a period of 30 years or more. Figure 208 shows a trailer containing magnetizing equipment being lowered to the turbine room floor. The spindle has been lifted out of the case and mounted on temporary bearings so that it can be rotated while blading and shaft are tested with portable magnetic particle equipment.

8. TESTING OF LARGE AND HEAVY ARTICLES AND COMPONENTS. One of the characteristics of magnetic particle testing which makes

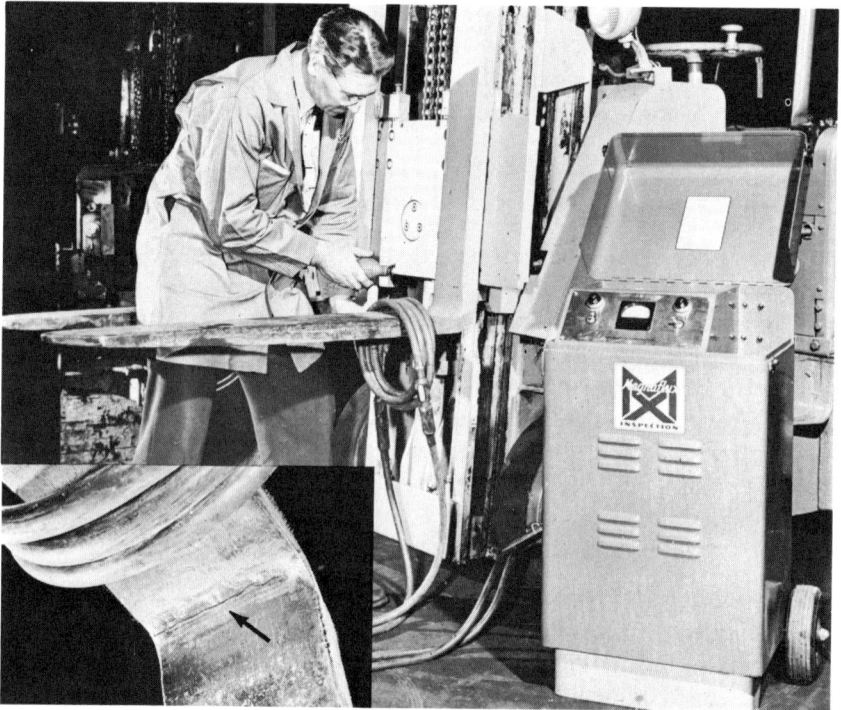

Fig. 207—Testing the Lifting Fingers of a Fork Truck.
Inset: Close-up of Fatigue Crack Indication.

Fig. 208—Preparations for Testing Steam Turbine
Spindle and Blading for Cracks.

it applicable to such a large variety of industrial testing problems, is the fact that size and shape are in no case a deterrent to its use. Tiny nuts and bolts can be inspected equally as well as large machines and structures. The automatic testing of huge billets weighing a number of tons on massive equipment has been described in Chapter 19.

Also described in Chapter 19 was the application of magnetic particle testing to the inspection of solid fuel rocket engines and motor cases. Figures 209 and 210 illustrate the equipment used for the testing of the Polaris motor cases. Figure 209 shows the unit for testing the case, and Fig. 210 shows the testing of the domes and nozzles.

Fig. 209—Special Unit Designed for the Inspection of Solid Fuel Missile Motor Cases.

In the case of liquid fuel rocket engines, thrust units, fuel tanks and all pumps, piping and tubing are carefully inspected with every possible kind of nondestructive test. The problem here is not one of guarding against fatigue cracks to ensure long life, since the rockets to date give only "one-shot" service. Here the object of the inspection is to find and eliminate any defect which could cause the assembly to malfunction in any way when fired. Magnetic particle testing is used on all ferromagnetic parts to make sure that no crack or other defect is present which would lower the strength of the part below that assumed by the design. Figure 211 is a view of three thrust units of a liquid fuel rocket engine, seen from below, mounted

Fig. 210—Special Unit for Inspection of Domes and Nozzles.

(Courtesy Rocketdyne Division, North American Aviation Corporation)
Fig. 211—Thrust Units of a Liquid Fuel Rocket Engine, Mounted on the Test-Firing Stand.

on the test-firing stand. And Figure 212 shows the magnetic particle test being applied to butt welds in a cylindrical motor chamber. Permanent magnet yokes with dry powder are being used. These are most convenient since there are no electric cables or power packs required; and effective because only a very small section of weld is inspected at one time, and the yoke gives a strong concentrated field for this purpose.

(Courtesy Rocketdyne Division, North American Aviation Corporation)
Fig. 212—Magnetic Particle Testing of Butt Welds on a Chamber of a Liquid Fuel Rocket Engine, Using Permanent Magnet Yoke.

9. SOME UNUSUAL SPECIAL APPLICATIONS. The versatility of the magnetic particle method has found for it some interesting and out-of-the-ordinary applications. One of these is the testing of racing car engines and car parts before the annual Memorial Day "500" race at Indianapolis. This year—1966—marks the 30th in which cars have been tested with magnetic particles at the track. Some critical parts, such as the steering parts, crankshafts, connecting rods, etc., must pass magnetic particle inspection before the cars are permitted on the track. In the early days of the application of magnetic particle testing, some drivers objected to the test even though many accidents had previously resulted from steering part failures. Their fear was that cracks would be found that would keep them out of the race. This attitude has changed completely, and the freedom from failure of parts during the race which the tests

Fig. 213—The "500" Race at Indianapolis Speedway, Start of the 1965 Race.

help to assure, with increased chances for completing the race, has now made such tests welcome.

In one recent race the record favoring such testing is impressive. Of 22 of the cars qualifying, only three had not had their engines and transmissions tested. Of these three, two dropped out of the race—one at 68 laps because of drive shaft troubles, and one at 80 laps due to gear failure. Nine of the first ten cars to finish had had their engines and other critical parts inspected, and in seven of these, cracked parts had been found and replaced.

Another interesting application was the inspection in the field of a 24 inch gas transmission pipe for a three mile crossing of the Mississippi River. Although the pipe had already been inspected at the mill, extraordinary steps were taken to make sure that no failure would occur in this critical service. In order that all sources of possible failure were eliminated, magnetic particle testing was used at the site to inspect the entire outer surface of the pipe. By means of magnetic leech contacts spaced ten feet apart along the pipe, 1500 amperes of direct current were passed to magnetize the pipe circularly. Figure 214 shows a close-up of this operation. Dry powder was used, and all indications were ground out with a small hand grinder. See Figure 189-b, Chapter 22. If the discontinuity could not be removed without going deeper than .060 inch (pipe wall was 0.500 inch thick) the section containing the defect was rejected and cut out of the line. Defects found consisted of cracks, laps, laminations or other breaks in the surface. A total of 13 sections, from three inches to twenty inches in length were found to contain rejectable defects. In addition to magnetic particles, radiography and ultrasonics were also used to check the welds and the wall thickness.

Fig. 214—Close-up of Inspection of 24 Inch Line Pipe,
Three-Mile Mississippi River Crossing.

Testing of steel billets for more accurate conditioning before they are rolled into finished sizes and shapes has already been described. An unusual application in one plant is the testing of 2″x2″ billets of a nickel-iron alloy, used for drawing into fine wire for the electronics industry. The wire, used for lead-wires into encapsulated devices, must be free of seams to avoid leakage of air into the capsules. The alloy was sufficiently magnetic that indications of seams could be obtained on the billets. By careful removal at the stage, the resulting wire when drawn down to very small diameter, was free of any seams.

In the laying of 2400 miles of telephone cable between the mainland and Hawaii, and also a trans-Atlantic cable, magnetic particle testing played a part in insuring trouble-free performance. In this length of cable, 114 repeater stations were required. These are complex electronic units, incorporated into the cable, required in long transmission lines to keep the signals clear and separate. The devices are protected with numerous layers of insulation, waterproofing and protection from physical damage. Structural strength without sacrifice of flexibility is given by a series of hardened steel collars. These rings were inspected with fluorescent magnetic particles for seams, splits, surface nicks or scratches, and any minor surface irregularities. Defects were found in six to eight percent of the rings!

Fg. 215—Testing of Protective Steel Collars on the First Trans-Pacific Telephone Cable. Collars are Magnetized on a Large Diameter Central Conductor.

Figure 215 shows the method of magnetizing, using direct current. Four rings at a time were placed on the central conductor for magnetizing and inspection with fluorescent magnetic particles. At the right are shown four rings with indications of surface seams or cracks.

One application involved the use of the method to detect defects in a non-magnetic material. Cast aluminum rotor bars for electric motors were inspected for internal voids using fluorescent magnetic particles. Direct current of moderate intensity was passed through the bars and the fluorescent magnetic particle suspension flowed over the surface. The magnetic particles arranged themselves in a uniform pattern on all surfaces of the bars. The pattern consisted of circumferential lines, corresponding to the lines of force around the bar, as in a magnetograph. Bars without voids showed an all-over uniform coating of fluorescent particles. The adherence of particles was, of course, produced by the *external* magnetic field due to the current the bars were carrying. When voids were present, the pattern of particles was distorted at the location of the voids, where the current density in the bars changed because of the reduction in the conductive cross-section of the bars at such location. Presence of voids was thus clearly indicated.

CHAPTER 24

TESTING OF WELDMENTS, LARGE CASTINGS
AND FORGINGS

1. INTRODUCTION. The testing of large objects and structures presents some special problems for which special techniques have been developed. The inspection of welds in moderate to heavy sections is included in this group, for the reason that special procedures are required.

In the case of weld testing, the article itself may be large, as for example a pressure vessel, or a 60 story building, but the inspection is normally confined to the local area of the weld itself. This has led to the use of yokes and prods, which are especially suited to such local magnetizations. The use of prods and the requirements of current values and contact spacing have been thoroughly discussed in Chapter 10, Sections 6 through 9. Some additional details of weld inspection techniques will be outlined in the succeeding sections of this chapter.

Large castings and forgings, too large to handle or be accommodated by standard, or fixtured magnetizing units, are also inspected with prods or yokes. However, in this case the local nature of the fields produced are a handicap, since, unlike welds where the location of sought-for defects is known, the entire large surfaces of castings and forgings must be painstakingly gone over, point by point—a laborious and time-consuming project. The introduction of high output multi-directional (Duovec®) power packs, delivering up to 20,000 amperes or more, has made possible overall magnetization of medium large articles. This system achieves an increased effectiveness in defect location, and a great reduction in time required for the inspection. It is described in detail in Chapter 19, Section 18.

2. WELD DEFECTS. Figures 216 and 217 show, respectively, the terms used in relation to fillet welds, and the defects and their designations which are commonly found in fillet and butt welds. In addition cracking may occur in the weld metal at the base of the Vee of a multiple-pass Vee butt weld in a heavy section, especially in the

440

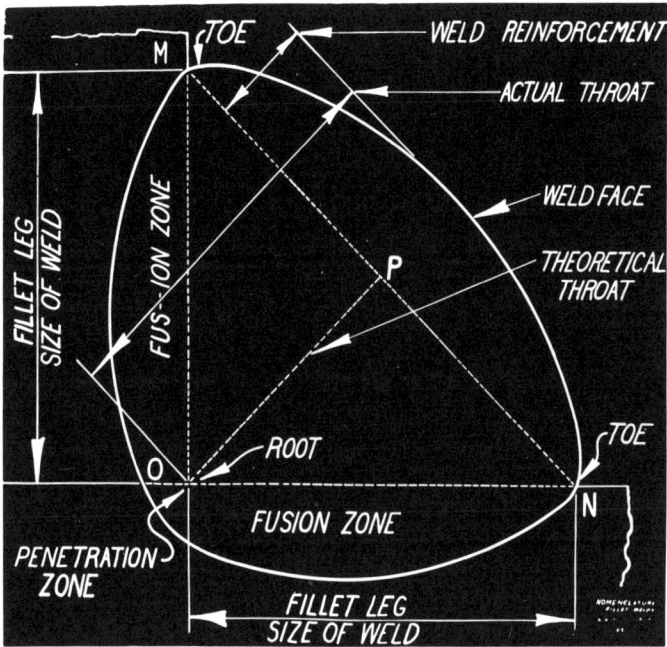

Fig. 216—Nomenclature of a Fillet Weld.

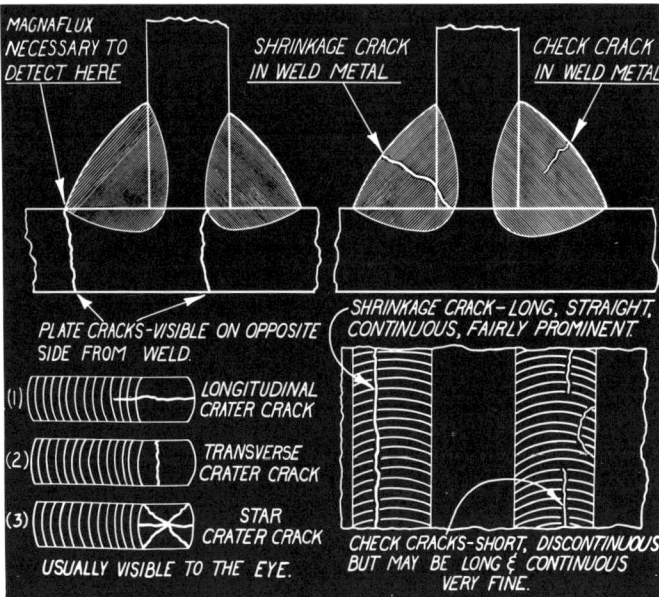

Fig. 217—Typical Weld Defects.

first two or three passes. These cracks, if not detected and chipped out, will often propagate into the subsequent layers of weld metal as they are layed on. Other defects are slag inclusions and voids, lack of penetration between weld and base metal, and cracks in the heat-affected zone in the base metal adjacent to the weld.

Many of these defects are open to the surface and are readily detected with magnetic particles, using prods or yokes. For detection of slag inclusions and voids, and lack of penetration at the root of the weld, which are below the surface, prod magnetization is best, using half wave D.C. and dry powder.

3. MAGNETIZING TECHNIQUES FOR WELD TESTING. The use of prods for producing circular magnetization in the inspection of welds, using half wave D.C. and dry powder, has long been standard practice for this test by the magnetic particle method. Today yokes are used extensively for the location of surface cracks, though they are not suited if discontinuities lying wholly below the surface, such as slag inclusions, are being sought. For this purpose prods should be used. Yokes, using either A.C. or D.C., or half wave D.C. —or even permanent magnet yokes—can introduce a strong field across or along the surface of the weld, and will readily locate all surface cracks. They are much more convenient to use than are prod contacts. In checking for cracks in the early beads of a mul-

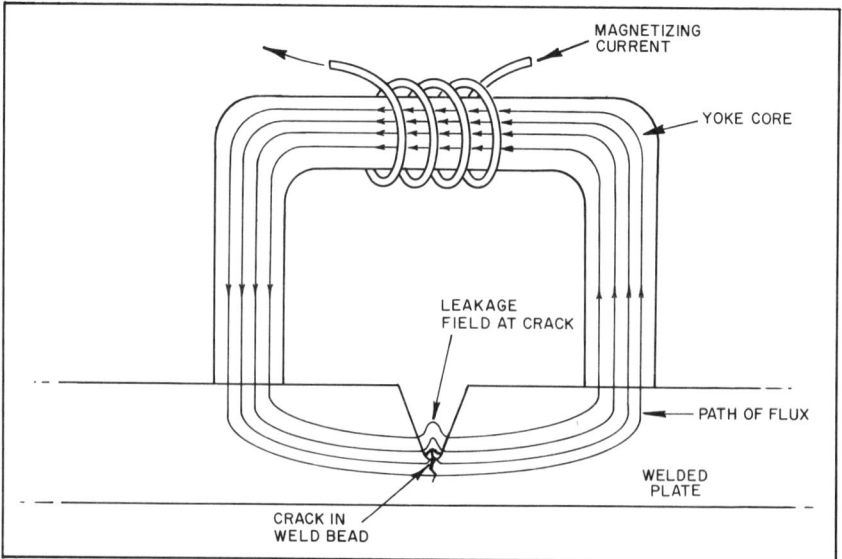

Fig. 218—Field Produced by Yoke Across a Vee Butt Weld.

tiple-pass weld in a thick section, the yoke is very effective.

4. OTHER NONDESTRUCTIVE TEST METHODS FOR WELD INSPECTION. Radiography was specified and used for weld inspection before the magnetic particle method was developed. Radiography excels in locating internal defects such as slag inclusions, gas pockets, lack of penetration or other voids, and is still the most effective method for this purpose. Radiography is not reliable for locating fine surface cracks, however. Today, ultrasonic methods are used extensively for the location of surface cracks, as well as such subsurface discontinuities as lack of fusion, lack of penetration, and cracks in the heat-affected zone.

Eddy Current methods also have applications in this field, either by a probe scanning the weld zone or by passing the welded product through an encircling coil. Other methods of detecting leakage fields which are used to some extent for weld inspection (other than magnetic particles) employ probes for picking up the field disturbances and indicating them on a meter, oscilloscope, or pen recorder, or sometimes the magnetic tape technique. The latter is rolled under pressure over the magnetized weld bead, and the leakage fields are transferred and fixed magnetically on the tape. The tape is subsequently scanned and read by electronic means. These techniques are limited to applications where the weld surface is quite smooth, as on electric resistance welded pipe, for example.

The magnetic particle method is however still the most effective in many circumstances, particularly in irregular sections where the contour of the part or the configuration of the surface at the welded junction makes both radiography and ultrasound inapplicable. Ultrasound has no advantage over magnetic particles in sensitivity for detecting surface cracks—in fact, the magnetic particle method is the *most* sensitive for very shallow cracks, and is more readily interpreted because of the easy read-out of surface crack patterns. Ultrasound, on the other hand, has advantages for high speed inspection on welds in regular-shaped parts, lending itself better to automation and recording.

5. PROD MAGNETIZATION AND DRY POWDER TECHNIQUE. The requirements for the proper application of dry magnetic particles using prod contacts for circular magnetization have been thoroughly discussed in Chapter 13. It is suggested that the reader refer back to that chapter, especially Sections 4 through 10. The requirements for surface preparation, the effect of position—vertical or overhead versus horizontal—the importance of how the powder is applied, etc., all apply specifically to weld inspection, and need not be repeated here. The important matter of prod spacing and amount of current is detailed in Chapter 10, Sections 6 through 9, and should also be reviewed.

6. YOKE MAGNETIZATION. If a yoke is to be used, the same requirements for surface preparation as for prods apply. Positioning of the yoke with respect to the direction of the defects sought, is, of course, different from that of the prods. Prods are spaced *along* the weld bead to locate cracks parallel to the bead; they are placed on *opposite sides* of the weld to locate transverse cracks. Since the *field* traverses a path between the poles of the yoke, these must be placed on *opposite sides* of the bead for parallel cracks, and *along* the weld for transverse cracks.

Powder application is similar to that used with prods. However, because of the strong external fields associated with the poles of the yoke, powder should be applied sparingly and be directed at the area between the poles. Since discontinuities lying wholly below the surface are not looked for when yokes are being used, the powder application is not critical and the indications of surface cracks are easily seen. Such tests are widely used on hull welds in nuclear submarines, for example.

7. TYPE OF CURRENT. It has been stated that if discontinuities lying wholly below the surface of a weldment are sought, the prod method using half wave D.C. is used. A.C. is satisfactory so long as surface cracks only are of interest. A.C. has no appreciable depth penetrating power, but produces a good field for surface cracks. In discussions of the location of discontinuities lying wholly below the surface the question is often asked "how deep a defect can be located with magnetic particles?" The reader is referred to Chapter 15 for a discussion in great detail of the concept of depth in relation to detectability of deep-lying defects with magnetic particle testing.

There is one set of circumstances in which the penetrating power of half wave results in confusion. This is the case of a double-tee fillet weld in which complete penetration is not specified and an open root is permissible and nearly always present. When half wave is used with prods, this open root will most probably be indicated on the weld surface as shown in Fig. 219. Mr. J. W. Owens pointed

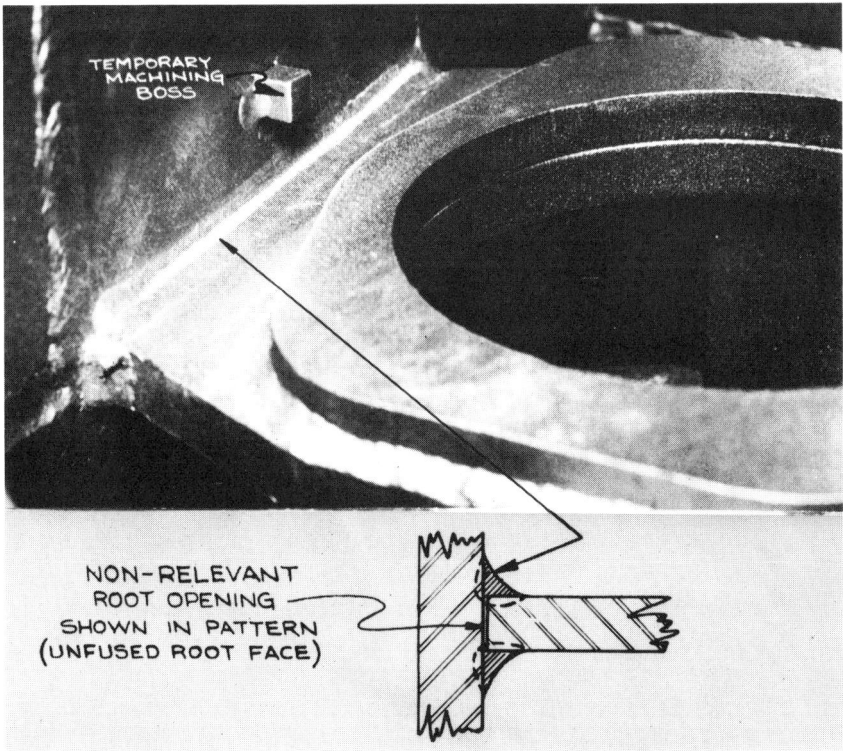

(Courtesy James W. Owens)

Fig. 219—Indication of Non-Relevant Open Root in Fillet Weld.

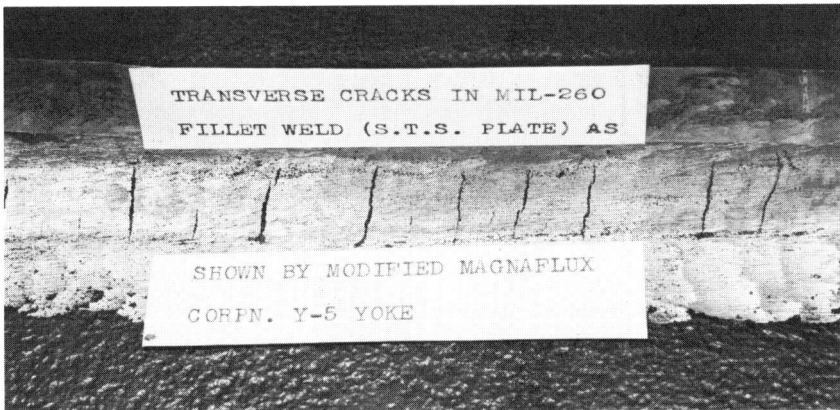

(Courtesy Newport News Shipbuilding Company)

Fig. 220—Magnetic Particle Indications of Transverse Cracks
in Mil 260 Fillet Weld, Shown with Yoke.

445

out in 1944 that this confusion could be overcome by using A.C. instead of half wave on such welds. A.C. will show surface cracks in the weld metal, but will not penetrate deeply enough to produce the confusing indication of the open root, which in this case is there by design and therefore is not a weld defect. Portable units for weld inspection often can deliver either half wave D.C. or A.C. so that each is available when needed in such a situation.

8. EXAMPLES OF WELD INSPECTION WITH MAGNETIC PARTICLE TESTING. A few examples of the application of magnetic particle testing to the inspection of welds will serve to illustrate some of the things that are being done in this area.

A. *Inspection of Structural Welds.* With welded construction taking over the fabrication of all sorts of steel structures, the need for shop and field inspection of welds to assure safety has become urgent, and today nearly all such construction calls for weld inspection. Any or all of the three standard nondestructive tests for welds are used—radiographic, ultrasonic and magnetic particle.

Fig. 221—Inspecting Welds with Magnetic Particles on the 32-Story Michigan Consolidated Gas Company Building in Detroit, Michigan.

Radiographic inspections of welds under the conditions shown in Fig. 221 are difficult, but not impossible, except where long welded box beams give no opening to place film properly. A major handicap is in radiation danger, requiring the area to be cleared of people. Costs are also quite high. Ultrasound also has many applications in this field. It has the advantage of easy portability and freedom from radiation hazard. Costs are in the medium range. But careful handling is required for a delicate instrument in precarious locations such as shown in Fig. 221.

Magnetic particle portable magnetizing units can be placed on a construction floor or platform, and cables of considerable length used to reach the actual point of inspection. Only the prods, which include the current-control switch, and the powder applicator need to be handled by the inspector. Light portable units with outputs up to 900 amperes were designed especially to facilitate these field inspections, and are in wide use today. Remote current-control can also be had at the point of inspection by means of a small dial on a light cable. The inspector can operate the infinitely variable self-regulating control as much as 100 ft. from the magnetizing unit. Under some circumstances, permanent magnet yokes can be used if it is not feasible to supply power to a standard portable unit.

Structural weld inspection with magnetic particles received its first impetus in California, where the State Department of Education began to use welded, single story school buildings as a safeguard against earthquake damage. The practice of all-welded fabrication spread slowly at first, but has become a standard practice today. Figure 222 shows another application of magnetic particle testing of a critical weld at the base of the 285-foot television mast being erected on top of one of the Marina City Towers in Chicago.

B. *Electric Resistance Welded Steel Pipe.* The welding of steel pipe by the electric resistance method provides an example of the automatic production testing of welds. Figure 223 shows an installation which is applying the magnetic particle test to the weld of a 10 inch pipe, within a few feet of the weld stand. The powerful electromagnetic yokes shown are used to produce a field transverse to the weld. A special coarse dry powder is applied in excess, and the surplus is recovered and

Fig. 222—Testing Weld at the Base of a 285-Foot Television Tower on the Roof of One of the Twin Towers of Marina City in Chicago.

returned to the storage hopper shown. The seam then passes under the eyes of the inspector, who marks the indication of cracks as they go by—either by hand or by paint-spray application which he controls.

Speeds are from 50 to 150 feet per minute, depending on pipe size, which may run as large as 36 inches in diameter. Different pole pieces for the electromagnets are provided to fit different sizes of pipe. The unit can be made fully automatic, although the cost of complete automation is difficult to justify. Defects found are chipped or ground out and repaired by hand. Most important is the ability to control the welder, so that, as soon as defects occur, their cause can be eliminated. Failures on final hydrostatic tests have almost been eliminated by this test.

C. *Pressure Vessels.* The earliest application of radiographic nondestructive testing was on welded pressure vessels—both

Fig. 223—Inspection at the Weld Stand of Resistance-Welded Pipe.

hammer- and fusion-welded. Magnetic Particle testing was early used on many types of pressure vessels because of its great reliability for the location of *surface* cracks—an area in which radiography is less than fully dependable. Welds around nozzles and man-ways, which could not be x-rayed because of shape and contour, were also magnetic particle tested.

Such tests continue today, with a trend toward greater use of magnetic particle testing on pressure vessels for missiles and nuclear equipment—where every precaution against any reduction in strength at a weld must be taken. The extreme sensitivity of magnetic particles for very fine and very shallow cracks is a large factor in its use for such purposes.

449

Fig. 224—Testing a Welded Oil Storage Tank.

9. STEEL CASTINGS. Magnetic particle testing is used extensively on castings of all kinds. Defects sought include all kinds of surface flaws such as shrink cracks, hot tears, porosity, blowholes and slag pockets. For small castings, standard units are generally used, with either the dry or the wet method. Defective castings are segregated and salvaged, if possible, by grinding out the defect, provided it does not penetrate too deeply into the section. Larger castings are inspected with prods, again with either dry or wet magnetic particles. Defects are cut or ground out, and repair by welding is often permissible.

The overall method of multi-directional magnetization already described in Chapter 19 is the most rapid and most satisfactory method for the testing of very large castings, some of which may weigh many tons. Although this application requires heavy and relatively expensive electrical equipment, sometimes with outputs of 20,000 amperes or more, the saving in inspection time per casting, and the increased effectiveness in reliably detecting important surface cracks, can justify the expense of the installation on a purely

450

economic basis. The savings are most pronounced when a large run of castings from the same general design are being made. Once the proper magnetizing pattern has been worked out, the process is purely repetitive. Inspection time per casting has usually been reduced to only 25% of that required for prod inspections, and in one application at least, to only 10% of the former prod inspection time. Prod inspection of large and complicated castings frequently did not locate all defects on the first test, and additional cracks would show up on re-inspection after repair. Because the all-over method usually finds all significant cracks on the first inspection, re-inspection time after repair is often cut in half.

With the overall method of magnetization, fluorescent magnetic particles in water suspension are invariably used, mainly because of the greater visibility of the indications. This advantage is especially important when cracks occur in holes or recesses difficult to illuminate and see into with white light, but for which the black light that can be projected into such places is adequate to make indications readable. Dry powder does not perform well at the high levels of D.C. magnetization usually employed for the overall method because it tends to form patterns of the external field, thus losing the mobility necessary to form indications of defects.

If defects lying wholly below the surface are considered important, the prod method with dry powder is still the most effective. Locations where porosity or internal shrinks are likely to occur can usually be predicted, and the prod inspection confined to those few areas.

10. NONDESTRUCTIVE TESTING AS A DESIGN TOOL. One of the most useful applications of magnetic particle testing methods (as well as for all other nondestructive testing methods) is in the checking of pilot casting procedures and mold design on a new article. Testing of pilot castings can indicate locations subject to shrink cracking, porosity, etc., and point to remedial changes in the design of the casting itself, or to corrections in chills, or to the mold with respect to locations of gates and risers.

11. GRAY IRON CASTINGS. Gray iron castings are inspected for many crack-type surface defects with the magnetic particle method. These are the most common defects occurring in gray iron. Because of the distribution of the graphite flakes throughout the metal the magnetic field is badly distorted, making location of below-the-surface defects difficult or impossible, especially on the lower grades. In the case of chilled iron which is extremely hard, the permeability is low and this again defeats deep defect detection, although surface cracks are readily found. In the case of nodular and malleable iron, somewhat better results can be expected for deep-lying defects, although as a practical matter the greatest number of applications and needs for iron castings lie in the detection of surface cracks only.

One area of saving in the gray iron foundry by use of 100% magnetic particle inspection is the detection of handling cracks. Gray iron is very brittle, and rough handling in shake-out and in subsequent sorting operations often produces more cracks than are present when the casting leaves the mold. Figure 225 shows some

Fig. 225—Some Typical Handling Cracks in Gray Iron Castings.

typical handling cracks indicated with magnetic particles. When the number of such cracks showing up on inspection becomes excessive, special handling precautions can be set up and such production losses largely eliminated. Some shake-out machines have been completely rebuilt to correct this tendency to cause cracks in castings.

In the production of malleable castings, the white iron castings are commonly inspected before malleableizing. A cracked white iron casting takes up valuable space in the furnace and can only result in a defective malleable casting which must be scrapped. Magnetic particle testing in the white iron stage assures that only sound castings go into the malleableizing furnace—a considerable economic saving.

12. FORGINGS. Many of the practices in testing large castings are duplicated in the case of large forgings, too large to handle by normal magnetizing and inspection techniques. The prod method is normally used on such articles, although in some cases the overall method is applicable. Fig. 226 shows inspection on overhaul of an 18 inch marine tail shaft, with fatigue cracks showing in the keyway. In such a shaft, the areas vulnerable to fatigue are at the

Fig. 226—Magnetic Particle Test of 18 inch Marine Tail Shaft, Showing Fatigue Cracks in the Keyway.

keyways where the drive gear and the propeller are attached, at the outer end of the outboard bearing, and at points adjacent to couplings and line bearings.

13. SUMMARY. The successful inspection of weldments and large castings and forgings requires the highest type of skill and knowledge of the method.

It is only in the case of volume-produced identical items and the simplest welded articles, or small castings and forgings, that procedures which can be carried out by the average inspector can be easily set up. For complex weldments and large parts of intricate shape, the inspector or nondestructive testing engineer must work out the proper magnetizing technique himself, which may be different for each case. Success here depends on a full knowledge of the theory and practice of magnetic particle testing as has been detailed in the pages of this book, supplemented with practical experience.

CHAPTER 25

STANDARDS AND SPECIFICATIONS FOR MAGNETIC PARTICLE TESTING

1. GENERAL. Because of the very large number of variables that may affect the end result of magnetic particle testing it is highly desirable that the procedures for conducting any given inspection be exactly specified and described. Also, such guides to rejectable conditions as it may be feasible to establish should be set up to assure, as far as possible, uniform interpretation of the indications that are produced. The former objective is relatively easy to accomplish and is independent of the human factor, provided that specified procedures are faithfully followed by the operator. The second objective, however, must in most cases rely on human judgment based on someone's knowledge and experience with interpretation of various magnetic particle indications of defects.

Since the effect of a given discontinuity on the usefulness of a part may differ widely in different parts intended for different types of service, it is obviously impossible to write *general* standards for acceptance or rejection. It is a curious fact, however, that only a few years ago industry was clamoring for the sponsors of the method to tell the users where to draw the accept-reject line. Clearly it is the function of all nondestructive testing methods to locate and if possible identify discontinuities. But the accept-reject decision must be made by the engineers who design and use the part being tested, for only they can judge accurately the in-service effect of a condition disclosed by nondestructive testing.

2. SPECIFICATIONS FOR MAGNETIC PARTICLE TESTING. For the purposes of this discussion we are concerned with specifications of only one type—namely, those which lay down rules for the method for specific applications, to insure reliable, uniform and reproducible results, and *not* acceptance standards.

The tendency to spell out every detail of procedure for these purposes certainly has some desirable objectives, but there are also some results of it that are less desirable. By reducing all activities with regard to the inspection either to automatic processes or to processes governed by rules and specifications, the following advantages are presumably achieved:

(a) Increased over-all reliability of the inspection.

(b) More uniform results.

(c) Reproducibility of results by different operators or inspectors.

(d) Inspectors can be used who do not have a high level of education or experience.

(e) Minimized training of operators and inspectors.

(f) Conservation of time of skilled and experienced inspectors.

Some less desirable results may be the following:

(a) The tendency to relieve operators of the necessity for thinking and developing skills, and feeling a personal responsibility for results.

(b) De-emphasizing the importance of watching for abnormalities or unusual occurrences in materials and processes that may affect the inspection, or product usefulness.

(c) The tendency to make the inspection perfunctory—which it should *never* be.

Still, the balance of weight is definitely on the desirable side and standards and specifications are certainly important and necessary to insure reliable tests. Therefore much has been written and proposed to provide guides and controls in all phases of magnetic particle testing practice.

3. TYPES OF SPECIFICATIONS. Specifications for magnetic particle testing fall into several groups, each with a somewhat different objective from the others. These groups are:

(a) Broad procedural guides—general.

(b) Company procedural guides.

(c) Procedural guides for specific types of products, or for industries.

(d) Procedures for testing specific articles, specified by the purchaser or by the company internally—process specifications.

(e) Procedures specified for user overhaul inspection of a company's products.

(f) Specifications for certification of operators.

(g) Limits for acceptance or rejection, set up by a buyer, or by a company itself for quality control.

(h) Repair Station requirements—especially for aircraft.

(i) Equipment specifications.

(j) Instructions for operating specific types of equipment or individual special units.

4. BROAD PROCEDURAL GUIDES. A number of technical societies have prepared broad procedural guides for the use of their members who wish to learn about or use magnetic particle testing. These consist for the most part of a very general description of magnetic particle testing. They are intended to be informative but not binding. They give the reader an idea of how the method works, and the steps by which it is properly applied. Usually examples of applications are given, often with some illustrations of types of indications of typical conditions.

The sections on magnetic particle testing in the Handbook of the Society for Nondestructive Testing constitute by far the most complete presentation of this type. Some of the other Societies whose handbooks include a section on magnetic particle testing are the American Welding Society, the American Society for Metals, the American Society of Mechanical Engineers, the Society of Automotive Engineers, the American Society of Tool and Manufacturing Engineers, and others. In addition, there are various books, published both here and abroad, dealing with nondestructive testing that include similar more or less complete descriptions of the method.

5. COMPANY PROCEDURAL GUIDES. Many large companies that make extensive use of magnetic particle testing have prepared their own versions of the procedures that are to be used in their operations. These are usually rather broad in their approach, except that they may be limited to those specific procedures applicable to their products. As an example, if the dry method is not used in their plant, no guide for the use of this technique need be included. Also the procedure specified may be written around the particular type of equipment used in that plant.

Such manuals are used for training new operators, and for the guidance of those regularly using the method. Included in this type of specification or manual are those issued by various facilities of the Armed Services, prescribing the materials and techniques that are approved for use. Such Technical Orders (T.O.'s) are issued by the U.S. Air Force, for example, for the use of inspectors engaged in testing aircraft parts on overhaul. These T.O.'s are usually very complete and cover all types and phases of magnetic particle testing which are used in their specific areas of application.

6. PRODUCT OR INDUSTRY SPECIFICATIONS. Some method specifications have been written from the point of view of a single industry or type of product. For example, the description of magnetic particle testing which appears in the Handbook of the American Welding Society is directed entirely toward its use for the testing of welds and weldments. The U.S. Air Force Technical Orders are, of course, written for use in the testing of aircraft and aircraft engine parts. A company that makes only castings would prepare a specification that would only cover the tests of castings and associated weldments, and this would be no broader than need be for this single purpose. The U.S. Air Force has issued a specification that is limited to tests of castings, as a class.

7. PROCESS SPECIFICATIONS. A much narrower type of specification is that drawn up to prescribe the exact method of testing of a single specific article. These are usually prepared by the quality control or engineering department, and in addition to calling for an exact testing procedure, usually also include a definition of indications which may be tolerated and of those for which the article must be rejected. Such process specifications may refer to, or make use of, the broad method description of a company-prepared or hand book type of procedural guide, but should be specific with regard to those factors which will insure proper inspection of the particular part to which the process specification refers.

Some of the factors that should be covered are:

(a) Cleaning or pre-inspection preparation (paint stripping or removal of plated coatings when required, degreasing, etc.).

(b) Specific method of magnetization and particle application to be used.

(1) Wet or dry method.

(2) Circular or longitudinal magnetization or both.

(3) A.C., D.C. or Half-Wave D.C. current.

(4) Continuous or residual method.

(5) Fluorescent or visible magnetic particles.

(6) Amperes or ampere turns of magnetizing force.

(c) A specific type of equipment is often called for.

(d) Procedures and limits for demagnetization if required.

(e) Post-cleaning when required for following processes.

(f) Post-inspection surface protection — corrosion prevention and packaging.

In addition, critical areas are usually defined and the indications described which may be accepted, and those which are immediate cause for rejection, so that disposition of what otherwise might be questionable parts can be expedited. Length, direction and frequency of indications and their location with regard to critical stress areas are usually the means for classifying acceptable and rejectable indications.

These process specifications and acceptance limits are an excellent device for insuring good inspection. They must be prepared with care by competent engineers who are fully aware of the service requirements of the part, the directions and magnitude of the stresses and the location of stress concentration areas. Those preparing the specification should also be thoroughly familiar with the various techniques of magnetic particle testing, and the degrees of sensitivity control obtainable.

Similar specifications and acceptance limits may be agreed upon in advance between buyer and manufacturer to cover the inspection and the acceptance and rejection limits for a *specific* purchased part.

8. MAINTENANCE OR OVERHAUL INSPECTION. Manufacturers of complex and critical products (as, for example, aircraft engines, diesel engines, etc.) frequently include a section in their maintenance manuals in which recommended testing procedures are given for use on overhaul. Such a guide is broad enough to cover general techniques and equipment, but is specific with respect to the tests of certain critical parts and components. Such guides are of great value, since they tend to assure in advance that the inspection for maintenance purposes that will be performed by the user will be suitable and proper for the specific items involved. The chance is minimized that the inspection may go wrong because the inspector responsible may not be sufficiently experienced to perform the inspection properly. Advice is often included in such manuals as to where to look for trouble and how to judge the significance of an indication when found; and, if possible, how to repair the part.

9. CERTIFICATION OF OPERATORS. The U.S. Air Force and other Federal and State Government departments that make regular use of these test methods long recognized the danger of poor results if operators and inspectors are not properly trained and experienced. For this reason a system of operator qualification was long in use, and specifications were issued under which operators were required to be qualified. Such qualification in some form was a pre-requisite for an operator who was going to conduct inspections—especially of Government-purchased materials, or on Government-used equipment such as aircraft or parts on overhaul.

Usually such certification was limited to certain types of inspection on certain types of equipment. If the operator changed jobs he needed new certification tests in order to qualify for the new work. Operators and inspectors in vendor's plants were required to be so certified as were those directly employed by Government agencies.

Responsibility for certification of operators under current Government rules now rests with the prime contractor for a given government project. Qualification tests are often given in accordance with Government specifications, and include written examinations on materials and techniques, as well as actual demonstration of ability to find and interpret defects in actual parts.

10. STANDARDS FOR ACCEPTANCE OR REJECTION. Over the years there has been an insistent demand for standards to govern acceptance and rejection of parts on the basis of indications produced by this (and other) nondestructive testing methods. Users have complained that they are told how to use a method of test but are not told what to do about indications of defects that are disclosed. A little reflection should make it readily apparent that any broad standards for this purpose are impossible.

Whether or not a given discontinuity warrants rejection of a given part depends entirely upon factors that have no relation whatever to the inspection method that produces the indication of its presence, nor to the size or type of defect, per se. The service requirements of the part are the controlling factors. The nature of the stresses in service, stress concentrations, notch sensitivity of the metal, the critical nature of the function of the part—even location and orientation of the defect on the specific part—must all enter into a decision as to acceptance or rejection. These considerations cannot be generalized, but must be applied by those possessing the information, to each specific case. A given defect in one part for one type of service may warrant rejection, but may have no significance at all if present in another part for a different type of service.

The preparation of standards for acceptance or rejection is the responsibility of the designers of the part and of the quality control department, and are most effectively included along with the test process specifications. If such standards are not available and the person judging the results of the inspection does not possess the necessary knowledge, he must seek it from those who are responsible for the design and service performance of the part. He *should not expect* broad standards to be available. However, knowledge of the material and its intended service, as well as of defects in general and their effect on strength and performance, often allow an intelligent, experienced inspector, with the use of common sense, to arrive at a logical decision regarding acceptance or rejection.

11. REPAIR STATION REQUIREMENTS. Another type of specification, bearing on the use of magnetic particle (and other) inspection is that setting forth the requirements for repair stations for aircraft structures and aircraft engines. The Federal Aeronautics Administration (F.A.A.), which has jurisdiction over the approval of shops or stations for repair or overhaul, specifies the extent and type of equipment and other facilities necessary to insure proper and satisfactory inspection of all parts before they may be re-used. Magnetic particle equipment is a requirement for a fully authorized repair and overhaul station for aircraft and aircraft engines.

12. EQUIPMENT SPECIFICATIONS. Specifications for standard equipment for applying the magnetic particle inspection method have been prepared and are used for procurement of such equipment by the Air Force and by other Government agencies. Such equipment or material specifications are not (necessarily) applicable to general manufacturers.

13. OPERATING INSTRUCTIONS. Manuals or operating instructions are prepared by responsible manufacturers of test equipment to guide the purchaser in its proper use. These are not specifications in the usual sense of the term, but may include method procedures and techniques as applied to the use of the specific piece of equipment.

For large pieces of equipment, where automatic handling is involved, these manuals become lengthy and complex. The function of all controls—each button, switch and relay—must be explained so as to be understood by the persons responsible for keeping the unit operating properly. Successful operation of one of the large steel billet testing units (described in Chapter 19) would be difficult if not impossible without the specific instructions furnished in such a manual.

14. GOVERNMENT SPECIFICATIONS. Following is a list of U.S. Government specifications currently in use, which have to do with magnetic particle testing:*

Mil-C-6021   Castings, classification and inspection of.

Mil-I-6870 USAF   Inspection Requirements, Nondestructive, for Aircraft Materials and Parts.

Mil-Std-23   Nondestructive Test Symbols.

Mil-Std-271 Bu-Ships Military Standard Nondestructive Testing Requirements for Metals.

Navships 250-1500   Standard for Welding of Reactor Conduit Coolant and Associated Systems.

Mil-M-19698   Magnetic Particle Unit, Portable.

Mil-M-11472   Magnetic Particle Inspection Process, for Ferromagnetic Materials.

Mil-M-11473   Magnetic Particle Inspection, Soundness Requirements for Weldments.

Mil-Std-410   Qualification of Inspection Personnel (Magnetic Particle and Penetrant).

Bu-Weps Mil-I-6868   Inspection Process, Magnetic Particle, (General Requirements for Magnetic Particle Inspection)

Nav Aer 00-15PC-503   Technical Inspection Manual, Volume 3, Section 4, Magnetic Particle Inspection.

462

USAF T033, B2-1-1	Inspection of Material, Magnetic Particle Method.
Bu-Weps Mil-I-18620	Inspection, Magnetic Particle, Requirements for (Applicable to Overhaul of Airframes, Engines, Propellers and Accessories).
Mil-M-68678	Magnetic Particle Inspection Unit.
Mil-M-23527	Magnetic Particle Inspection Unit—Lightweight.
MS-17980	Magnetic Particle Indications on Steel Nuts.

Following is a list of some of the Technical Society Specifications which are currently in use:*

SNT-TC-1	NDT Personnel Qualification, Recommended Practice.
ASTM E-109-63	Dry Powder, Magnetic Particle Inspection.
ASTM A456-61T	Method and Specifications for Magnetic Particle Inspection of Large Crankshaft Forgings.
ASTM E125-63	Reference Photographs for Magnetic Particle Testing of Ferrous Castings.
ASTM A275-61	Steel Forgings, Heavy, Magnetic Particle Testing and Inspection of.
ASTM E138-63	Wet Magnetic Particle Inspection.
ASTM A340-61	Definition of Terms, Symbols and Conversion Factors Relating to Magnetic Testing.
AWS	Fourth Edition Handbook 8.24 to 8.36.
SAE AMS-2300	Magnetic Particle Inspection, Premium Aircraft Quality Steel, Cleanliness.
SAE AMS-2301	Magnetic Particle Inspection, Aircraft Quality Steel, Cleanliness.
SAE AMS-2640	Magnetic Particle Inspection.

*Taken from Materials Evaluation, March 1966, Vol. XXIV, Number 3, pp 158 and 159.

15. OTHER SPECIFICATIONS OF INTEREST. Among the many specifications on the subject of magnetic particle testing which have been put into use by industry and public works offices, a few are of interest as examples.

The State of California, Department of Public Works, Division of Highways, Materials and Research Department, have adopted specifications covering the inspection of welds in steel structures such as bridges and buildings. The designation of this specification is "Test Method No. Calif. 601C, Jan. 1959—Test Method and Techniques for Control of Welding Procedures Performed During Fabrication of Welded Steel Structures". Part VIII of this specification defines the procedures to be followed in the magnetic particle testing of welds—the currents to be used for different plate thicknesses per inch of prod spacings, precautions to be observed and method of reporting results. The State Road Commission of West Virginia has a similar specification, which gives a description of defects which are cause for rejection or which may be repaired.

The Interstate Commerce Commission has issued an order, #59-A, (10/1/63) which requires that all cargo compressed gas tank trucks must be completely re-inspected with magnetic particle testing methods at the rate of 10% of a total fleet every 6 months. All welds, internal and external must be tested.

An example of more highly specialized industry specifications is one issued by the Manufacturers Standardization Society of the Valve and Fittings Industry. Their MSS Standard Practice SP-53, "Quality Standards for Steel Castings for Valves, Flanges and Fittings" covers dry powder magnetic particle testing techniques as approved for this area of inspection.

16. SUMMARY. It is evident from this review that specifications which have been promulgated in the area of magnetic particle testing have come into being as the result of a need to make sure that proper and uniform procedures be followed in any given case of magnetic particle inspection, rather than to leave the selection of method and technique to different inspectors who may or may not be fully qualified to make the selection, and who are almost certain to have different ideas and come up with different method variations. In an operation such as magnetic particle testing in which there is reliance on the human factor, the careful consideration of method variations and selection of the most suitable one, by persons fully qualified to do so, gives the highest assurance that the greatest possible value will be obtained from the tests.

CHAPTER 26

TESTS FOR EVALUATION AND CONTROL OF EQUIPMENT AND PROCESSES

1. GENERAL. In applying magnetic particle testing (or any other non-destructive testing process) guides for the operation of the equipment and the details of the process assume that the operator and the equipment he is using will be capable of carrying out —and will faithfully carry out—the instructions and the procedures specified. However, operators and supervisors have need to verify that the critical requirements for magnetization and bath application are actually being met as the inspection proceeds.

Examples of things that might go wrong and lead to completely abortive results are the following:

(a) Meter reading magnetizing amperes not registering the *actual* current being passed through the coil or part.

(b) Wrong magnetic particles being used, or wrong suspending liquid for the bath.

(c) Bath concentration incorrect.

(d) For fluorescent particles, separation of fluorescent pigment from the magnetic particles during use.

(e) Insufficient black light intensity for the reading of fluorescent particle indications.

2. MALFUNCTIONING OF EQUIPMENT. If an operator, following procedural specifications, reads the called-for current on the ammeter, he assumes that the current is doing the specified magnetizing job. This is a fair assumption with equipment that has been properly designed and constructed on quality lines, and been properly maintained. However, ammeters can and do go out of calibration, and the supervisor should check the ammeter for accuracy at intervals.

Transformer coils can, with long use (or abuse) burn out, or connectors can loosen or corrode and break. When this happens the meter will usually not read at all, and operators should make a practice of watching the ammeter while magnetizing parts. In some installations a test is run at the start of a shift by placing a one inch copper bar between the heads of the unit and passing the maximum current through it. The meter is checked to make sure that it is reading up to the rated maximum output of the unit.

An example of an unusual failure of equipment was the case where the copper bar test appeared to pass the rated output current according to the meter reading. However, no current was flowing through the bar (or parts). It was found that a short circuit had developed between the head contact plate and the frame of the unit and the current was entirely by-passing parts clamped between the heads. In addition to such an electrical check on the meter, therefore, some overall performance test is desirable. (See Section 8, this Chapter.)

3. PROPER MAGNETIC PARTICLES AND BATH LIQUID. When an operator requisitions magnetic particles for his process from the stock room, he may assume that he has been issued the proper material. This is not necessarily the case, since the stock room attendant may pick up a container of material that is intended for use elsewhere in the plant. The operator should, of course, verify that the container label is correct, for the material he is using. Normally, if the material is not correct, the fact will be detected by its appearance and its behavior in use. However, with the unfortunate tendency to relieve the operator from the necessity of thinking, by issuing detailed rules for him to follow, he may not observe that an error has been made.

An instance when such an error led to an absurd failure of the method came to light some years ago when the stock room issued red lead paste instead of red magnetic particle paste, and the operator went ahead in carefree fashion with a nice red bath, until somebody noticed that for some time no defects had been found in parts tested on this particular unit. Fortunately such errors do not happen often, but the incident points up the need for alertness to make sure that all factors in the process are working properly.

In another case, in which an oil-base bath was being used, the stock room issued naphtha instead of the proper high-flash distillate. The result was a fire in the equipment when an arc at the contact heads ignited the naphtha bath. Again, alertness on the part of the operator or supervisor might have averted this loss. Specifications for a suitable oil for wet bath purposes are given later in this Chapter (Section 6).

4. BATH CONCENTRATION INCORRECT. In the course of use a certain amount of magnetic particle material is removed from the wet bath by clinging either magnetically or mechanically to the surfaces of parts. In time—depending on the number of parts tested—this results in a bath with insufficient magnetic particles and more testing material must be added. On the other hand, carry-out of bath liquid on the surface of parts, as well as evaporation of liquid when the unit is not in use, also acts to change bath concentration. Frequent tests for bath concentration are a most important requirement. A weak bath results in faint indications and thus tends to reduce the apparent severity of a discontinuity; or a faint indication which might be produced by a bath of proper concentration, might be missed entirely, if the bath is below strength. Sections 8 and 9 describe available bath-strength tests.

5. DETACHMENT OF FLUORESCENT PIGMENT. Modern fluorescent magnetic materials of the best quality are made by a process resulting in a dye-magnetic-particle combination which is permanent, and does not come apart even after prolonged, vigorous agitation. However, some types of fluorescent magnetic particles tend, in the course of use, to lose their attached fluorescent pigment under the action of the vigorous agitation in an inspection unit. If this occurs, fluorescent indications will become weaker in proportion to the number of particles which have lost their fluorescent color. In such a bath, the free dye tends to adhere over the entire surface of the part, not by magnetic attraction, and creates a confusing background. The non-fluorescent magnetic particles are attracted to leakage fields along with those still holding their fluorescent pigments, and act further to dilute or obscure the fluorescence of an indication.

Such a situation is quite different from the weakening of indications due merely to a lack of *magnetic* particles. Again, alertness on the part of the operator enables him to sense that something is wrong with the appearance of indications, as well as to observe an increase in fluorescent background.

6. SPECIFICATIONS FOR SUITABLE PETROLEUM BASE LIQUIDS FOR OIL TYPE WET BATH. The important characteristics for petroleum

468

distillate for the wet-bath suspensoid for magnetic particle testing were discussed in Chapter 24, Section 5. The tests and how to make them are listed below:

(a) Viscosity—Kinematic at 100°F (38°C)
   3 Centistokes, max.
   ASTM test method D-445-65

(b) Flash Point. Closed cup. 135°F (57°C) Min.
   ASTM test method D-93-62

(c) Initial Boiling Point. 390°F (199°C) Min.
   ASTM test method D-86-62

(d) End Point. 500°F (260°C) Max.
   ASTM test method D-86-62

(e) Color. Saybolt plus 25.
   ASTM test method D-156-64

(f) Sulphur Available. Must pass copper test.
   ASTM test method D-130-65

7. SETTLING TEST FOR BATH STRENGTH. The settling test for bath strength was devised over 30 years ago when the wet method of magnetic particle testing was first introduced. With some notable exceptions, it is still in use today, in spite of the fact that it lacks much in accuracy, and there are several inherent sources of error. It has the advantage of being simple and easily applied by the operator at the unit. It also gives information as to the condition of the bath from the point of view of contamination with dirt and lint. Other, newer methods are available for checking bath strength, although all have some draw-backs of their own. Some of these will also be described.

The settling test was discussed in general terms in Chapter 14, Section 8. Details of the test are given below:

Equipment: ASTM pear-shaped centrifuge tube, Goetz, 1.5 ml stem, graduated in 0.1 ml.

Stand for supporting the tube in a vertical position.

Procedure:

(1) Let pump run for several minutes to agitate the suspension thoroughly.

(2) Flow the bath mixture through hose and nozzle to clear hose.

(3) Fill the centrifuge tube to the 100 ml line.

(4) Place the tube and stand in a location free from vibration and let stand for 30 minutes for the particles to settle out.

(5) Read the volume of the *oxide* portion of the sediment, and refer to the table for the correct volume for the type of magnetic particle concentrate being used. If reading is too high, add liquid to the bath. If too low, add concentrate.

The sediment in the tube will consist of two layers, and sometimes three. The upper layer consists of light dust and lint, which being the lightest takes longest to settle. Below it is the oxide layer, and in taking the reading the upper layer should be disregarded. Its amount, however is a clue to the condition of the bath. If heavy iron dust from grinders or mill scale is present, this material, being heavier and of larger particle size, will settle out the fastest and appear as a layer under the oxide. It should also be disregarded.

The settling test worked well enough on the original magnetic particle paste materials, but the introduction of new materials varying in specific gravity, has made the test less definitive for use with such materials. Other methods of bath checking are coming into use.

Table VIII gives the settling readings for the several magnetic particle materials, for a new bath made up with the quantity of magnetic particle material recommended in the table.

As has been said, this method is not an absolute one, and the results may vary among different operators and with different bath liquids. Its main usefulness lies in day to day comparative readings for a given unit. The settling values in Table VIII should not be accepted as the standard for day to day checking. Rather, the settling volume for a new bath in a given unit should be determined and used for the standard for that unit. This eliminates the variables of bath liquids and individual preferences for concentrations differing from those shown in the Table. Build-up of mill scale or heavy dust, and of light dust and lint, can be observed, and when it becomes excessive corrective action can be taken. Usually this means that the bath must be discarded and a new one made up. As was said above, the middle layer of oxide should be read as the

TABLE VIII
Standardization Guide for Settling Tests

Type	Fluorescent Particles						Visible Particles				
Magnaflux Corporation Designation	14 AM	10 A	14 A	15	20 A	24	9 CM	7 C	9 C	27 A	29 A
Oz. per Gallon Recommended Concentration	(2)	0.2	1/6	1/6	1⅓	2	(2)	1¼	1¼	2⅓	2⅓
Settling* Volume Range-ml.	(2)	0.30 to 0.40	0.20 to 0.30	0.10 to 0.15 **	0.18 to 0.22	0.75 to 1.00 ***	(2)	1.3 to 1.5	1.7 to 1.9	1.1 to 1.2	1.4 to 1.5
Suspensoid											
Oil	(2)	X	X	X			(2)	X	X		
Water		(1)	(1)	(3)	X	X		(1)	(1)	X	X

1. Can be dispersed in water with the aid of WA-2A water conditioner.
2. Prepared Mix.
3. Can be dispersed in water with the aid of WA-3 water conditioner.
 *Settling time is 30 minutes in 100 ml. centrifuge tube, except as noted below:
 **Settling time is 5 minutes.
 ***Settling time is 45 minutes in 1 liter Imhof sedimentation cone.

Magnetic Particle Concentrates

measure of bath concentration of useful magnetic particles.

8. OTHER BATH STRENGTH TESTS. A number of other tests for checking bath strength have been proposed and some have been used. Most of these are based on measuring the amount of *magnetic material* in the bath. This can be done by means of a coil around the hose or pipe through which the bath is flowing, and measuring continuously, with read-out on a meter, the magnetic properties of the stream. This can be converted into the weight of magnetic particles in the bath per gallon.

Such a device might have some value in convenience, but actually would probably be no more accurate or satisfactory than the settling test. It would make no distinction between magnetic dust (common in foundries and steel mills) and magnetic testing particles. It would, of course, have to be calibrated separately for every type of magnetic particle used.

Various test blocks containing a series of artificial defects, just below the surface and having graduated depths, have been proposed and some are in use. Some of these are permanently magnetized, and a series of milled slots of the same width but increasing depth are used as indicators. Others use a single slot of varying depth, from shallow to deep. One version of such a device, shown in Fig. 227a, consists of a piece of soft iron having a hole drilled

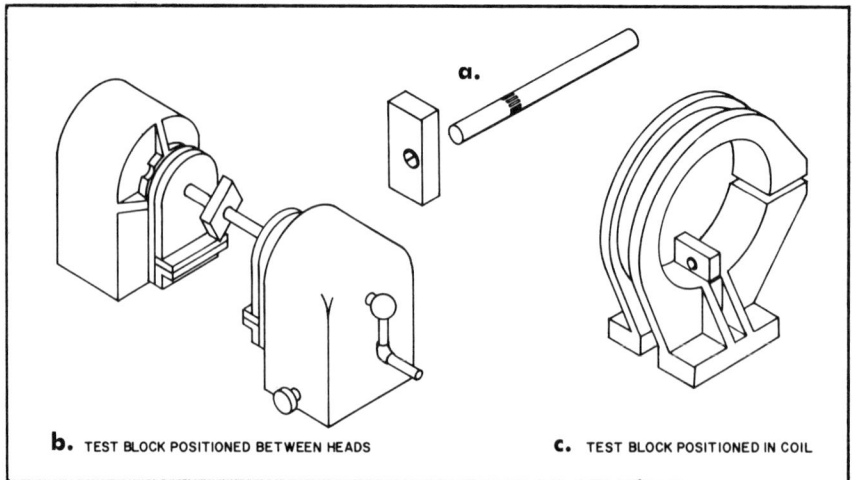

b. TEST BLOCK POSITIONED BETWEEN HEADS   **c.** TEST BLOCK POSITIONED IN COIL

Fig. 227. (a) Test Block for Measuring Bath Strength.
(b) Test Block in Use Between Heads of Unit.
(c) Test Block in Use in Coil.

472

through it, through which a copper rod is passed. A tapered slot, cut into the upper surface of the block and filled with a nonmagnetic material constitutes an artificial defect of varying depth. The block can be used either in a coil for longitudinal magnetism check, or by passing current through the copper rod passed through the hole. Figure 227 shows the block in use for both tests. Bath strength is indicated by the length of the indication when a current of a specified strength is passed.

Following is the test procedure:

Process the block by the wet continuous method. If the bath is good, and the circuits are operating correctly, the following results should be obtained:

*Head or Coil Current Amperes D.C.	Approximate Length of Indication % of defect length
500	Zero
1000	50%
1500	100%

*Five turn coil.

Failure to obtain the above results could be due to one or more of the following:

(1) Quality of the bath is poor—too strong or too weak.

(2) Head and/or coil circuit is not operating correctly.

(3) Operator's "wet continuous" technique is incorrect.

This device gives only qualitative information regarding bath strength, but does give an overall check on the equipment and the operator's technique. Other variations of this test device are in use and give similar results.

Development work is proceeding, and it is probable that a fully satisfactory quantitative bath strength testing device will be available in the near future.

9. TESTS FOR BLACK LIGHT INTENSITY. When fluorescent particles are being used frequent checking of black light intensity is of great importance, if reliable and reproducible results are to be obtained. A detailed discussion of the causes of black light variations is given in Chapter 15, Sections 16 to 19, and in Chapter 16, sections 14 to 18. Black light intensity is measured by exposing a photo-voltaic cell light meter to the radiation of the black light source held at a specified distance from the meter, with white light excluded.

Light meters used are:

Weston Model 703 Type 3, Sight Light Meter.

General Electric Light Meter, Model 8DW 40Y 16.

The meters are calibrated directly in foot-candles of white light. The reading is therefore not truly in foot-candles for black light. It is, however, quantitative and reproducible, and therefore gives reliable comparative results. Meter readings of not less than 90 foot-candles *at the point of inspection* are considered minimum for satisfactory brightness of fine fluorescent indications.

APPENDIX

The article which is printed below is the last bit of technical writing produced by A. V. de Forest, Founder (with F. B. Doane) of Magnaflux Corporation, before his death in April of 1945. The paper was to have been given by him at a Society meeting in February of that year. However, due to illness, he was unable to do so, and the author of this book read the paper in his stead. It has not been published to date, and is reproduced here in its entirety in A. V.'s original language.

The subject matter is basic to our interests in the detection of flaws and the causes of failure in metals. The discussion is clear and concise. It would be difficult to find a more comprehensive dissertation on this subject, which at the same time is as interestingly written and as understandable. After reading it, the various phases in which defects affect the strength of materials—and in which the properties of metals in their various aspects influence the effect of defects—are much more clearly understood.

The prime reason, however, for the inclusion of this paper in a book such as this, is the fact that it is a truly classical bit of technical exposition. The substance of the paper, written over twenty years ago, is still as pertinent today as it was in 1945, notwithstanding the great increase in our knowledge in these areas.

We feel that reading of this paper will be well worth while for anyone interested in the subject matter of this book.

Carl E. Betz

DEFECTS AND THE STRENGTH OF MATERIALS
by Alfred Victor de Forest. 1945.

Before it is possible to discuss defects and their relation to strength, it is necessary to understand clearly what is meant by these two simple words, defects and strength. As discussed here, both are related solely and directly to use. In this sense, therefore, a part is defective if it will not perform the function for which it is intended, quite regardless as to whether or not it meets an arbitrary set of specifications. Strength, likewise, will be used to mean ability to resist successfully the forces for which the part was designed under the circumstances under which it was intended to serve a useful purpose. We will have to limit the discussion to this comparatively narrow field or we shall be led far from the known paths of engineering.

Strength in metals is measured in many terms; tensile, compression, shear, torsion, and in brittle materials, bending as well. All these manifestations of strength are likewise defined by the rate at which loads are applied. If loads are applied for months and years, creep strength may be the all important factor. This is especially true when metals are used at elevated temperatures as in steam turbines, fired pressure vessels and especially gas turbines. Rates of loading for test purposes which require a few minutes to a few hours are misnamed "static" tests, and are those most commonly referred to, again erroneously, as physical tests, while in truth they are mechanical tests devised and used by mechanical engineers, and not by physicists.

Tests made at high rates of loading, where the metal is broken in less than a second, and sometimes in tenths of thousandths of seconds, are termed impact tests. The higher rates of loading when applied to armor and projectiles are also termed ballistic tests. The highest rates are those caused by the action of a detonating explosive applied directly to a metal as when a block of T.N.T. rests directly on the surface and is exploded by a primer. The rate of travel of the pressure wave in the explosive is approximately 17,000 ft. per second, about the speed of sound in steel, and this rate of application of load is far higher than that reached in normal ballistic tests. Ordinary impact tests, such as the Charpy and Izod, apply a blow at 15 ft. per second. Ballistic tests run from 800 ft. per second for revolver ammunition to 2500 ft. per second for rifles of normal variety. Velocities up to 5000 ft. per second are obtainable under special conditions. Explosions or detonations may operate up to 20,000 ft. per second.

The most important aspect of strength as applied to moving parts is termed endurance or fatigue strength. About one hundred years ago it became evident to engineers that metals did not stand un-limited repetitions of load even though this load was below the elastic limit. Very many investigators using a great diversity of testing machines have studied the effects of load reversals or load repetitions both from the point of view of learning the fatigue strength of the different metals and alloys and for the purpose of investigating the fatigue strength of full-sized parts. An excellent review of the subject is given in the book "Prevention of Fatigue of Metals" prepared by the Battelle Memorial Institute and pub-lished by John Wiley & Sons in 1941. As Magnaflux inspection is applied very frequently to moving parts, it is important to consider the various aspects of fatigue strength, because the factors which influence this are by no means the same factors which influence static or impact strength.

In this discussion we will confine ourselves to the magnetic ma-terials, that is, the ferritic steels.

1. GENERAL CONSIDERATIONS

The fatigue limit is the maximum stress a metal can withstand when the load is applied an indefinitely large number of times.

The fatigue strength of smooth specimens without notches or fillets is closely proportional to the ultimate tensile strength and under conditions of complete reversal of stress as produced in a rotating bend machine, the fatigue limit is found at about 45% to 50% of the ultimate; meaning that a 100,000 lb. per square inch steel will have a fatigue limit between 45,000 and 50,000 lbs. per square inch, regardless of the position of the yield point or the elastic limit. In extremely hard and strong steels containing mar-tensite and with tensile strengths above 250,000 lbs. per square inch, this proportion drops off and the fatigue limit lies around 35% to 40% of the ultimate.

The significant factor is that fatigue strength has no relation whatever to ductility, for the growth of fatigue cracks takes place with extremely small amounts of local deformation, in most cases too small for observation. The fact is well known but has not been completely explained. The exact mechanism by which the fatigue crack is produced is as yet unknown and so far there is no known method of inspecting a partially fatigued specimen in which an actual fatigue crack has not yet begun to grow. Very small fatigue cracks have been observed, of a length of only five thousandths of an inch, and their microscopic appearance is exactly the same as that of large cracks. There is undoubtedly a change in the metal at the location at which the fatigue crack will ultimately occur but what happens has as yet escaped positive proof.

The subject is complicated, as is shown by experiments where the specimen is loaded to a high fiber stress for a small number of repetitions of load and then subsequently tested at loads close to the fatigue limit. It is certain that in many cases the fatigue strength, as shown by the final test, is improved by not too many cycles of high stress, possibly by local yielding at high stress areas. When the high stressed level is repeated sufficiently, often the final fatigue strength is lowered.

It is also known that if the specimen is repeatedly loaded below the fatigue limit, the final endurance strength is in many cases raised, so that it would appear that the final fatigue limit is shifted one way or the other by previous strain histories. As there is no method of measuring directly what state the specimen is in, no engineering advantage can be taken of this situation and it is not considered that an old crankshaft is better than a new one by virtue of its previous history. The possible improvement is, in most cases, small, less than 10%, and the same conditions which produced the improvement might also be detrimental if the number of cycles and the stress limits had been above the damage line. As far as ordinary engineering practice is concerned, a part which has long been in use but contains no fatigue cracks may be considered as sound as it was originally.

All experiments show that all fatigue cracks grow slowly, especially so in steel. It may require an additional 10% to 40% repetition of load to break a fatigue specimen after the appearance of the first fatigue crack. The rate of growth of these cracks has not been adequately investigated but there is some evidence to show that different steels have different rates of growth and that these rates of growth are independent of the actual fatigue limit. From theoretical considerations it might be supposed that this rate of growth is in some way connected with the damping capacity or hysteresis loss in the metal, (i.e. the ability of the steel to absorb energy). At any rate, the materials which owe much of their strength to cold work, that is, wire and cold-rolled products, generally have a higher damping capacity and a slower rate of growth of fatigue cracks. Materials which are highly elastic, such as fully heat treated steels, and precipitation hardened alloys, show a more rapid rate of growth of fatigue cracks but in this field experiments are very limited. Further discussion of this point, and much detailed information on fatigue can be found in the book "Prevention of Fatigue of Metals", (previously referred to). The low rate of growth is a very important factor for the engineer, for it has been found that on repeated inspection at overhaul periods many small fatigue cracks are located before they have reached dangerous proportions.

One of the largest applications of Magnaflux inspection is in this field, where locomotive and railroad rolling stock, automotive and airplane parts are regularly inspected after a reasonable number of miles or hours of operation. It has been the universal experience on the railroads that such periodic inspection is well worth while, for in addition to preventing a possible accident, the small fatigue cracks when found may be machined, chipped or ground out and a smooth fillet provided instead of a sharp-bottom crack. The part is then returned to service and frequently operates for very long periods, if not indefinitely.

2. FREQUENCY OF LOADING

Many experiments indicate that the frequency of load application is unimportant, at least until the part reaches some 40,000 repetitions of load per minute. At these very high rates of loading, most materials tend to become hot, even though below the fatigue limit, and the fatigue limit tends to increase. These high frequencies are not met with in large parts but are primarily interesting when an attempt is made to build very high-frequency fatigue testing equipment in order to get through the test in a shorter time. This may be done on small specimens and some testing equipment has been built to operate above 30,000 cycles a minute. As it is the number of cycles that counts, in many cases of slowly repeated loads the total life of the structure may not be sufficient to reach the actual fatigue limit and in this case the permissible stress may be higher than would be the case in high-speed machinery.

3. CYCLES TO REACH THE FATIGUE LIMIT

In the steels under discussion the fatigue limit is reached after ten million cycles, provided the part is kept free from corrosion. That is, for any given loading, failure will occur within ten million cycles if the loading is at or above the fatigue limit for the specimen. This rather definite and limited number of cycles is of great advantage, and fatigue tests can be discontinued if the part has not failed after this number of loads. In the case of the non-ferrous materials, such as the aluminum and copper alloys, this limitation does not exist and the fatigue limit decreases continuously as the number of cycles is increased up to a billon or perhaps more. That is, loading, to be under the fatigue limit, must take into account the expected number of stress cycles the part may have to sustain throughout its life. The reason for this situation has not been explained.

4. VARIATIONS IN FATIGUE STRENGTH

It is the universal experience that individual specimens show far more variations among themselves when tested in fatigue than when tested under ordinary static conditions. There is, therefore, more uncertainty as to the life of a part under fatigue than can be explained by the extent to which the steel is not uniform as shown by static loading. The reason for this undoubtedly is that the fatigue crack is initiated by very small variations in surface conditions or local metallurgical circumstances, such as particularly oriented grain boundaries. It is, therefore, to be expected that if one hundred specimens are tested in exactly the same manner, the weakest specimen may easily be 10% below the average. The stress level between the weakest and the strongest in any set of tests is called the scatter band and, quite obviously, good engineering requires that the width of this scatter band be taken into consideration when designing a part. The width of this scatter band may be particularly great in complicated components, such as ball bearings in which fatigue may take place anywhere on the surface of any ball or on the mating surface of raceways. This occurs even though the ball bearings are more carefully heat treated and sized and made from cleaner steels than any other engineering components.

5. SURFACE CONDITIONS

Under almost all conditions of fatigue loading, cracks start from the surface of the parts unless some particular treatment has been used to raise the strength of the surface. Smooth specimens are stronger than rough ones but the degree of roughness is difficult to measure. It appears probable that the sharpness of the notch in the surface is the most important factor. A ground surface may be smoother than a machined surface but the sharpness of the fine notches or scratches left by the abrasive grit may be greater than those left by the machine tool. The direction of the notch is likewise important and rather coarse grinding which lies parallel to the direction of stress is far less detrimental than a smooth appearing grind in which the grinding marks are at right angles to the stress. It is for this reason that airplane engine parts are, whenever possible, finished so that the grinding marks run parallel to the major stress direction.

Corroded or etched surfaces are particularly detrimental and where the fatigue limit is important, surfaces produced by casting or forging are smoothed by machining or grinding.

6. METALLURGY OF SURFACES

In addition to the smoothness of finish, any metallurgical change between the surface and the underlying metal is particularly important. A decarburized layer a thousandths or two deep on the surface of strong steel very greatly reduces the fatigue strength, for a small fatigue crack forms in the weak decarbonized layer which then proceeds through into the underlying metal. This condition is true even though the surface is protected by a hot-dip coating or an electro plate. Fatigue cracks may start easily in the electro plate layer and proceed from there into the strong steel underneath.

Usually all electro plating is detrimental to fatigue life as is likewise hot-dipped galvanizing. In the latter case, fatigue cracks occur in the brittle iron-zinc alloy which can propagate into the steel even though the thickness of this layer is less than a thousandth of an inch. These protective coatings, however, may serve a very useful purpose in preventing direct corrosion.

There are many methods of improving the strength of the surface of steel, for instance, by carburizing, nitriding, cyaniding and flame or induction hardening. In all of those instances the fatigue strength of the surface of the part may be very materially increased, sometimes to such an extent that the fatigue crack forms at the junction between the hardened surface and the normal underlying steel. In this case, while the fatigue crack actually begins below the surface, it progresses rapidly outward through the strong layer of the external surface, and in any event the crack lies close enough to the surface to be readily found by Magnaflux inspection.

Another method of strengthening surfaces has become extremely widely used—the improvement by means of shot peening. Surfaces treated in this manner are somewhat roughened by the shot blast but the surface is hardened by cold work and, therefore, strengthened; and at the same time the surface is put into initial compression so that under repeated loads the maximum tension is less than would have been the case in the absence of such treatment. It is as yet impossible to say what proportion of improvement is due to the changed stress condition and what to the increase in strength and hardness due to cold work, but in any event shot peening can markedly improve the fatigue strength of parts whether of steel or of other metals. It is current practice in many cases to shot blast instead of grind as a method of improving forgings, for it has been found that a rough forging, shot blasted, may be as good as or better than the same forging after a careful surface grind. Quite obviously, in parts of complicated shape, shot blasting is far more economical. It must be remembered that this shot blasting must be done with proper care and control, for there is a very distinct maximum in the results obtained by variations in the size of shot, the velocity of the shot and the length of time of blasting.

7. CORROSION

All metals show a great reduction in fatigue strength if the part is subjected to corrosion during the application of load. It would appear that surfaces under repeated stress are much more readily attacked than surfaces under static load. Furthermore, the attack is continuous and the fatigue limit continues to decrease way below the normal fatigue limits. One can, therefore, say that there is no real fatigue limit if a part is tested under water. The strength will be reduced continuously during the time of operation. If the part is steel, operating in salt water, the reduction is very rapid and good engineering demands absolute protection against such conditions. Many engineering parts have to operate in an unprotected manner, for instance, hollow drill steel in which water is circulated to remove stone dust from the bottom of the hole. After prolonged operation, fatigue cracks start from the inside diameter of the drill steel. Sucker rods for pumping oil wells likewise have to withstand corrosion fatigue and have a rather limited life. Tail shafts of ships are protected from direct contact with the water inasfar as possible by bronze liners and rubber gaskets of many different types. In spite of the fact that such shafting operates at a very low fiber stress, failure is inevitable if the water tightness of the packing is not perfect.

In the same way turbine blades operating in steam have a lower fatigue limit than they would have in air and the fatigue resistance of these blades must be studied in relation to corrosion rather than as to stress cycles. There is strong evidence that fatigue testing in air results in lower fatigue limits than though such tests were conducted in a vacuum, brass, for instance, showing a 20% improvement in some cases when tested in the absence of air. While this is interesting, fatigue testing is usually carried on in an air atmosphere without humidity control, although perhaps humidity control would reduce some of the variations which are normally found.

8. Notches and Stress Raisers

Another great source of variation between static testing and fatigue testing is in the very important effect of notches in relation to fatigue. A notch produced by a change in section on the surface serves to concentrate the stress and in this manner produces a very definite lowering of the fatigue strength. For instance, a small hole in a comparatively large plate under tension produces a three to one stress concentration. This means that the stress immediately adjacent to the hole is three times the average stress in the plate and in calculating the fatigue strength of a part with an oil hole, such factors must be taken into consideration. A stress concentration at a fillet will, of course, depend on the radius of the fillet and the relative diameters of the two parts connected by the fillet. A square notch is, therefore, more dangerous than a large radius at such a transition.

In most bolts there is a square notch under the head, unprotected by any fillet. Such a condition is extremely weakening to the fatigue strength if the bolt is in repeated tension. A fillet at this point will markedly increase the strength of the bolt and such fillets are provided in proper designing of connecting rod bolts. The thread end of a bolt is likewise a series of notches and has a very detrimental influence on fatigue strength. A "V" bottom thread, especially if produced by grinding, is far more detrimental than a rounded root as in the Whitworth system. The stress raising effect of such mechanical notches may easily be measured by photo-elastic means which accurately measures the stress distribution at notches of all types where the stress can be considered in a plane.

Inasmuch as stresses computed from the theory of elasticity or measured by means of photo-elasticity assume the material to be perfectly elastic, it is found on fatigue testing that the notch severity is not quite as pronounced as theory indicates, because the metals themselves are not perfectly elastic. The less elastic the material, the greater is the discrepancy between the theory and practice. This means that the harder, stronger and more elastic metals are more injured by notches than the softer and less perfectly elastic metals. Fatigue tests are, therefore, conducted to determine the notch sensitivity of different materials and a knowledge of this factor is important for proper selection. It frequently happens that a less notch-sensitive carbon steel is superior in the notched condition to a stronger and more elastic alloy steel on account of this notch effect. It is important to realize that the shallow notches, extremely sharp at their roots, are more detrimental than deeper and more rounded notches and the defects found by Magnaflux are primarily detrimental to fatigue because of their extreme sharpness rather than on account of their depth.

The notches caused by grinding cracks and heat-treating cracks are, by their nature, extremely sharp regardless of their depth and, therefore, constitute the worst form of stress raiser. Notches caused by cold shuts in forgings or laps in rolled bars may also be sharp and, if they run perpendicular to the direction of applied load, constitute a very serious weakness. The lap resulting from an over fill in rolling bar stock lies parallel to the direction of rolling. If this bar stock is made into a bolt which is primarily loaded only in tension or shear, the presence of the notch may be quite unimportant, merely on account of the direction of stress. If, for any reason, this same stock were used in repeated torsion, as is the case in a helical spring, the notch is then a serious source of weakness. In order to decide whether or not any particular discontinuity in the metal is important and constitutes a defect, it must be considered in relation to the direction of loading.

Notches are likewise caused by the presence of non-metallics in steel. These non-metallics are distributed in the direction of rolling or forging and, if they are parallel to the stress, may be relatively unimportant, becoming more important as the stress is applied at right angles. For instance, seamless tubing used in tension or compression may contain longitudinal non-metallics which are quite harmless but if this tubing is loaded by hoop tension, as is the case when fluctuating hydraulic pressure is applied, the exact same non-metallics may constitute serious weakness. The same argument holds in regard to slight imperfections in welded tubing. These may be extremely serious under internal-pressure conditions in the tubes, as for instance, in Diesel engine oil injection lines, but if the tubing is used for airplane construction, where loads are distinctly longitudinal, the same discontinuities may be harmless. It is to be noted that bars or tubing in torsion are seriously weakened by longitudinal discontinuities, for in torsion the maximum tension stress lies at an angle of 45° to the axis. Helical springs must, therefore, be particularly carefully inspected and all serious longitudinal discontinuities removed.

Non-metallics in steel, particularly the manganese suphide type, present rounded surfaces rather than sharp discontinuities and are, therefore, realtively harmless as compared with cracks, cold shuts, or deep seams. In almost all cases the stress raising effect of the non-metallics is far less than the stress raising effect of notches put in by improper design. If a part, such as a bolt or crankshaft, already contains mechanical notches of considerable severity, there will be no weakening produced by non-metallics whose effect is considerably less than that of the mechanical notch unless the non-metallic occurs at a position of maximum load. It is, therefore, customary to ignore non-metallics which occur elsewhere than at known high stressed positions. A severe non-metallic stringer at an oil hole may be dangerous, whereas in the metal a short distance from the oil hole it may be harmless.

There is a relationship between grain direction and stress under fatigue loading, cross-grain metal being weaker than with a longitudinal grain, primarily because the distribution of the non-metallics follows the flow lines of the steel in complicated forgings. The direction of the flow lines is frequently important. These flow lines may be shown by Magnaflux inspection where a heavy current and the wet continuous method is used. The same information can also be found by deep etching. Inasmuch as the flow lines in the steel are always present in forgings, Magnaflux indications of the flow lines must not be considered a source of weakness but rather as an indication of the structure of the metal. Investigation of flow lines should be of interest to the metallurgist rather than the inspector.

While non-metallics themselves may not be of importance in many cases, they may lead to serious difficulty. For instance, steels which are quenched outright may be cracked during heat treatment by the stress raising effects of non-metallics and in this case the effect of the non-metallic may be serious as an indication of steel quality. The same is true of ball-bearing steels in which the presence of non-metallics at the working surface of the ball may lead to early failure.

Magnaflux inspection is much used in determining the quality of steel in regard to its non-metallic content. It is quite customary to inspect aircraft quality steel and tool quality steel by this means, not because the non-metallics are a source of weakness in the bar which is inspected, but purely because such an inspection is quicker and cheaper than examination by means of microscope. Such tests are again of a metallurgical rather than of an engineering importance.

9. INFLUENCE OF A SERIES OF NOTCHES

It is well known that a series of notches is less detrimental to fatigue strength than a single notch of the same shape and depth. A part which is threaded throughout its length is, in fatigue, stronger than a part containing a single notch or which is threaded only for half its length. A single non-metallic of considerable size in an otherwise clean steel is, therefore, a more severe stress raiser than a well distributed system of non-metallics throughout the material. In the same way, a single blow-hole in otherwise sound metal is more detrimental than a large area of porosity caused by small blow-holes. In general, blow-holes have rounded contours and do not constitute a severe source of weakness. Porosity and blow holes in welds are, therefore, relatively harmless although they may appear to be very pronounced under x-ray examination. Sand pockets act in the same way in castings. Shrinkage cracks in castings, however, are usually intergranular and constitute sharp notches, injuring the material far more than porosity and sand. Such shrinkage cracks lying near the surface are usually removed and repairs are made by welding.

It has long been recognized that gray iron is remarkably free from notch sensitivity. This behavior is usually explained by the fact that the material already contains many interior notches in its grain structure at the ends of the graphite flakes. As these notches are more severe than the external notch caused by suitable fillets, the effect of the mechanical notch is obscured by the original weakness of the metal. Here again we have a system of a very large number of small notches, the behavior of which is included in the characteristics of the metal.

The same may be said of free cutting steels in which non-metallics are introduced for the purpose of breaking the chip during machining. Such steels, although extremely dirty in the usual sense, are in fact quite trustworthy, for their weakness is generally known and considered in the design. If, however, a part is designed for clean steel and a substitute is then made to free machining steel, the results may be unsatisfactory, particularly if the load is not directly parallel to the rolling direction. Free machining steels are, therefore, substantially as good as clean steels in tension and compression but are weaker under transverse loads, or torsion.

10. EFFECT OF NOTCHES UNDER IMPACT LOAD

It is now known that there is a relationship between rate of loading, temperature of steel, and severity of a notch, all steels becoming more brittle as the temperature is lowered. The transition from the ductile to the brittle behavior is very clearly marked and is called the drop-off point. The temperature at which this occurs is characteristic of the type of steel used, its alloy content and its method of deoxidation in the furnace. In general, Bessemer steel has a higher drop-off point than open hearth steel and rimmed steel is also more sensitive to lowered temperatures than fully killed steels. The addition of alloys, particularly nickel, lowers the drop-off point.

Whatever the inherent temperature sensitivity of the steel, if it breaks in a brittle manner at a higher temperature, the more severe is the notch condition, and exactly as in the case of fatigue, the sharpness of the notch rather than its depth, is important. Parts which are to operate under impact, particularly under low temperatures, must, therefore, be free of stress raisers, both those produced by design and those produced by accident. Non-metallics may, therefore, be important as in the case of fatigue, if the stresses are not parallel to the rolling direction, particularly at low temperatures. Very much less information is available here than in the case of fatigue but direct experiments on full-sized parts may always be used to substantiate any questionable conditions.

Increased brittleness due to low temperature, leading to failure under impact, is characteristic of ferritic steel only. Non-magnetic steels do not show this effect nor do any other engineering alloys, and containers for liquid oxygen and other low-temperature chemical equipment are not made of magnetic steel.

All metals exhibit two forms of strength, one the shearing strength in which different elements of the structure slide over each other without immediate failure thereby allowing a change of shape under load. When the shearing strength is relatively low, a metal is said to be ductile. Metals also have a definite cleavage strength, meaning the force required to pull the structure apart under direct tension. All brittle materials, both metals and such substances as glass and rock, fail by cleavage. Whether a metal fails in a ductile or brittle manner depends on the relation between the two types of strength. In all magnetic steels the shearing strength increases as the temperature is lowered, more rapidly than the cleavage strength, and when the two strengths are equal, or the shear strength is the higher, a brittle failure results.

The effect of rate of loading is likewise to increase the shear strength so that the rate of loading required influences the drop-off point in the same way as a lowering of temperature. This characteristic of steel has long been known. Steel rails, for instance, are more prone to break the lower the temperature at which they are operated. Low carbon steel, such as used in chain, becomes undependable at low temperatures, and chain is frequently heated before being used on wrecking cranes or other equipment operated in severe winter weather. A careful workman likewise warms his cold chisel before chipping when the weather is severe. It is generally recognized that welded steel ships are more apt to sustain severe damage from cracks in cold weather.

From all that has been said it is evident that before it can be determined whether or not a discontinuity constitutes a defect, it is necessary to know the conditions of service, including both the amount and the direction of those stresses for which the part is intended. As far as the stress distribution in service parts is concerned, this subject is in the hands of a relatively new profession, that of stress analysts. The new engineering tools of Stresscoat and the wire strain gage have made it possible to determine accurately what stresses are encountered in service. The more accurately these factors are known, the more fully the distinction may be made between harmless and harmful conditions. This art is rapidly advancing and should form a part of the education of all those dealing with the strength of materials.

It is customary for engineers to design parts with a certain factor of safety. When a failure may be extremely costly it is customary to either use a large factor or inspect the part with additional care. For instance, the failure of an airplane engine part may result in very great damage as compared with a corresponding failure in an automobile. It is, therefore, good engineering to ask for a more rigorous inspection of airplane parts than of automotive parts, particularly if the failure of the automotive part does not endanger the vehicle. Springs may, therefore, be used at a lower factor of safety and with less rigorous inspection than parts of the steering gear or of the brake system. Much skill and judgement is required to reach the best compromise between weight, cost and safety.

A. V. de Forest
Feb. 1945

APPENDIX

BIBLIOGRAPHY

1. SAXBY, S. M., *Magnetic Testing of Iron.* Engineering, Vol. 5, P. 297, 1868.

2. BURROWS, C. W., *On the Best Method of Demagnetizing Iron in Magnetic Testing.* Bulletin 78, United States Bureau of Standards, Pp. 205-274, 1908.

3. STEINHAUS, W. & GUMLICH, E., *Experimental Investigations in the Theory of Ferro-Magnetism.* Verhandlungen der Deutschen Physikalischen Gesellschaft, No. 21, 1915.

4. STEINMETZ, CHARLES P., *Transient Electric Phenomena and Oscillation.* 1920.

5. MITRA, S. K., *On the Demagnetization of Iron by Electromagnetic Oscillation.* J. de Physique, P. 259, 1923.

6. SPOONER, THOMAS, *Properties and Testing of Magnetic Materials,* 1927.

7. TSCHETVERIKOVA, M., *Demagnetization of Iron Compounds by Electric Oscillation.* Ztschr, f. Physik, 44, No. 139, 1927.

8. CULVER, CHAS. A., *Electricity and Magnetism.* October, 1930.

9. WATTS, T. R., *Magnetic Testing of Butt Welds.* Journal of the American Welding Society, Vol 9, No. 9, P. 49 f., September, 1930.

10. WILLIAMS, SAMUEL R., *Magnetic Phenomena.* 1931.

11. GEROLD, E., *Kritische Betrachtung der Magnetischen Verfahren fuer Werkstoffpruefung.* Stahl und Eisen, Vol. 51, No. 14, P. 428, April 2, 1931.

12. MEHL, R. F., *Nondestructive Testing Methods in the Iron and Steel Industry.* Yearbook American Iron and Steel Institute, Pp. 126-184, 1932.

13. DE FOREST, A. V., *Magnetic Testing of Iron and Steel.* Iron and Steel Engineer, Vol. 9, No. 4, P. 170, April, 1932.

14. DE FOREST, A. V., *Magnetic Test Locates Flaws in Valve Springs*. Iron Age, Vol. 130, No. 3, P. 107, July 21, 1932.

15. DOANE, F. B., *Magnetographic and X-ray Tests of Pipe Welds Compared*. Iron Age, Feb. 2, 1933.

16. JACOBS, F. C., *A Crack Sleuth—Magnetic Powder Inspects Turbine Blading*. Power, May, 1933.

17. DOANE, F. B., *Magnaflux Inspection for Cracks and Seams*. Iron Age, September 26, 1933.

18. McCUNE, C. A., *Magnaflux in the Transit Field*. Transit Journal, September 20, 1933.

19. VON SCHWARZ, M. & KRAUSE, J., *Magnetische Untersuchung zum Fehlernachweis in Ferromagnetischen Werkstoffen*. Maschinenschaden, Vol. 11, No. 7, P. 107, 1934.

20. HATFIELD, H. S., *The Action of Alternating and Moving Magnetic Fields Upon Particles of Magnetic Substances*. Physics Society (London), 46 (5), P. 604, 1934.

21. DE FOREST, A. V., *Magnaflux Testing for Discontinuities in Magnetic Materials*. U.S. War Dept., Air Corps Technical Report, 1935.

22. DE FOREST, A. V., *Dynamic Strength of Machine Parts Affected by Quality of Surface*. Iron Age, Feb. 21, 1935.

23. DAVIS, C. W., *Some Observations on the Movement and Demagnetization of Ferromagnetic Particles in Alternating Current Magnetic Fields*. Physics, Vol. 6, June, 1935.

24. WARREN, A. G., *Flash Magnetization for Examining Castings and Forgings*. Engineering, Vol. 140, No. 3638, Pp. 353-354, Oct. 4, 1935.

25. DE FOREST, A. V., *The Rate of Growth of Fatigue Cracks*. Journal of Applied Mechanics. March, 1936.

26. METROPOLITAN VICKERS ELECTRIC COMPANY, *Fault Detection in Metals by Magnetic Methods*. Engineering, Vol. 141, No. 3669, P. 504, May 8, 1936.

27. NOBLE, H. J., *Magnaflux Inspection Methods in Airplane Engines*. Metals and Alloys, Vol. 7, No. 7, P. 167, July, 1936.

28. YANT, J. W., *The Magnaflux Inspection of Pressure Vessel Welds.* Paper presented Oct. 23, 1936, at ASME Symposium, Cleveland, Ohio.

29. BERTHOLD, R., *Fault Detection by the Magnetic Powder Process.* Engineering Progress, Berlin, Nov., 1936.

30. WILHELM, H. & ELSASSER, H., *A Method of Detecting Circumferential Cracks in Hollow Cylinders.* Mitteilungen Verein Grosskessel-Bezitzers, December 30, 1936.

31. HODGE, JAMES CAMPBELL, *Welding of Pressure Vessels.* Journal of American Society for Naval Engineers, Pp. 519, 521, Nov., 1936.

32. BITTER, FRANCIS, *Introduction to Ferromagnetism.* 1937.

33. SKILLING, HUGH H., *Transient Electric Currents.* 1937.

34. BERTHOLD, R. & GOTTFELD, F., *A New Aid in Testing Welds.* Stahlbau, Vol. 10, No. 4, P. 31, Feb. 12, 1937.

35. CAVANAGH, R. F., *Magnaflux Inspection of Boiler Drums and Unfired Pressure Vessels.* Mechanical Engineering, Vol. 59, P. 153, March, 1937.

36. RATHBONE, T. C., *Detection of Fatigue Cracks by Magnaflux Methods.* Mechanical Engineering, Vol. 59, P. 147, March, 1937.

37. DIXON, E. O., *Metallurgical Qualities in Forgings.* Aero Digest, July, 1937.

38. ADAMS, C. A., *The Nondestructive Tests of Welded Joints.* Paper presented at the Eleventh General Meeting of the National Board of Boiler and Pressure Vessel Inspectors, at New York. May 25, 1937.

39. YAMIN, I. V., *New Method of Detecting Cracks in Welded Seams.* Zavodskaya Laboratoriya, Vol. 6, P. 1284, 1937.

40. DE FOREST, DOANE, MCCUNE, *Nondestructive Tests of Welds.* American Welding Society Handbook, 1938.

41. WEVER, F. & HAENSEL, H., *Beitrag zur Magnetischen Werkstoffpruefung.* Mitteilungen aus dem Kaiser-Wilhelm Institut, Vol. 20, No. 8, 1938.

42. DE FOREST, TABER, *Detection of Fatigue Cracks.* California Oil World, March 26, 1938.

43. HAENSEL, H., *Die Fehlererkennbarkeit bei der Magnetischen Zerstoerungsfreien Pruefung*. Archiv fuer das Eisenhuettenwesen, Vol. 11, No. 10, April, 1938.

44. GRIGOROV, K. V., *Fixation of the Precipitate of Magnetic Inspection*. Zavodskaya Laboratoriya, Vol. 7, Pp. 737-738, 1938. Abstract in Chemical Abstracts, Vol. 33, Pp. 101-109, 1939.

45. MCBRIAN, RAY, *Magnetic Testing of Car and Locomotive Parts*. Railway Electrical Engineer, July, 1938.

46. MCCUNE, C. A., *Magnaflux Inspection of Gas Cylinders*. Iron Age, July 14, 1938.

47. MCBRIAN, RAY, *D. & R. G. W. Installs Magnaflux Inspection*. Railway Electrical Engineer, September, 1938.

48. BERTHOLD, R. & SCHIRP, W., *The Physical Bases of the Magnetic Powder Process*. Atlas der Zerstoerungsfreien Prufverfahren, published by Johann Ambrosius Barth, Leipzig, 1938.

49. BERTHOLD, R., *Technical Aids of the Magnetic Powder Process*. Ibid, Leipzig, 1938.

50. LEE, JOHN G., *Airplane Demagnetization and Neutralization*. Journal of the Aeronautical Sciences, Vol. 2, No. 12, Oct., 1939.

51. PORTEVIN, ALBERT M., *Interpretation of Magnetic Patterns*. Metal Progress, April, 1939.

52. LESTER, H. H., SANFORD, R. L. & MOCHEL, N. L., *Nondestructive Testing in the United States of America*. I.E.E. Journal, Vol. 84, No. 509, May, 1939.

53. BERTHOLD, R., *Nondestructive Testing Based on Magnetic and Electrical Principles*. I.E.E. Journal, Vol. 84, No. 509, May, 1939.

54. WHITE, A. E., *Changes in High-Pressure Drum to Eliminate Recurrence of Cracks Due to Corrosion Fatigue*. Paper presented at the Annual Meeting of the ASME at New York, Dec. 5, 1938. Published in the Transactions, ASME, Aug., 1939.

55. MCCUNE, C. A., *Detection of Flaws in Steel and Steel Parts*. National Safety Council News, October, 1939.

56. GRAF, S. H., *Industrially Significant Nondestructive Testing Methods for Engineering Materials and Machine Parts*. Paper pre-

sented to SAE, Seattle, Washington, March 10, 1939 and ASM, Portland, May 12, 1939.

57. TIMMONS, J. J., *Magnaflux Inspection of the Power Plant.* Power Plant Engineering, April, 1940.

58. DE FOREST, A. V., *Nondestructive Testing of Metals.* Mining and Metallurgy, July, 1940.

59. MCBRIAN, RAY, *New Denver and Rio Grande Laboratory in Step with Modern Progress.* Railway Purchases and Stores, P. 457, Oct., 1940.

60. JAEKEL, W., *Magnetic Powder Method.* Die Giesserei, Vol. 27, Pp. 262-265, 1940.

61. SCHRADER, H., *Practical Experiences with Magnetic Powder Method of Detecting Cracks.* Stahl und Eisen, Vol. 60, Pp. 634-645 and 655-660, 1940.

62. MCCUNE, C. A., *Magnaflux Inspection of Railroad Steel Parts.* Brotherhood of Locomotive Fireman & Enginemen's Magazine, Vol. 109, No. 6, P. 397, Dec., 1940, Southern and Southwestern Railway Club Proceedings, July, 1940.

63. *Principles of Magnaflux Inspection.* First Edition, DOANE, F. B., Magnaflux Corporation, January, 1940.

64. DE FOREST, A. V., *Magnaflux on the Railroads.* New England Railroad Club, April 8, 1941.

65. *Recommended Practice for Magnaflux Rating.* SAE Journal, Vol. 48, Pp. 18-19, January, 1941. SAE Handbook, P. 335.

66. *Prevention of the Failure of Metals Under Repeated Stress.* Battelle Memorial Institute, 1941.

67. *Airplane Welding and Materials.* Second Edition, JOHNSON, J. B., 1941.

68. *Inspection of Metals.* PULSIFER, HARRY B., 1941.

69. *Magnaflux Aircraft Inspection Manual.* DOANE, F. B. & THOMAS, W. E., Magnaflux Corporation, August, 1941.

70. DAVIS, C. W., *A Rapid Practical Method of Demagnetization, Involving High Frequency.* Nature.

71. HASTINGS, CARLTON, *The Magnetic Powder Method of Inspecting Weldments and Castings for Sub-surface Defects.* Welding Journal, January, 1943.

72. THOMAS, W. E. & DE FOREST, T., *Bolt and Nut Inspection by the Magnetic Particle Method.* Published by Magnaflux Corporation, July, 1943.

73. BETZ, C. E., *Use of Magnaflux by the Railroads.* Proceedings, Southern and Southwestern Railway Club, September, 1943.

74. BETZ, C. E., *Magnetic Particle Inspection.* Canadian Metals and Metal Industries, May, 1944.

75. MAGES, M. L., *Demagnetization During Magnetic Particle Inspection.* Aeronautical Engineering Review, Vol. 3, No. 10, Oct., 1944.

76. OWENS, JAMES E., *Routine Inspection and Salvage of Machinery Weldments.* The Welding Journal, October, 1944.

77. *Tentative Method for Magnetic Particle Testing and Inspection of Heavy Steel Forgings.* American Society for Testing Materials, Standards, 1944, Part I, A 275-44 T.

78. *Tentative Method of Magnetic Particle Testing and Inspection of Commercial Steel Castings.* American Society for Testing Materials, Standards, 1944, Part I, A 272-44 T.

79. FFIELD, PAUL, *Conditioning of Steel Castings to Standards of Quality.* Transactions, American Foundrymen's Association, Vol. 52, Pp. 173-204, 1944.

80. *Tentative Recommended Practice for Determining the Inclusion Content of Steel.* American Society for Testing Materials, Standards, 1944, Part I, E 45.

81. COTTON, JOHN F., *Magnetic Powder Inspection of Large Castings.* Transactions, American Foundrymen's Association, 1944.

82. *Magnetic Particle Inspection.* Inspection Handbook for Manual Arc Welding; American Welding Society, 1945.

83. BROWN, A. L. & SMITH, J. B., *Failure of Spherical Hydrogen Storage Tank.* The Welding Journal, March, 1945.

84. EVANS, SIDLEY O., *Quality Control During Production of Electric Resistance Welded Tubing.* The Welding Journal, September, 1945.

85. *Symposium on Magnetic Particle Testing.* Published by American Society for Testing Materials, 1945.

86. *Transcript of First Railroad Magnaflux Conference.* Published by Magnaflux Corporation, February, 1946.

87. MCMASTER, R. C. & BANTA, H. M., *Progress Report on Drill String Research Nondestructive Testing of Drill Pipe.* Battelle Memorial Institute, Drilling Contractor, August 15, 1946.

88. FREAR & LYONS, *Nondestructive Testing of Castings.* Industrial Radiography, Winter, 1946-1947.

89. WALSH, D. P., *Practical Tool Inspection and Quality Control Methods.* The Tool Engineer, December, 1946.

90. MCBRIAN, RAY, *Testing with Magnaflux on the D. & R. G. W. RY.* Railway Engineering and Maintenance, February, 1947.

91. MUELLER, H. M. & YEAST, W. E., *Magnetic Particle Inspection of Chromium Plated Tools.* Metal Progress, March, 1947.

92. *Transcript of Conference on Weld Inspection with Magnaflux.* Published by Magnaflux Corporation, May, 1947.

93. FICK, N. C., *Present Methods and Trends with Notes on Literature.* Metals Review, May, 1947.

94. THOMAS, W. E., *Inspection Methods Using Magnaflux and Zyglo in Production Industries.* Nondestructive Testing, Fall, 1947.

95. MARADUDIN, A. P., *Inspection of Pressure Vessels and Tanks.* Western Metals, April, 1948.

96. MCCUTCHEON, D. M., *Application of Nondestructive Testing to Automotive Parts.* Paper presented at S.A.E. Summer Meeting, June, 1948.

97. RODA, DONALD E., *Magnetic Particle Inspection in Engineering.* Iron Age, Vol. 162, No. 6, August 5, 1948.

98. RODA, DONALD E., *Standards of Magnetic Particle Inspection.* Iron Age, Vol. 162, No. 7, August 12, 1948.

99. HITT, WM., *The Value of Scientific Inspection to Industry.* Western Machinery and Steel World, July, 1948.

100. *Introduction to Complex Variables.* CHINCHILL, R. V.; McGraw-Hill, 1948.

101. *Principles of Electricity.* PAGE, L. & ADAMS, Y. I.; Van Nostrand, 1949.

102. *Electromagnetic Felds, Vol. I, Mapping of Fields.* WEBER, ERNST; John Wiley & Sons, 1950.

103. *Static and Dynamic Electricity.* SMYTHE, W. R.; McGraw-Hill, 1950.

104. *Handbook of Experimental Stress Analysis.* HETENYI, M.; John Wiley & Sons, 1950.

105. CAINE, JOHN B., *What's Wrong with Castings?* Foundry, June, 1951.

106. SWEET, JOHN W., *Weldment Inspection in Aircraft Construction.* Nondestructive Testing, Fall, 1951.

107. GILL, STANLEY A., *Various Inspection Methods Used in Heat Treating Shops.* Metal Treating, May-June, 1952.

108. *Dictionary of Conformal Representations.* KOBER, H.; Dover, 1952.

109. *Conformal Mapping,* NEHARI, Z.; McGraw-Hill, 1952.

110. *Electromagnetics.* KRAUS, JOHN D.; McGraw-Hill, 1953.

111. THOMAS, W. E., *Economic Factors of Nondestructive Testing.* Nondestructive Testing, March, 1953.

112. ALLEN, ARTHUR H., *Improved Tools Expedite Magnetic Particle and Penetrant Inspection.* Metal Progress, January, 1954.

113. OYE, LLOYD J., *Nondestructive Testing of Structures.* The Welding Journal, March, 1954.

114. MCCLURG, G. O., *Theory and Application of Coil Magnetization.* Journal Society for Nondestructive Testing, Jan.-Feb., 1955, 11 23-25.

115. MIGEL, HAMILTON, *Magnetic Particle, Penetrant and Related Inspection Methods as Production Tools for Process Control.* Steel Processing, February, 1955.

116. WILSON, T. C., *Nondestructive Testing of Refinery Equipment*. The Petroleum Engineer, July, 1955.

117. SMITH, O. G., *Magnetic Particle Technique Makes Billet Inspection Positive and Efficient*. Iron Age, May 5, 1955.

118. DEVRIES, A. J., *Productive Inspection Assures Quality and Reduces Scrap*. Chicago Midwest Metal Worker, January, 1956.

119. CATLIN, FRANKLIN S., *Nondestructive Sample Testing for Cracks Aids Heat Treating*. Metal Treating, March-April, 1956.

120. MCCABE, J. L. & HIRST, B., *Use of Tap Water in Magnetic Inspection*. Metal Progress, July, 1956.

121. HARRER, JOHN R., *Greater Acceptance of Welding Through the Use of Inspection Methods*. The Welding Journal, March, 1957.

122. *Standardization Aids Weld Inspection of Tractors*. Editorial, The Welding Engineer, April, 1957.

123. *Steel Hunts Harder for Flaws*. Business Week, May 4, 1957.

124. CATLIN, F. S., *Test Methods Valuable in Improving Casting Design*. Foundry, March, 1957.

125. *The Making, Shaping and Treating of Steel*. United States Steel Corporation, Seventh Edition, 1957.

126. *Brittle Behavior of Engineering Materials*. PARKER, EARL R.; John Wiley & Sons, 1957.

127. MARSCHALL, H. G., *Magnetic Particle Testing and Its Latest Means of Indication*. Werkstatt und Bertrieb, Vol. 91, Pp. 640-653, Nov., 1958.

128. *Magnetic Particle Inspection*. The Welding Handbook, Fourth Edition, 1958; American Welding Society.

129. *Nondestructive Testing*. HINSLEY, J. F.; McDonald & Evans Ltd., London, England, 1959.

130. *Progress in Nondestructive Testing, Vol. I*. STANFORD, E. G. & FEARON, J. H.; The McMillan Co., New York (Heywood & Co., London), 1959.

131. ARROTT, A. & GOLDMAN, J. E., *Fundamentals of Ferro-Magnetism*. Electrical Manufacturing, Pp. 109-140, March, 1959.

132. *Handbook, Society for Nondestructive Testing, Vol. II*, Sections 30 to 34, 1959.

133. *Ferro-Magnetism.* RICHARD M. BOZORTH; D. Van Nostrand Co., Inc., 1959.

134. *Physics of Magnetism.* SÓSHIN CHIKAZUMI; John Wiley & Sons, 1959.

135. *Black Light Pin Points Flaws in Tube-Conditioning Line.* Iron Age, Editorial, Nov. 24, 1960.

136. CHRISTENSEN, A. E., *Method Control in Testing—Key to Reliability.* Tool and Manufacturing Engineer, Dec., 1960.

137. STARR, ROY C., *Shop Quality Control Safeguards Automotive Spring Manufacture.* Canadian Machinery, May, 1960.

138. *Allowable Stresses in Flash-Welded Joints.* Product Engineering, Editorial, Oct. 31, 1960.

139. WALSH, D. P., *Reducing Maintenance Cost Through Inspection.* Mining Congress Journal, August, 1960.

140. BETZ, CARL E., *The Nondestructive Testing Engineer—Today's Career Opportunity.* Nondestructive Testing, Jan.-Feb., 1960.

141. *Progress in Nondestructive Testing, Vol. II.* STANFORD, E. G. & FEARON, J. H.; The McMillan Co., New York (Heywood & Co., London), 1960.

142. *Symposium on Fatigue of Aircraft Structures.* ASTM, 1960.

143. *Magnetic Transitions.* BELOV, K. P.; Consultants Bureau, N. Y., 1961.

144. VANCE, SAM, *Magnetic Particle Inspection at Regular Intervals Can Prevent Major Troubles.* Pipe Line Industry, February, 1961.

145. NIXON, FRANK, *Simple Fatigue Testing as an Essential Corollary to Nondestructive Testing.* Nondestructive Testing, March-April, 1961.

146. BOGART, HENRY G., *The Place for Nondestructive Tests in the Field of Plant and Equipment Overhaul.* Presented, Winter Annual Meeting, American Society of Mechanical Engineers, Dec. 1, 1961.

147. *Magnaglo Takes a Close Look.* Wireco Life, July-August, 1961, American Steel & Wire Div., U.S. Steel Corp.

148. *Missile Parts Get Fast Check.* Iron Age, Editorial, July 13, 1961.

149. EUBANKS, PAUL E., *Making Efficient Use of Magnetic Particle Inspection.* Foundry, July, 1961.

150. *Engineering and Technology Report—Nondestructive Testing.* Factory Magazine, Jan., 1961.

151. EMISH, C. F., *Tubing Quality is Built In—Tests Cannot Produce It.* Steel, June 5, 1961.

152. MULLER, E. A. W., *Material-Pruefung Nach dem Magnet-Pulver Verfahren.* Akademische Verlagsgesellschaft, Leipzig, 1961.

153. *Metals Handbook.* 9th Edition, Vol. I. American Society for Metals, 1961.

154. *Analysis of Service Failures.* Republic Alloy Steels Handbook, Republic Steel Corporation, 1961.

155. SITARAIN, R. V., *A Useful Engineering Theory of High Stress Level Fatigue.* Aerospace Engineering, Oct., 1961, P. 18.

156. *Glowing Lines Outline Seams in Steel Billets and Blooms.* Iron Age, Editorial, Oct. 4, 1962.

157. *Magnetic Particle Testing of Malleable Castings.* Report, AFS Malleable Division, Finishing and Inspection Committee (6 G), American Foundry Society Transactions, Vol. 70, 1962.

158. THOMAS, W. E., *Conditioning Practices and Their Effect on Product Quality and Cost.* Iron and Steel Engineer, Nov., 1962.

159. DALE, DON, *The Crack of Doom.* Sports Car Graphic, May, 1962.

160. MCPARLAN, JOS. L., *How Nondestructive Testing Aids Power Station Maintenance.* Power Engineering, March, 1957, P. 57.

161. HARZ, JOHN J., *How to Find Trouble Before It Starts.* National Engineer, August, 1962.

162. *Billets Handled Automatically in Nondestructive Test Unit.* Automation, November, 1962.

163. *Progress in Nondestructive Testing, Vol. III.* STANFORD, E. G. & FEARON, J. H.; The McMillan Co., New York (Heywood & Co., London), 1962.

164. *Principles and Practice of Nondestructive Testing.* LAMBLE, J. H.; John Wiley & Sons, 1963.

165. WALORDY, ALEX, *Magnaflux—The Key to Engine Reliability.* Cars, April, 1963.

166. HENDRON, J. A., *Automated Magnetic Particle Testing of Pressure Vessels.* Nondestructive Testing, July-August, 1963.

167. BRIGGS, CHAS. W., *Significance of Discontinuities in Steel Castings on the Basis of Destructive Testing.* Materials Research and Standards, ASTM, June, 1963, Pp. 472-479.

168. GOHIN, G. R., *The Mechanism of Fatigue.* Materials Research and Standards, AST & M, Vol. 3, No. 2, Feb., 1963, P. 106. Fatigue of Metals.

169. HARDRATH, H. F., *Crack Propagation and Final Failure.* Ibid., P. 116. Fatigue of Metals.

170. PETERSON, R. E., *Engineering and Design Aspects.* Ibid., P. 122. Fatigue of Metals.

171. PARK, F. R., *Fracture.* International Science and Technology, March, 1963, P. 24.

172. *Micromagnetics.* WM. FULLER BROWN; Interscience Publication, 1963.

173. BEZER, H. J., *Magnetic Methods of Nondestructive Testing.* The British Journal of Nondestructive Testing, Vol. 6, No. 4, Dec., 1964, P. 109.

174. *Metals Handbook, Vol. II,* 8th Edition, 1964. American Society for Metals.

175. *Sonics Unite with Magnetic Glow for Automatic Billet Testing.* Iron Age, Editorial, Oct. 15, 1964.

176. *Merry-go-Round Test System.* Tooling and Production, Editorial, September, 1964.

177. MIGEL, HAMILTON, *Acceptance Standards for Nondestructive Tests.* Mechanical Engineering, April, 1964.

178. *Symposium on Nondestructive Testing in the Missile Industry.* AST&M, 1964.

179. VLAHOS, C. J., *Nondestructive Testing Pays Its Way in Maintenance.* Mill and Factory, August, 1964.

180. HAYES, D. A., *The Use of Mechanized Nondestructive Testing in Quality Control.* Given at Chicago Technical Meeting of American Iron & Steel Institute, October 20, 1965.

181. WULPI, D. J., *How Components Fail—Modes of Fracture.* Metal Progress, Vol. 88, No. 3, Sept., 1965, P. 72.

182. WULPI, D. J., *How Components Fail—Types of Loading.* Ibid., No. 5, Nov., 1965, P. 83.

183. WULPI, D. J., *How Components Fail—Effects of Variables.* Ibid., No. 6, Dec., 1965, P. 65.

184. *Defects and Failures in Pressure Vessels and Piping.* HELMUT THIELSCH; Reinhold Publishing Corp., 1965.

185. *Table of Specifications and Standards for Nondestructive Testing.* Materials Evaluation, Vol. XXIV, No. 3, March, 1966, P. 158.

186. HOMER, RICHARD G., *Assuring Quality Forgings by Nondestructive Testing.* Metal Progress, March, 1966.

187. THOMAS, W. E., *Development and Application of Nondestructive Testing in the Aerospace Industry.* Third Aerospace Conference on Nondestructive Testing, Georgia Institute of Technology, Atlanta, May 2, 1966.

188. *Testing Combination Sets High Production Standards.* Tooling & Production, June, 1966.

189. BETZ, C. E., *Nondestructive Testing—Recent Achievements.* Industrial Quality Control, Vol. 23, No. 1, July, 1966, P. 19.

190. *Manual of Experimental Stress Analysis, Second Edition.* Edited by W. H. Tuppeny, Jr. and A. S. Kobayashi, SESA, 1966.

SUBJECT INDEX

Abrasive Wheel, for sectioning parts, 403
Acceptance limits, 459
Acceptance standards, 460
Accept-Reject Decision, 64, 417
Accept-Reject Standards, 338, 460, 461
Aircraft overhaul, 428, 429, 461
F.A.A. Requirements, 461
Air Force, U.S., 57, 438
Equipment Specs, 461
Air gap, defined, 125
All-over patterns, 383
Alternating current (A.C.)
Defined, 120
Demagnetizing with, 314
First use of, for magnetizing, 54
Loops, for demagnetizing, 319
Magnetizing with, 54, 157, 193
Permanent Magnetization with, 157
Rectified, 154
Sensitivity comparisons, with D.C. & Half Wave, 234, 236
Skin effect of, 152, 159, 195, 196
Sources of, 157
Vs. D.C., factor in Equipment Design, 347
Vs. D.C., for magnetizing, 151, 231
Vs. D.C., for surface and subsurface cracks, 231, 232, 233, 234
Yokes, 145
American Society of Mechanical Engineers, 457
American Society for Metals, 457
American Society for Testing and Materials, 457
Recommended Prod Spacings and current values of, 205, 206
Specifications of, 463
American Society of Tool Engrs., 457
American Welding Society, 457, 463

Ammonium Persulphate Etch, 404
Ampere, defined, 120
Magnetizing, half-wave, 155
Ampere turns, defined, 124, 145
Rule for determining, for longitudinal magnetizing, 181
Amperes
Rule for determining, for Circular magnetizing, 198
Prod magnetizing, 206
Per inch of prod spacing, 206
Analog methods, for field distribution, 175
Analysis, Experimental stress, 414
Appearance of Indications
Affected by spread of emergent field, 376
Factor in interpretation, 343, 344
Applications of Mag. Part. Testing
Classification of, 419
Industrial, 419
Applying Magnetic Particles
Dry Powders, 250
Factor in design, 345
Wet Bath, 266
Arc, Eye damage from, 153
Parts damaged by, 151
Quenching of, on A.C. break, 157, 158
Austentic steels, 102
Automatic equipment, 333, 342
Advantages of, 343
Analysis of testing problems for, 344
Bolts, testing with, 335
Compromises in design of, 341, 348
Current, method of applying, factor in design of, 348
Current, type of, factor in design of, 347
Defined, 334
Early, 56, 57
Future trends in, 365
Magnetic particles, method of applying, factor in design of, 345